Multi-objective Optimization Techniques

The book establishes how to design, develop, and test different hybrids of multi-objective optimization algorithms. It presents several application areas of multi-objective optimization algorithms.

- Presents a thorough analysis of equations, mathematical models, and mechanisms of multi-objective optimization algorithms.
- Explores different alternatives of multi-objective optimization algorithms to solve binary, multi-objective, noisy, dynamic, and combinatorial optimization problems.
- Illustrates how to design, develop, and test different hybrids of multi-objective optimization algorithms.
- Discusses multi-objective optimization techniques for cloud, fog, and edge computing.
- Highlights applications of multi-objective optimization in diverse sectors such as engineering, e-healthcare, and scheduling.

The book is primarily written for senior undergraduates, graduate students, and academic researchers in the fields of electrical engineering, electronics, communications engineering, computer science and engineering, and mathematics.

Multi-objective Optimization Techniques

Variants, Hybrids, Improvements, and Applications

Edited by
Tarik A. Rashid, Aram M. Ahmed, Bryar A. Hassan, Zaher Mundher Yaseen, Seyedali Mirjalili, Nebojsa Bacanin, and Sinan Q. Salih

CRC Press is an imprint of the
Taylor & Francis Group, an informa business

Designed cover image: Shutterstock

First edition published 2025
by CRC Press
2385 NW Executive Center Drive, Suite 320, Boca Raton FL 33431

and by CRC Press
4 Park Square, Milton Park, Abingdon, Oxon, OX14 4RN

CRC Press is an imprint of Taylor & Francis Group, LLC

ISBN: 978-1-032-58998-5 (hbk)
ISBN: 978-1-032-98949-5 (pbk)
ISBN: 978-1-003-60155-5 (ebk)

DOI: 10.1201/9781003601555

Typeset in Time
by Newgen Publishing UK

Contents

Biography

Tarik A. Rashid

Prof. Dr. Tarik Ahmed Rashid is a Principal Fellow for the Higher Education Authority and a professor in the Department of Computer Science and Engineering at the University of Kurdistan Hewlêr, Iraq. He pursued his Post-Doctoral Fellowship at the Computer Science and Informatics School, College of Engineering, Mathematical and Physical Sciences, University College Dublin, Ireland. His areas of research cover the fields of artificial intelligence, nature inspired algorithms, swarm intelligence, computational intelligence, machine learning, and data mining. He is a member of IEEE, Machine Intelligence Research Labs. He has journal editorial experience as an editor/board member and acted as a keynote conference speaker at several conferences, conference chair, conference program committee member, etc. Overall, he has authored and edited 140 Web of Science (WoS) and Scopus documents, including books and book chapters published by CRC, Springer, Elsevier, and IET.

Professor Rashid is among the Top 4 researchers in Iraq by number of WoS-indexed published documents in the Engineering research field over a five-year period (2019–2023). It is noteworthy that Professor Rashid was on the prestigious Stanford University list of the World's Top 2% of Scientists for the years 2021, 2022, and 2023. The ranking was performed under the condition of 44 criteria. Professor Rashid was also on the list of Top 10 researchers in the Al-Ayen Iraqi Researchers Ranking (2022). AIR-Ranking 2022 is a national ranking organized by Al-Ayen University. The ranking was performed under the condition of 24 criteria.

It is worth mentioning that Professor Rashid's team has designed a number of single and multi-objective optimization algorithms, such as FDO, CDDO, DSO, ANA, FOX, LPB, Goose, Leo, SHOA, ECA*, and iECA*. His team has also designed several multi-objective optimization algorithms, such as MOFDO, MOLPB, and GMOCSO.

Aram M. Ahmed

Assistant Professor Dr. Aram M. Ahmed received his B.Sc. degree in Software Engineering from Koya University/Iraq in 2010–2011 and his master's degree in Computer Network Technology from Northumbria University/UK in 2012–2013. He received his Ph.D. degree in Optimization and Metaheuristic Algorithms from Sulaimani Polytechnic University/Iraq in 2020–2021.

Dr. Ahmed is interested in modeling biological and natural systems using computational techniques. He is an active researcher who has published books and papers in peer-reviewed academic journals with high impact factors. He is currently working as a lecturer and researcher at the University of Kurdistan Hewler, Iraq.

Bryar A. Hassan

Bryar A. Hassan pursued bachelor's and master's degrees in Software Engineering from the University of Southampton, and the joint Ph.D. degree in Information Technology. He is a renowned Assistant Professor in Computer Science at the University of Kurdistan Hewler. He is on the list of the World's Top 2% Scientists

Rankings 2023 (by Stanford University and Elsevier). He has worked as a software developer and entrepreneur for several years. He is currently a renowned Assistant Professor; he inspires students through innovative teaching and cutting-edge research in Artificial Intelligence and optimization algorithms. His research interests include artificial intelligence, optimization algorithms, medical computing, semantic web, and NLP.

Zaher Mundher Yaseen

Dr. Zaher Mundher Yaseen is an Assistant Professor and research scientist in the field of civil and environmental engineering. He obtained his master's and Ph.D. degrees from the National University of Malaysia in 2014 and 2017, respectively. Currently, he is working at King Fahd University of Petroleum and Minerals, Saudi Arabia. The scope of his research is quite broad, covering water resources engineering, environmental engineering, knowledge-based system development, climate change and the implementation of data analytic and artificial intelligence models. He has published over 480 research articles in international journals and his total number of citations is over 20,000 (Google Scholar H-Index = 79). He has collaborated with more than 900 researchers in over 50 countries. He has served as a reviewer for more than 140 international journals and has been an academic editor of several Clarivate journals.

Seyedali Mirjalili

Seyedali Mirjalili is a Professor at the Center for Artificial Intelligence Research and Optimization at Torrens University. He gained international recognition for his contributions to nature-inspired artificial intelligence (AI) techniques, with over 600 published works that have received more than 110,000 citations and an H-index of 110. He has been on the list of the Top 1% of highly cited researchers since 2019, and Web of Science named him one of the most influential researchers in the world. In 2022 and 2023, *The Australian* newspaper recognized him as a global leader in AI and a national leader in the Evolutionary Computation and Fuzzy Systems fields. He serves as a senior member of IEEE and holds editorial positions at several top AI journals, including *Engineering Applications of Artificial Intelligence*, *Applied Soft Computing*, *Neurocomputing*, *Advances in Engineering Software*, *Computers in Biology and Medicine*, *Healthcare Analytics*, *Applied Intelligence*, and *Decision Analytics*.

Nebojsa Bacanin

Nebojsa Bacanin received his Ph.D. degree from the Faculty of Mathematics, University of Belgrade, in 2015 (study program Computer Science, average grade 10.00). He started his university career in Serbia 18 years ago at the Graduate School of Computer Science in Belgrade. He currently works as Full Professor, Vice-Rector for Scientific Research and Head of the Applied Artificial Intelligence study program at Singidunum University, Belgrade, Serbia. He is involved in scientific research in the field of Computer Science, and his specialty includes stochastic optimization algorithms, swarm intelligence, soft-computing, optimization, and modeling, as well as artificial intelligence algorithms, swarm intelligence, machine learning, image processing, and cloud and distributed computing. He has published more than 420 scientific papers (more than 180 SCIE papers) in high-quality journals and international conference proceedings indexed in Clarivate Analytics JCR, Scopus, WoS, IEEExplore, and other scientific databases. As a member of numerous editorial boards of cutting-edge international journals and committee boards of international conferences, he regularly performs editing and review activities. He has also been included in the prestigious Stanford University list of the World's Top 2% of Researchers, when his whole career is considered, in the field of Artificial Intelligence. Additionally, according to the AD scientific index, he is currently listed as the best researcher in the Computer Science area in Serbia.

Sinan Q. Salih

Sinan Q. Salih received the B.Sc. degree in information systems from the University of Anbar, in 2010, the M.Sc. degree in computer science from the College of Information Technology, Universiti Tenaga Nasional, in 2012, and the Ph.D. degree in soft computing and intelligent systems from the Faculty of Software Engineering, University of Malaysia Pahang, in 2019. He is currently a lecturer at the Technical College of Engineering, Al-Bayan University. He has published more than 70 research papers and co-authored papers with more than 50 researchers. His current research interests include optimization algorithms, nature-inspired metaheuristics, machine learning, and feature selection problems for real-world problems.

Contributors

Hanan K. AbdulKarim
Software and Informatics Engineering Department, College of Engineering
Salahaddin University – Erbil, Erbil, Iraq

Mzhda S. Abdulkarim
Department of Computer Science, College of Science
Charmo University, 46023, Chamchamal, Sulaimani, Iraq

Abdulhady A. Abdulla
Artificial Intelligence and Innovation Centre,
University of Kurdistan Hewler, Erbil, Iraq

Parastoo Afrasyabi
Faculty of Geodesy and Geomatics Engineering
K.N. Toosi University of Technology, Tehran, Iran

Aram M. Ahmed
Computer Science and Engineering Department, School of Science and Engineering
University of Kurdistan Hewler, Erbil, Iraq

Omed H. Ahmed
Department of Information Technology
University of Human Development, Sulaymaniyah, Iraq

Aso M. Aladdin
Department of Computer Science, College of Science
Charmo University, 46023, Chamchamal, Sulaimani, Iraq

Yusra H. Ali
Computer sciences department
University of Technology, Iraq

Nawzad K. Al-Salihi
Computer Science and Engineering Department, School of Science and Engineering
University of Kurdistan Hewler, Erbil, Iraq

Milos Antonijevic
Faculty of Informatics and Computing Singidunum University, Belgrade, Serbia

Nebojsa Bacanin
Faculty of Informatics and Computing
Singidunum University, Belgrade, Serbia

Milos Dobrojevic
Faculty of Informatics and Computing
Singidunum University, Belgrade, Serbia

Salar Farahmand-Tabar
Department of Civil Engineering, Faculty of Engineering
University of Zanian, Iran

Dana Rasul Hamad
Computer Science Department, Faculty of Science
Soran University, Soran, Erbil, KRG, Iraq

Rebwar Kh. Hamad
Department of Information Technology, Koya Technical Institute
Erbil Polytechnic University, Erbil, Iraq

Mahmood Yashar Hamza
Computer Engineering Department,
Tishk International University,
Erbil, Iraq

Alla A. Hassan
Department of Database Technology,
Computer Science Institute
Sulaimani Polytechnic University,
Sulaimani, 46001, Iraq

Bryar A. Hassan
Computer Science and Engineering
Department, School of Science and
Engineering
University of Kurdistan Hewler,
Erbil, Iraq;
Department of Computer Science,
College of Science
Charmo University, 46023,
Chamchamal, Sulaimani, KRI, Iraq

Dler O. Hasan
Department of Computer Science,
College of Science, Charmo University,
46023, Chamchamal, Sulaimani, Iraq

Abubakr S. Issa
Computer sciences department
University of Technology, Iraq

Sajjad Shamkhi Jaber
Computer sciences department
University of Technology, Iraq

Luka Jovanovic
Faculty of Informatics and Computing
Singidunum University,
Belgrade, Serbia

Jafar Majidpour
Computer Science Department
University of Raparin, Rania
46012, Iraq

Seyedali Mirjalili
Center for Artificial Intelligence
Research and Optimization
Torrens University Australia, Fortitude
Valley, Brisbane, QLD 4006,
Australia

Mokhtar Mohammadi
Department of Information Technology
Lebanese French University, Erbil
44001, Kurdistan Region, Iraq

Rebaz Mohammed
Computer Science and Engineering
Department, School of Science and
Engineering
University of Kurdistan Hewler,
Erbil, Iraq

Soran R. Mohammed-Taha
Department of Computer Science,
College of Science
Charmo University, 46023,
Chamchamal, Sulaimani, Iraq

Araz Ibrahim Mustafa
Department of Computer Science,
College of Science
Charmo University, 46023,
Chamchamal, Sulaimani, Iraq

Kaniaw A. Noori
Database Technology Department,
Technical College of Informatics
Sulaimani Polytechnic University,
Sulaimani, Iraq

Tarik A. Rashid
Computer Science and Engineering
Department, School of Science and
Engineering, AIIC
University of Kurdistan Hewler,
Erbil, Iraq

Soran Ab. M. Saeed
Database Technology Department, Technical College of Informatics
Sulaimani Polytechnic University, Sulaimani, Iraq

Mohamed Salb
Faculty of Informatics and Computing
Singidunum University, Belgrade, Serbia

Sinan Q. Salih
Technical College of Engineering
Al-Bayan University, Baghdad, Iraq

Yasameen Sami
Department of Computer Engineering
Tishk International University, Kurdistan Region, Erbil, Iraq

Esra Muhannad Shawkat
Computer Science and Engineering Department, School of Science and Engineering
University of Kurdistan Hewler, Erbil, Iraq

Sina Shirgir
Department of Structural Engineering
Faculty of Civil Engineering, University of Tabriz, Iran

Shahla U. Umar
Network Department, College of Computer Science and Information Technology
Kirkuk University, Kirkuk, Iraq

Zaher Mundher Yaseen
Civil and Environmental Engineering Department
King Fahd University of Petroleum and Minerals, Saudi Arabia

Miodrag Zivkovic
Faculty of Informatics and Computing
Singidunum University, Belgrade, Serbia

1 Optimum Distribution of MR Dampers in Structures

A Multi-objective Approach Using the Fuzzy-PSO Algorithm

Salar Farahmand-Tabar and Sina Shirgir

1.1 INTRODUCTION

Beside gradient-based methods [1], metaheuristics represent potent optimization strategies utilized for tackling intricate engineering dilemmas beyond the reach of conventional optimization techniques [2–5]. These algorithms leverage diverse mathematical and computational principles, facilitating the efficient exploration of solution spaces, identification of promising solutions, and eventual convergence to the optimal solution. Despite their efficacy, metaheuristics confront limitations in terms of efficiency, accuracy, and scalability. Some may be trapped in local optima or exhibit slow convergence, while others require excessive fitness evaluations to attain optimal solutions. Moreover, by increasing the complexity relating to the problem, metaheuristic algorithms may witness performance degradation, diminishing their effectiveness in real-world engineering problem-solving [6–9].

Consequently, there exists an imperative to ameliorate the performance of metaheuristic algorithms to improve their effectiveness in resolving complex engineering challenges. This improvement can be achieved through the development of novel algorithms addressing extant limitations or by integrating diverse techniques like machine learning, opposition-based learning, and adaptive parameter control into existing algorithms to enhance their performance. Advancing the performance of metaheuristic algorithms has profound implications across engineering domains such as mechanical, civil, and electrical engineering, pivotal in solving intricate problems, thereby fostering the development of more efficient, cost-effective, and dependable systems while enhancing productivity and safety [10–13].

Fuzzy logic, an approach based on degrees of truth rather than strict binary (true/false) values, has been integrated into metaheuristic algorithms to tackle complex optimization problems. This integration introduces a level of uncertainty and ambiguity,

DOI: 10.1201/9781003601555-1

allowing for more flexible decision-making processes. In fuzzy metaheuristics, each solution is represented not as a precise value, but rather as a fuzzy set with membership functions indicating the degree to which the solution satisfies certain criteria. By incorporating fuzzy techniques into metaheuristic algorithms, such as Particle Swarm Optimization (PSO), the algorithms gain the ability to handle imprecise and vague information more effectively. This enables them to navigate solution spaces with greater adaptability and robustness, thereby enhancing their overall performance in finding optimal solutions. The utilization of fuzzy logic in metaheuristics represents a powerful approach to address optimization challenges by introducing a more nuanced and flexible framework for problem-solving.

Numerous studies have employed the fuzzy technique to enhance various optimization algorithms, demonstrating its efficacy in boosting performance [14]. For instance, Cuevas et al. [15] investigated fuzzy-based metaheuristics and Yalaoui et al. [16] utilized fuzzy-assisted metaheuristics to solve the scheduling problem. In a distinct context, Arora et al. [17] applied fuzzy metaheuristics to ANFIS-based prediction models. Additionally, Patel [18] integrated fuzzified metaheuristics into control applications.

Control devices are categorized into passive, active, and semi-active types based on their mechanisms. Semi-active response control systems that employ a Magnetorheological (MR) Fluid Damper (MRD) have particularly gained prominence across engineering applications owing to their unique advantages, especially on building structures [19–23]. MRDs, comprising magnetically polarizable particles dispersed within a carrying liquid, exhibit a rapid response to magnetic field fluctuations while maintaining rheological stability [24–28]. MRDs with this behavior offer benefits like continuous adjustment, compact size, low energy consumption, and convenient control [29, 30], which lead to finding extensive utility in shock absorption and engineering applications across various domains, such as automobiles, military applications, ships, and bridges [31–36].

The literature predominantly focuses on tuned mass dampers [37–45], with recent studies delving into modeling and control strategies for MR devices [46–48]. However, there exists an increasing demand for design methodologies with enhanced performance that reduce manufacturing time and costs. Optimizing the designs of MRDs necessitates considering different factors, posing challenges for traditional optimization methods.

In this chapter, a fuzzy version of the PSO algorithm is applied to optimize the distribution of structural control systems employing MRDs. Optimal MR damper placement crucially impacts structural performance, thus necessitating the use of enhanced methods for optimum design. Performance evaluation involves selecting a benchmark 40-story tall shear frame, comparing results across various optimization methods.

1.2 FUZZY PSO ALGORITHM

PSO has emerged as a powerful metaheuristic algorithm for solving optimization problems. In recent years, there has been a growing demand to tackle optimization problems involving multiple conflicting objectives, leading to the development of Multi-objective PSO (MOPSO) algorithms. However, while MOPSO techniques have shown promise in handling such multi-dimensional optimization challenges, they can be improved to maintain a balance between convergence and diversity in the search space. To address this, an enhanced approach is suggested: an MOPSO algorithm that incorporates fuzzy logic to adaptively adjust the exploration-exploitation trade-off. To provide a comprehensive understanding, first the fundamentals of single-objective PSO is presented, followed by an exploration of the MOPSO technique with the integration of fuzzy logic into the framework to enhance its effectiveness in solving multi-objective optimization problems.

1.2.1 Single-objective Particle Swarm Optimization

The PSO algorithm [49] is improved through a swarm of particles that continually adjust their positions in each iteration to optimize the search process. The versatility of PSO is demonstrated through hybridization, enhancements, and various variants, leading to its application in real-world scenarios across several domains. These areas include industrial, environmental, smart city, commercial, healthcare, and general aspect applications [50]. In PSO, each particle within the swarm aims to obtain the best possible solution by moving toward its own best previous position (p_{best}) and the global best position (g_{best}). This movement is particularly effective in problems where the objective is minimization, as expressed by the following equation:

$$p^t_{best_i} = x^*_i \mid \mathrm{f}\left(x^*_i\right) = \min_{\substack{k=\{1,2,\ldots,l\} \\ i=\{1,2,\ldots,N\}}} \left(\left\{\mathrm{f}\left(x^k_i\right)\right\}\right)$$

$$g^l_{best_i} = x^l_* \mid \mathrm{f}\left(x^l_*\right) = \min_{\substack{i=\{1,2,\ldots,N\} \\ k=\{1,2,\ldots,t\}}} \left(\left\{\mathrm{f}\left(x^k_i\right)\right\}\right) \tag{1}$$

where i denotes the particle's index and t denotes the ongoing iteration count. Furthermore, f represents the objective function being either optimized or minimized and x denotes the position vector of the particle, which can be considered as a potential solution. Lastly, N denotes the total number of particles within the swarm. The position (x) and velocity (v) of each individual particle i undergo updates according to the subsequent equations at every iteration $t+1$:

$$V^{l+1}_i = \omega V^{l+1}_i + c_1 r_1 \left(p^l_{best_i} - x^l_i\right) c_2 r_2 \left(g^l_{best_i} - x^l_i\right)$$

$$x^{t+1}_i = x^t_i + V^{t+1}_i \tag{2}$$

The update equations rely on the velocity vector (V), which is greatly influenced by the inertia weight (ω). This weight acts as a crucial factor in balancing between exploiting local search and exploring global search. To incorporate randomness into the process, random vectors r_1 and r_2 are uniformly generated within the range $[0,1]^D$, spanning across the dimensions of the search space (D) or the size of the problem being addressed. Furthermore, the equations integrate positive constants c_1 and c_2, known as acceleration coefficients. These coefficients play a key role in guiding the algorithm toward optimal solutions. The PSO algorithm operates through a series of steps aimed at uncovering the most optimal solution:

Step 1: For each particle, initial vectors for velocity and position are generated randomly.
Step 2: The algorithm evaluates the objective function while considering predefined constraints.
Step 3: The current fitness value for each particle is compared and replaced with its previous best, if it is superior.
Step 4: The global best is updated by comparing its value with the fitness value associated with each particle.
Step 5: Updating is done on the velocity and position vectors of each particle.
Step 6: Steps (2)–(5) are reiterated until the stopping criteria are met or the predetermined maximum number of iterations is reached.

Throughout each iteration, there are predetermined maximum values for velocity and position, which regulate the changes in velocity and position for each particle.

1.2.2 Multi-objective Fuzzy Particle Swarm Optimization

Numerous methods exist and are applied for the multi-objective optimization of real-world problems [51–53]. The Multi-objective Fuzzy PSO (MOFPSO) technique draws inspiration from the "sub-population" strategy. Its core concept involves breaking down the original problem into several less complex parts. By collecting solutions (particles) from these smaller problems, an improved overall solution can be achieved. The search space is divided into a dynamic number of major spaces, with our experiments utilizing five such spaces. Each major space is then subdivided into a variable number of minor spaces, also set to five in our trials. Population-based and Pareto-based methods have been combined to improve results. The priority of choosing a new search space for the subsequent iteration is determined by an algorithm assessing areas through a fuzzy controller [54–56]. For calculating the position and velocity of a new particle, we employ the core PSO algorithm:

$$V(i+1) = V_i c_1 r_1 \left(p_{best_i} - x_i\right) + c_2 r_2 \left(g_{best_i} - x_i\right);$$

$$x_{i+1} = x_i + V_{i+1} \tag{3}$$

Utilizing fuzzy logic in MOPSO involves setting parameters such as $c_1 = 1.4$, $c_2 = 1.5$, and generating random values r_1, r_2 from the range [0, 1]. The maximum velocity limit is determined as half the range of the variable. When a particle's velocity approaches zero, the particle is shifted to a different dimension within the n-dimensional search area, facilitating escape from local optima. Furthermore, the algorithm employs two accumulators, namely repository and deleted-repository. These accumulators enable the fuzzy controller to assess the densities of both dominated and non-dominated particles. The fuzzy controller plays a crucial role in deciding which search area should be chosen for choosing the best global solution, denoted by g_{best}.

1.2.2.1 Steps and Structure of MOFPSO

Algorithm 1 outlines the MOFPSO in a concise manner [57]. The initial step involves parameter initialization such as: getting ranges, initializing particles and their fitness, and initializing the major and minor parts. In computing the velocity phase, three steps are performed, each employing a diverse strategy for selecting the global best solution, g_{best}: (1) choosing g_{best} from all previously achieved candidates (if multiple candidates exist, one is randomly chosen), (2) the fuzzy controller seeking the best search space to obtain g_{best} based on particle density in both the repository and deleted-repository, and (3) searching for g_{best} according to the major part. The "Evaluate fitness" step calculates the fitness value of the particle, considering the objective functions. The "Update p_{best}" step continuously updates the best position obtained by the particle under consideration. The Evaluate *Non-dominated* method is utilized to select the non-dominated particle.

Algorithm 1. MOFPSO algorithm

Initialize parameters ();
Update $pbest$ ();
Evaluate $Non\text{-}dominated$ ();
while $i <= max.iteration$ do
  Compute $Velocity$ ();
  Evaluate $fitness$ ();
  Update $pbest$ ();
  Evaluate $Non\text{-}dominated$ ();
$i = i + 1$;
end while
Make $report$ ();

1.2.2.2 Fuzzy Logic Controller and Membership Function

The entire search space is subdivided into five divisions and each of these subspaces is then further divided into five subsections. The objective is to identify the most suitable subsection for selection in the next iteration. This decision is based on assessing the presence of non-dominated particles within a given search space. The fuzzy controller evaluates the particle density in both the repository and deleted-repository. The focal region of interest is where the density of dominated particles is medium or low, and the density of non-dominated particles is medium to high. For this reason, the considered fuzzy controller examines densities in both repositories.

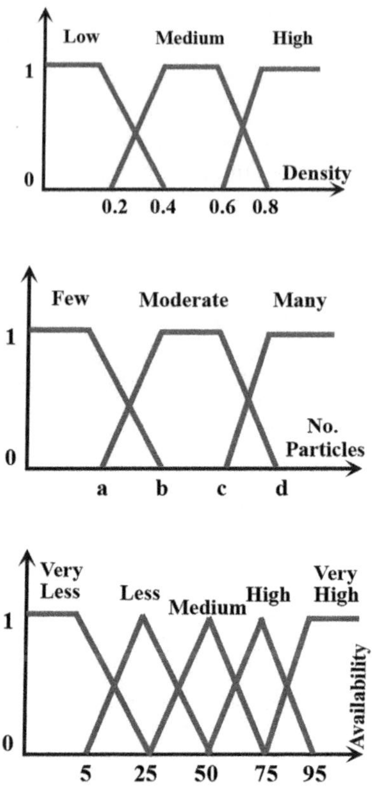

FIGURE 1.1 Graphical representation of the membership function.
a) Input fuzzy variable density of non-dominated particles.
b) Input fuzzy variable number of particles.
c) Output fuzzy variable availability of non-dominated particles.

A fuzzy controller is utilized that employs three fuzzy sets for each of the input variables: number of particles and its density, for both the repository and deleted-repository accumulators. For the variable density, the sets are High (H), Medium (M), and Low (L). Regarding the number of variable particles, the fuzzy sets are Many (MA), Moderate (MO), and Few (F). Additionally, the output variable availability of non-dominated particles is characterized by five fuzzy sets: Very Less Availability (VLSP), Less Availability (LSP), Medium Availability (MP), High Availability (HP), and Very High Availability (VHP).

The membership functions are illustrated in Figure 1.1. It's worth noting that the boundaries of fuzzy sets for the variable number of particles vary as iterations progress. The related values of points *a*, *b*, *c*, and *d* in the fuzzy sets are computed through a coefficient ($coefficient = 0.1 \times population.size \times current.iteration$). The calculation is performed as follows:

TABLE 1.1
Three-dimensional matrix defining the fuzzy rules: availability depending on densities in the both accumulators

Repository →	L			M			H		
Del-rep↓ No.P→	F	MO	MA	F	MO	MA	F	MO	MA
L	HP	MP	LSP	HP	HP	VHP	HP	VHP	VHP
M	HP	MP	LSP	MP	MP	MP	MP	HP	HP
H	MP	LSP	VLSP	MP	LS	VLSP	MP	MP	LSP

$$a = coefficient$$

$$b = coefficient + \frac{coefficient}{2}$$

$$c = 2 \times coefficient + \frac{coefficient}{2}$$

$$d = 3 \times coefficient$$

Eighteen specific fuzzy rules have been formulated for the controller, presented in Table 1.1. Activating the pertinent rules, the fuzzy controller employs the Mamdani (max-min) implication method to compute the fuzzy output. Extracting a crisp value from all activated rules, the weighted-average defuzzification method is utilized.

1.3 PROPOSED OPTIMAL PLACEMENT STRATEGY FOR MR DAMPERS

This section introduces an approach to design a semi-active control system for tall buildings by strategically placing MR dampers using metaheuristic algorithms. To explore the efficacy of this approach, a 40-story shear building is modeled to analyze the performance of control systems. Key structural parameters, such as stiffness, damping, mass, moment of inertia, and elevations are detailed in Table 1.2. The placement of MR dampers, including their location and quantity on each story, is considered as a design variable in this optimization process. The objective function is defined as the sum of the ratio between the maximum controlled and uncontrolled drift across all stories. This function serves as the optimization target, aiming to minimize the overall structural response to seismic events:

$$Objective\ Functions \begin{cases} OF1 = \sum_{i=1}^{n} \left(\frac{Max.\ Controlled\ Drift_i}{Max.\ Uncontrolled\ Drift_i} \right) \\ OF2 = \sum_{i=1}^{n} N_{damper}(i) \end{cases}$$

TABLE 1.2
Story-related parameters of the 40-story frame

Story	m_i (t)	k_i (MN/m)	c_i (MNs/m)	I_i (kgm^2)	z_i (m)
Base	1960	-	-	1.96×10^8	0
1	980	2130.00	42.60	1.31×10^8	4
2	980	2100.97	42.02	1.31×10^8	8
3	980	2071.95	41.44	1.31×10^8	12
4	980	2042.92	40.86	1.31×10^8	16
5	980	2013.90	40.28	1.31×10^8	20
6	980	1984.87	39.70	1.31×10^8	24
7	980	1955.85	39.12	1.31×10^8	28
8	980	1926.82	38.54	1.31×10^8	32
9	980	1897.79	37.96	1.31×10^8	36
10	980	1868.77	37.38	1.31×10^8	40
11	980	1839.74	36.79	1.31×10^8	44
12	980	1810.72	36.21	1.31×10^8	48
13	980	1781.69	35.63	1.31×10^8	52
14	980	1752.67	35.05	1.31×10^8	56
15	980	1723.64	34.47	1.31×10^8	60
16	980	1694.62	33.89	1.31×10^8	64
17	980	1665.59	33.31	1.31×10^8	68
18	980	1636.56	32.73	1.31×10^8	72
19	980	1607.54	32.15	1.31×10^8	76
20	980	1578.51	31.57	1.31×10^8	80
21	980	1549.49	30.99	1.31×10^8	84
22	980	1520.46	30.41	1.31×10^8	88
23	980	1491.44	29.83	1.31×10^8	92
24	980	1462.41	29.25	1.31×10^8	96
25	980	1433.38	28.67	1.31×10^8	100
26	980	1404.36	28.09	1.31×10^8	104
27	980	1375.33	27.51	1.31×10^8	108
28	980	1346.31	26.93	1.31×10^8	112
29	980	1317.28	26.35	1.31×10^8	116
30	980	1288.26	25.77	1.31×10^8	120
31	980	1259.23	25.18	1.31×10^8	124
32	980	1230.21	24.60	1.31×10^8	128
33	980	1201.18	24.02	1.31×10^8	132
34	980	1172.15	23.44	1.31×10^8	136
35	980	1143.13	22.86	1.31×10^8	140
36	980	1114.10	22.28	1.31×10^8	144
37	980	1085.08	21.70	1.31×10^8	148
38	980	1056.05	21.12	1.31×10^8	152
39	980	1027.03	20.54	1.31×10^8	156
40	980	998.00	19.96	1.31×10^8	160

$$subjected\ to \begin{cases} 0 \le N_{damper}(i) \le 5 \\ 40 \le N_{damper-total} \le 80 \end{cases} \tag{4}$$

The optimization problem sets constraints on the number of dampers per story, where the number of stories is denoted by n. Each story can have between 0 and 5 dampers, with 0 indicating no damper use and 5 representing the maximum allowable dampers per story. Another constraint is the total number of dampers installed throughout the building, set to a minimum of 40 dampers corresponding to the number of stories in the structural model and a maximum of 80 dampers. The optimization algorithms are limited to a maximum of 500 iterations as the stopping criteria.

To ensure reliability, statistical analysis is conducted based on 30 independent experiments, with the best solution being reported. The structural model is subjected to both controlled and uncontrolled scenarios using numerical time-history methods. A white noise ground acceleration with a peak ground acceleration (PGA) of 0.4 g, depicted as $W(t)$ in Figure 1.2, is utilized as an illustrative example. Additionally, the impact of different distributions of MR dampers on tall building behavior is investigated through seismic analyses using historical earthquake records. Several earthquake events are chosen, and their details are summarized in Table 1.3 for analysis purposes.

The assessment of control performance relies on various performance indicators, comparing the responses of the structure under uncontrolled and controlled conditions. Six criteria are utilized, with the initial three focusing on maximum values, while the latter three pertain to the structural norm response $(.) \to (\|.\|)$. These criteria are outlined in Table 1.4 and include the following:

- δ^{max} : Max. uncontrolled inter-story drift.
- $\ddot{x}_u^{max}$: Absolute value of roof acceleration.
- F_b^{max} : Max. force of the base shear.
- $d_i(t)$: Controlled inter-story drift of the i th level.
- $\ddot{x}_{ai}(t)$: Absolute acceleration of the i th level.
- m_i : Seismic mass of the i th level.

The proposed method's flowchart is illustrated in Figure 1.3.

TABLE 1.3
Characteristic data of ground motion

Event	Station	Fault	Year	Magnitude	PGA (g)
El Centro	El Centro	Strike slip	1940	7	0.349
Northridge	Beverly Hills	Reverse	1994	6.7	0.344
Kobe	Shin-Osaka	Strike slip	1995	6.9	0.693

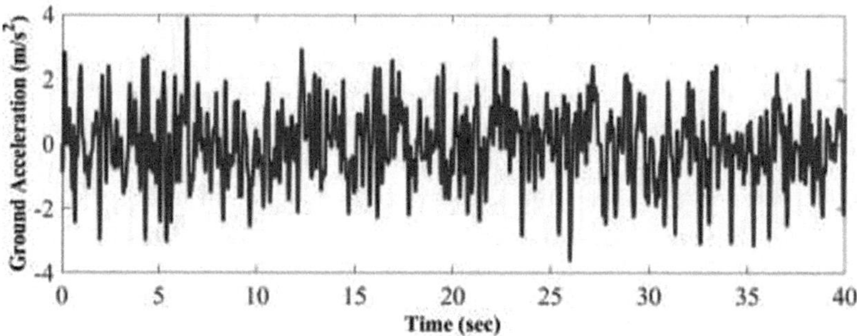

FIGURE 1.2 Time history of white noise ground acceleration, $W(t)$, $PGA = 0.4$ g.

TABLE 1.4
Performance criteria for the building with the control system

- Peak inter-story drift:

$$J_1 = max\left\{\frac{\max\limits_{t.i}\left(\left|d_i(t)\right|\right)}{\delta^{max}}\right\}$$

- Peak level acceleration:

$$J_2 = max\left\{\frac{\max\limits_{t.i}\left\{\ddot{x}_{ai}(t)\right\}}{\ddot{x}_u^{max}}\right\}$$

- Peak base shear force:

$$J_3 = max\left\{\frac{\max\limits_{t}\left|\sum_i m_i \ddot{x}_{ai}(t)\right|}{F_b^{max}}\right\}$$

- Normed inter-story drift:

$$J_4 = max\left\{\frac{\max\limits_{t.i}\left(\left\|d_i(t)\right\|\right)}{\left\|\delta^{max}\right\|}\right\}$$

- Normed level acceleration:

$$J_5 = max\left\{\frac{\max\limits_{t.i}\left\|\ddot{x}_{ai}(t)\right\|}{\left\|\ddot{x}_u^{max}\right\|}\right\}$$

- Normed base shear force:

$$J_6 = max\left\{\frac{\max\limits_{t}\sum_i\left\|m_i \ddot{x}_{ai}(t)\right\|}{\left\|F_b^{max}\right\|}\right\}$$

1.4 RESULTS AND DISCUSSION OF THE CASE STUDY

The design optimization is based on objective functions related to the number of dampers and relative displacement responses. Through multi-objective optimization, a Pareto front that showed a range of optimal solutions was successfully identified, each offering a unique trade-off between our considered design objectives (Figure 1.4). Among this diverse set of solutions, we pinpointed the one that stood out as the most desirable, marked in the figure on the Pareto front. This particular solution represents an exceptional balance between the reduction of seismic responses and the number of MR dampers. Furthermore, to provide a wider perspective on the design space, we also obtained two additional solutions considering the minimum and maximum number of dampers, one on each side of the Pareto front. These solutions, while not considered Pareto-optimal, serve as valuable reference points, showcasing the extremes of performance with respect to each individual objective. Such a comprehensive analysis empowers engineers to make decisions with a holistic view of the available design options, facilitating choices that align with project-specific priorities and constraints.

Figure 1.5 showcases the optimal allocation of the quantity and placement of MR dampers throughout the 40-story frame. The illustration highlights that several stories, including 10, 14–16, 22, 27, and 29, do not need any dampers, while others require multiple dampers, with a maximum of 5 dampers allocated to stories 1, 12,

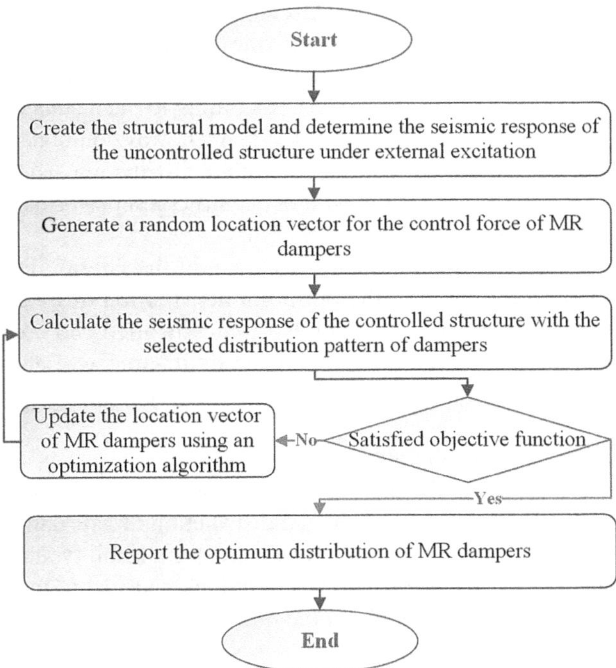

FIGURE 1.3 Flowchart for the optimal distribution of the MRDs.

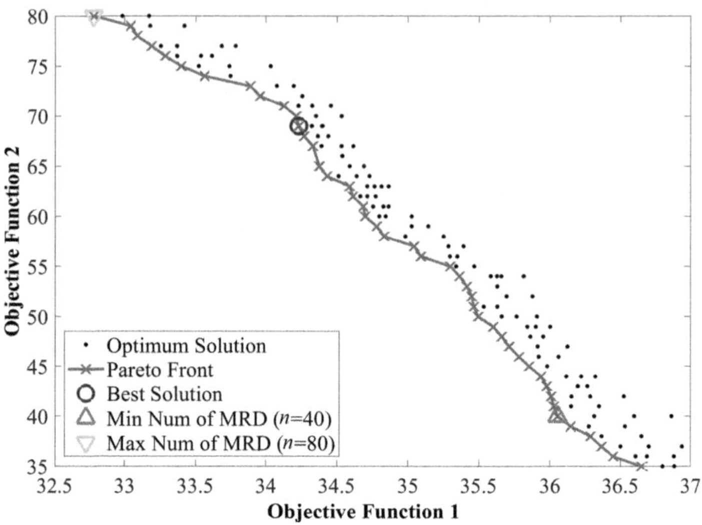

FIGURE 1.4 Pareto solutions of the structure with the optimum design of MR dampers.

36, and 37. Certain stories exhibit a demand for more than 4 dampers, indicating susceptible stories in the structure.

By analyzing the performance indicators for controlling the seismic response of the MR-equipped structure under three different realistic earthquakes (Table 1.3), it becomes evident that designing structures with the optimum distribution of dampers in the multi-objective mode exhibits a remarkable ability to reduce seismic responses by considering the cost of dampers (Table 1.5). According to the results in Table 1.5, the structure with the optimum distribution of 40 dampers was dominated by the results of the Pareto-optimal solution, with the optimum distribution using 69 dampers. Considering 80 dampers gives results with better structural performance with higher costs of the control system, which is not desired.

In Figures 1.6 and 1.7, the maximum response of the lateral displacement and drift of the structural model with the optimum distribution of 69 MR dampers is presented under different earthquakes in both uncontrolled and controlled states. This comparison confirms the efficiency of the multi-objective approach in simultaneously reducing both the number of dampers (cost) and the drift responses of the structure.

Integrating the fuzzy technique into the optimization process yields notable improvements, evident in both the convergence and results. These enhancements play a pivotal role in achieving an optimized distribution of MR dampers within tall buildings, offering the potential for heightened structural stability and improved performance during seismic occurrences. These findings underscore the importance of incorporating the fuzzy technique within the multi-objective optimization algorithm, offering engineers and researchers a valuable tool for efficiently addressing complex structural optimization challenges with reliable solutions.

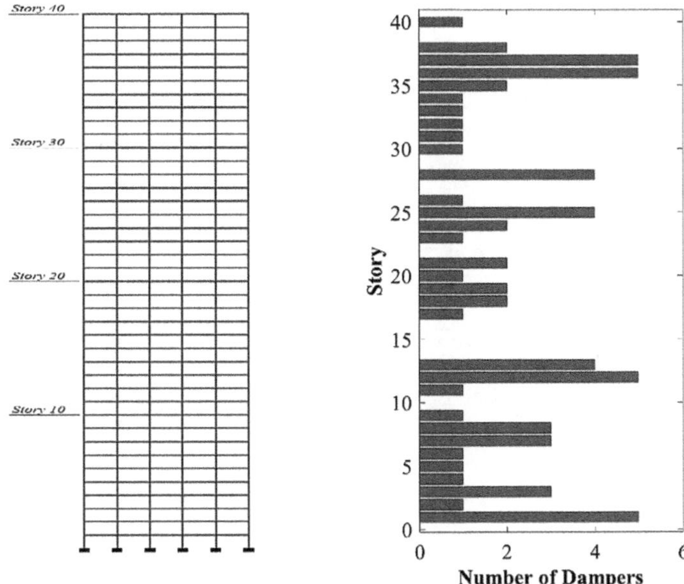

FIGURE 1.5 Damper distribution for the optimal structure.

TABLE 1.5
Performance responses of the structure with the optimum distribution of dampers

Event		N_{damper}		
		40 (Single-Obj.)	69 (Multi-Obj.)	80 (Single-Obj.)
El Centro	J_1	0.8674	0.8257	0.7857
	J_2	0.8172	0.7845	0.7672
	J_3	0.8692	0.8409	0.8051
	J_4	0.8408	0.8017	0.7729
	J_5	0.7753	0.7539	0.7284
	J_6	0.8344	0.8198	0.7873
Kobe	J_1	0.8237	0.7932	0.7718
	J_2	0.8846	0.8542	0.8168
	J_3	0.9014	0.8819	0.8591
	J_4	0.7679	0.7361	0.7096
	J_5	0.8260	0.7847	0.7511
	J_6	0.8942	0.8752	0.8376
Northridge	J_1	0.7926	0.7662	0.7445
	J_2	0.8176	0.7934	0.7698
	J_3	0.8624	0.8452	0.7975
	J_4	0.7382	0.7062	0.6843
	J_5	0.7782	0.7244	0.6911
	J_6	0.8465	0.8261	0.7585

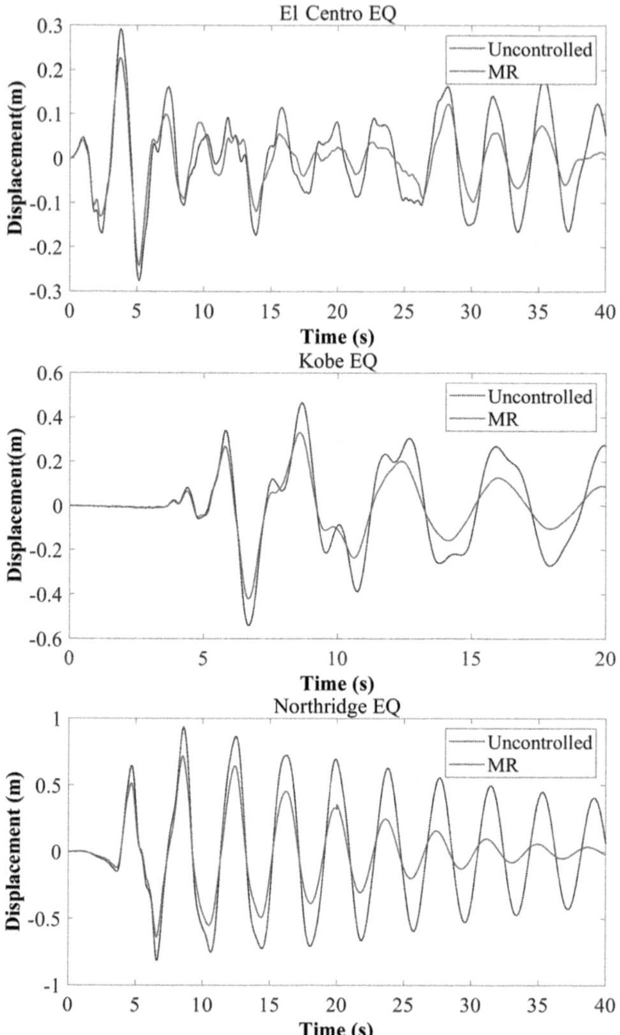

FIGURE 1.6 Results of the displacement response of the structure.

1.5 CONCLUSIONS

In this chapter, the effectiveness of the MOFPSO algorithm was focused on to tackle structural optimization challenges. To address common issues encountered in metaheuristic algorithms, such as slow convergence rates and susceptibility to local optima, the fuzzy technique was integrated into the PSO algorithm to enhance its exploration and exploitation capabilities. The investigation focused on the performance of the proposed algorithm within the context of a 40-story tall building framework equipped with MR dampers, aiming to determine the optimal number and placement of control devices. The primary objective of the optimization efforts was

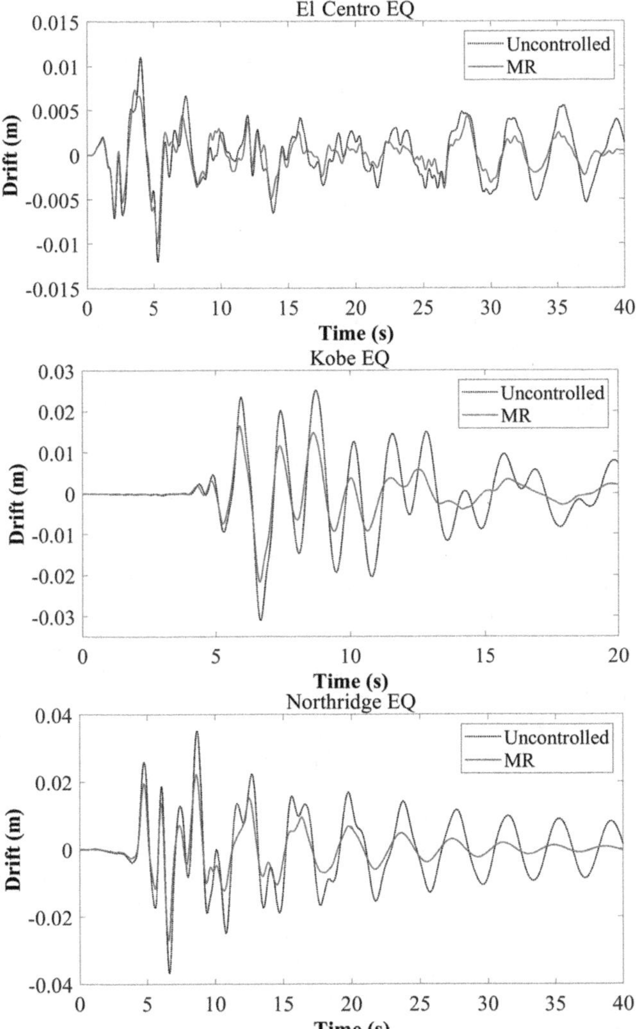

FIGURE 1.7 Results of the drift response of the structure.

to minimize the structural response under earthquake excitations (specifically, the El Centro, Northridge, and Kobe earthquakes) by strategically locating MR dampers. Various structural responses, including maximum inter-story drift, story acceleration, base shear, and their respective norm values, were analyzed. The efficacy of the system's performance was assessed based on the calculated J indices, indicating the efficiency of the damper distribution patterns in mitigating structural responses. Overall, the results demonstrate that MR dampers distributed using the approach that considers multiple objectives with Fuzzy-PSO exhibit appropriate performance in reducing seismic responses, including drift, lateral displacement, acceleration, and base shear.

REFERENCES

1. Farahmand-Tabar S, Abdollahi F, Fatemi M. (2023). Robust conjugate gradient methods for non-smooth convex optimization and image processing problems. In: Kulkarni AJ, Gandomi AH (eds) *Handbook of formal optimization*. Springer, Singapore. https://doi.org/10.1007/978-981-19-8851-6_42-1
2. Farahmand-Tabar S, Ashtari P. (2020). Simultaneous size and topology optimization of 3D outrigger-braced tall buildings with inclined belt truss using genetic algorithm. *The Structural Design of Tall and Special Buildings* 29(13): e1776. https://doi.org/10.1002/tal.1776
3. Farahmand-Tabar S, Shirgir S. (2023). Synergistic collaboration of motion-based metaheuristics for the strength prediction of cement-based mortar materials using TSK model. In: Kulkarni AJ, Gandomi AH (eds) *Handbook of formal optimization*. Springer, Singapore. https://doi.org/10.1007/978-981-19-8851-6_43-1
4. Farahmand-Tabar S, Rashid TA. (2023). Steel plate fault detection using the fitness dependent optimizer and neural networks. In: Kulkarni AJ, Gandomi AH (eds) *Handbook of formal optimization*. Springer, Singapore. https://doi.org/10.1007/978-981-19-8851-6_41-1
5. Farahmand-Tabar S. (2023). Memory-driven metaheuristics: improving optimization perfor-mance. In: Kulkarni AJ, Gandomi AH (eds) *Handbook of formal optimization*. Springer, Singapore. https://doi.org/10.1007/978-981-19-8851-6_38-1
6. Farahmand-Tabar S, Shirgir S. (2023). Boosting the efficiency of metaheuristics through opposition-based learning in optimum locating of control systems in tall buildings. In: Kulkarni AJ, Gandomi AH (eds) *Handbook of formal optimization*. Springer, Singapore. https://doi.org/10.1007/978-981-19-8851-6_37-1
7. Farahmand-Tabar S, Sadrekarimi N. (2023). Overcoming constraints: the critical role of penalty functions as constraint handling methods in structural optimization. In: Kulkarni AJ, Gandomi AH (eds) *Handbook of formal optimization*. Springer, Singapore. https://doi.org/10.1007/978-981-19-8851-6_40-1
8. Farahmand-Tabar S, Shirgir S. (2023). Incorporating nelder mead simplex as an accelerating operator to improve the performance of metaheuristics in nonlinear system identification. In: Kulkarni AJ, Gandomi AH (eds) *Handbook of formal optimization*. Springer, Singapore. https://doi.org/10.1007/978-981-19-8851-6_39-1
9. Farahmand-Tabar S, Shirgir S. (2023). Opposed pheromone ant colony optimization for property identification of nonlinear structures. In: Dey N (eds) *Applications of ant colony optimization and its variants*. Springer, Singapore. https://doi.org/10.1007/978-981-99-7227-2_5
10. Farahmand-Tabar S, Babaei M. (2023). Memory-assisted adaptive multiverse optimizer and its application in structural shape and size optimization. *Soft Computing*, 27(16):11505–27. https://doi.org/10.1007/s00 500-023-08349-9
11. Farahmand-Tabar S, Shirgir S. (2023). Positron-enabled atomic orbital search algorithm for improved reliability-based design optimization. In: Kulkarni AJ, Gandomi AH (eds) *Handbook of formal optimization*. Springer, Singapore. https://doi.org/10.1007/978-981-19-8851-6_44-1
12. Farahmand-Tabar S, Ashtari P, Babaei M. (2023). Dynamic intelligence of self-organized map in the frequency-based optimum design of structures. In: Kulkarni AJ, Gandomi AH (eds) *Handbook of formal optimization*. Springer, Singapore. https://doi.org/10.1007/978-981-19-8851-6_45-1
13. Farahmand-Tabar S. (2023). Frequency-based optimization of truss dome structures using ant colony optimization (ACOR) with multi-trail pheromone memory. In: Dey

N (ed) *Applications of ant colony optimization and its variants*. Springer, Singapore. https://doi.org/10.1007/978-981-99-7227-2_11

14. Alkan N, Kahraman C. (2021). Fuzzy Metaheuristics: A State-of-the-Art Review. In: Kahraman, C, Cevik Onar S, Oztaysi B, Sari I, Cebi S, Tolga A (eds) *Intelligent and Fuzzy Techniques: Smart and Innovative Solutions. INFUS 2020. Advances in Intelligent Systems and Computing,* 1197. Springer, Cham. https://doi.org/10.1007/978-3-030-51156-2_168
15. Cuevas E, Zaldívar D, Pérez-Cisneros M. (2018). Metaheuristic Algorithms Based on Fuzzy Logic. In: *Advances in Metaheuristics Algorithms: Methods and Applications. Studies in Computational Intelligence*, 775. Springer, Cham. https://doi.org/10.1007/978-3-319-89309-9_8
16. Yalaoui N, Ouazene Y, Yalaoui F, Amodeo L, Mahdi H. (2013). Fuzzy-metaheuristic methods to solve a hybrid flow shop scheduling problem with pre-assignment. *International Journal of Production Research* 51(12): 3609–3624. https://doi.org/10.1080/00207543.2012.754964
17. Arora A, Arabameri A, Pandey M, Siddiqui MA, Shukla UK, Bui DT, Mishra VN, Bhardwaj A. (2021). Optimization of state-of-the-art fuzzy-metaheuristic ANFIS-based machine learning models for flood susceptibility prediction mapping in the Middle Ganga Plain, India. *Science of the Total Environment* 750: 141565. https://doi.org/10.1016/j.scitotenv.2020.141565
18. Patel, HR. (2022). Fuzzy-based metaheuristic algorithm for optimization of fuzzy controller: fault-tolerant control application. *International Journal of Intelligent Computing and Cybernetics* 15(4): 599–624. https://doi.org/10.1108/IJICC-09-2021-0204
19. Soong T. (1988). State-of-the-art review: Active structural control in civil engineering. *Engineering Structures* 10(2): 74–84. https://doi.org/10.1016/0141-0296(88)90033-8
20. Soong TT, Spencer BF. (2002). Supplemental energy dissipation: State-of-the-art and state-of-the-practice. *Engineering Structures* 24(3): 243–259. https://doi.org/10.1016/S0141-0296(01)00092-X
21. Spencer Jr B, Nagarajaiah S. (2003). State of the art of structural control. *Journal of Structural Engineering* 129(7): 845–856. https://doi.org/10.1061/(ASCE)0733-9445(2003)129:7(845)
22. Hurlebaus S, Gaul L. (2006). Smart structure dynamics. *Mechanical Systems and Signal Processing* 20(2): 255–281. https://doi.org/10.1016/j.ymssp.2005.08.025
23. Younespour A, Ghaffarzadeh H. (2016). Semi-active control of seismically excited structures with variable orifice damper using block pulse functions. *Smart Structures and Systems* 18(6): 1111–1123. https://doi.org/10.1177/1077546313519285
24. Kumar JS, Paul PS, Raghunathan G, et al. (2019). A review of challenges and solutions in the preparation and use of magnetorheological fluids. *International Journal of Mechanical and Materials Engineering* 14(1): 1–18.
25. Zhang Y, Li D, Cui H, et al. (2020). A new modified model for the rheological properties of magnetorheological fluids based on different magnetic field. *Journal Magnetism and Magnetic Materials* 500: 166377.
26. Jolly MR, Bender JW, Carlson JD. (1999). Properties and applications of commercial magnetorheological fluids. *Journal Intelligent Material Systems and Structures* 10(1): 5–13.
27. Zhao D, Shi X, Liu S, Wang F. (2020). Theoretical and experimental investigation on wave propagation in the periodic impedance layered structure modulated by magnetorheological fluid. *Journal Intelligent Material Systems and Structures* 31(6): 882–896.

28. Li DD, Keogh DF, Huang K, Chan QN, Yuen ACY, Menictas C, Timchenko V, Yeoh GH. (2019). Modeling the response of magnetorheological fluid dampers under seismic conditions. *Applied Science* 9(19): 4189. https://doi.org/10.3390/app9194189
29. Yang J, Ning D, Sun SS, et al. (2021). A semi-active suspension using a magnetorheological damper with nonlinear negative-stiffness component. *Mechanical Systems and Signal Processing* 147: 107071.
30. Yoon DS, Kim GW, Choi SB. (2021). Response time of magnetorheological dampers to current inputs in a semi-active suspension system: Modeling, control and sensitivity analysis. *Mechanical Systems Signal Processing* 146: 106999. https://doi.org/10.1016/j.ymssp.2020.106999
31. Song X, Dong X, Yan M, X Li. (2020). Investigation of an automobile magnetorheological damper with asymmetric mechanical characteristics. *Journal Physics: Conference Series* 1678(1): 012012. https://doi.org/10.1088/1742-6596/1678/1/012012
32. Hu G, Liu Q, Ding R, Li G. (2017). Vibration control of semi-active suspension system with magnetorheological damper based on hyperbolic tangent model. *Advance in Mechanical Engineering* 9(5): 1687814017694581. https://doi.org/10.1177/1687814017694581
33. Luu M, Martinez-Rodrigo MD, Zabel V, Kénke C. (2014). Semi-active magnetorheological dampers for reducing response of high-speed railway bridges. *Control Engineering Practice* 32: 147–160.
34. Heo G, Kim C, Lee C. (2014). Experimental test of asymmetrical cable-stayed bridges using MR-damper for vibration control. *Soil Dynamics and Earthquake Engineering* 57: 78–85.
35. Deng ZC, Yao XL, Zhang DG. (2009). Research on the dynamic performance of ship isolator systems that use magnetorheological dampers. *Journal of Marine Science and Application* 8(4): 291–297.
36. Willey CL, Chen VW, Scalzi KJ, Buskohl PR, Juhl AT. (2020). A reconfigurable magnetorheological elastomer acoustic metamaterial. *Application Physics Letters* 117(10): 104102.
37. Bekdaş G, Nigdeli SM. (2011). Estimating optimum parameters of tuned mass dampers using harmony search. *Engineering Structures* 33: 2716–2723. https://doi.org/10.1016/j.engstruct.2011.05.024
38. Bekdaş G, Nigdeli SM. (2013). Response of discussion "Estimating optimum parameters of tuned mass dampers using harmony search. *Engineering Structures* 54: 265–267. https://doi.org/10.1016/j.engstruct.2013.08.015
39. Chey MH, Kim JU. (2012). Parametric control of structural responses using an optimal passive tuned mass damper under stationary Gaussian white noise excitations. *Frontiers of Structural and Civil Engineering* 6: 267–280. https://doi.org/10.1007/s11709-012-0170-x
40. Han B, Yan WT, Cu VH, Zhu L, Xie HB. (2019). H-TMD with hybrid control method for vibration control of long span cable-stayed bridge. *Earthquakes and Structures* 16: 349–358. https://doi.org/10.12989/eas.2019.16.3.349
41. Kaveh A, Mohammadi S, Hosseini OK, Keyhani A, Kalatjari V. (2015). Optimum parameters of tuned mass dampers for seismic applications using charged system search. *Iranian Journal of Science and Technology Transactions of Civil Engineering* 39: 21–40. https://doi.org/10.22099/IJSTC.2015.2739
42. Lee CL, Chen YT, Chung LL, Wang YP. (2006). Optimal design theories and applications of tuned mass dampers. *Engineering Structures* 28: 43–53. https://doi.org/10.1016/j.engstruct.2005.06.023

43. Li C. (2002). Optimum multiple tuned mass dampers for structures under the ground acceleration based on DDMF and ADMF. *Earthquake Engineering & Structural Dynamics* 31: 897–919. https://doi.org/10.1002/eqe.128
44. Li C, Liu Y. (2003). Optimum multiple tuned mass dampers for structures under the ground acceleration based on the uniform distribution of system parameters. *Earthquake Engineering & Structural Dynamics* 32: 671–690. https://doi.org/10.1002/eqe.239
45. Li C, Qu W. (2006). Optimum properties of multiple tuned mass dampers for reduction of translational and torsional response of structures subject to ground acceleration. *Engineering Structures* 28: 472–494. https://doi.org/10.1016/j.engstruct.2005.09.003
46. Azar BF, Veladi H, Talatahari S, Raeesi F. (2020). Optimal design of magnetorheological damper based on tuning Bouc-Wen Model parameters using hybrid algorithms. *KSCE Journal of Civil Engineering* 24: 867–878. https://doi.org/10.1007/s12205-020-0988-z
47. Hadidi A, Azar BF, Shirgir S. (2019). Reliability assessment of semi-active control of structures with MR damper. *Earthquakes and Structures* 17: 131–141. https://doi.org/10.12989/eas.2019.17.2.131
48. Farahmand-Tabar S, Shirgir S. (2023). Antlion-facing ant colony optimization in parameter identification of the MR damper as a semi-active control device. In: Dey N (eds) *Applications of ant colony optimization and its variants*. Springer, Singapore. https://doi.org/10.1007/978-981-99-7227-2_8
49. Kennedy J, Eberhart R. (1995). Particle Swarm Optimization. *Proceedings of the IEEE International Conference on Neural Networks* 4: 1942–1948. https://doi.org/10.1109/ICNN.1995.488968
50. Gad AG. (2022). Particle swarm optimization algorithm and its applications: A systematic review. *Archives of Computational Methods in Engineering* 29: 2531–2561. https://doi.org/10.1007/s11831-021-09694-4
51. Farahmand-Tabar S. (2024). Multi-objective Lichtenberg algorithm for the optimum design of truss structures. In: Dey, N. (eds) *Applied multi-objective optimization. Springer tracts in nature-inspired computing*. Springer, Singapore. https://doi.org/10.1007/978-981-97-0353-1_5
52. Farahmand-Tabar S, Afrasyabi P. (2024). Multi-modal routing in urban transportation network using multi-objective quantum particle swarm optimization. In: Dey, N. (eds) *Applied multi-objective optimization. Springer tracts in nature-inspired computing*. Springer, Singapore. https://doi.org/10.1007/978-981-97-0353-1_7
53. Farahmand-Tabar S, Shirgir S. (2024). Multi-objective adaptive guided differential evolution for passively controlled structures equipped with a tunned mass damper. In: Dey, N. (eds) *Applied multi-objective optimization. Springer tracts in nature-inspired computing*. Springer, Singapore. https://doi.org/10.1007/978-981-97-0353-1_3
54. Farahmand-Tabar S, Ashtari P. (2023). Bilinear Fuzzy Genetic algorithm and its application on the optimum design of steel structures with semi-rigid connections. In: Kulkarni AJ, Gandomi AH (eds) *Handbook of formal optimization*. Springer, Singapore. https://doi.org/10.1007/978-981-19-8851-6_36-1
55. Ashtari P, Karami R, Farahmand-Tabar S. (2021). Optimum geometrical pattern and design of real-size diagrid structures using accelerated fuzzy-genetic algorithm with bilinear membership function. *Applied Soft Computing* 110: 107646. https://doi.org/10.1016/j.asoc.2021.107646
56. Farahmand-Tabar S. (2023). Genetic algorithm and accelerating fuzzification for optimum sizing and topology design of real-size tall building systems. In: Dey N (ed)

Applied genetic algorithm and its variants. Springer tracts in nature-inspired computing. Springer, Singapore. https://doi.org/10.1007/978-981-99-3428-7_9

57. Yazdani H, Kwasnicka H, Ortiz-Arroyo D. (2011). Multiobjective particle swarm optimization using Fuzzy logic. In: Jędrzejowicz P, Nguyen NT, Hoang K (eds) *Computational collective intelligence. Technologies and applications. ICCCI 2011. Lecture notes in computer science*, vol 6922. Springer, Berlin, Heidelberg. https://doi.org/10.1007/978-3-642-23935-9_22

2 Multi-modal Routing in Urban Public Transportation with the Multi-objective Passing Vehicle Search Algorithm Incorporating Pandemic Protocols

Salar Farahmand-Tabar and Parastoo Afrasyabi

2.1 INTRODUCTION

Engineers generally aim to achieve their goals while minimizing computational resources and costs, encompassing diverse outcomes such as providing housing options, electrical appliances, and mechanical tools. Presently, two main optimization approaches serve these objectives: local optimizers, utilizing mathematical programming techniques and gradient data to explore nearby points from an initial position [1], and global optimizers, known as metaheuristics (MH), designed for expansive searches independent of continuous cost functions and variables [2–5]. These global optimizers find broad application across all domains, especially engineering [6–12], where they have the potential for performance enhancement and problem-solving across various types [13–20]. Among these challenges, the routing problem stands out for its intricate logistics and diverse constraints, making it particularly suited to benefit from the capabilities of these optimizers.

The declaration of the Covid-19 pandemic by the World Health Organization (WHO) on March 11, 2020, fundamentally altered daily life worldwide. This global crisis significantly impacted urban public transportation [21–24], as evidenced by a 40% decrease in usage in Hong Kong alone [25]. The repercussions extended beyond mere inconvenience, affecting air quality, traffic patterns, and viral transmission dynamics [26, 27]. Consequently, a tailored approach to managing urban transportation during the pandemic became imperative.

DOI: 10.1201/9781003601555-2

Metropolitan areas rely on diverse transportation modalities to facilitate daily commutes, with individuals making choices based on personal criteria such as congestion levels [28–30]. Since the onset of the pandemic, preferences shifted toward less crowded routes. Typically, urban transportation networks encompass metro, buses, taxis, and pedestrian infrastructure, each with its unique characteristics regarding congestion and interpersonal contact. The rise in private vehicle usage amid the pandemic exacerbated traffic congestion and environmental pollution [22].

Addressing the multi-modal routing [31–37] challenge posed by the pandemic necessitated innovative solutions. Conventional optimization methods proved inadequate due to the complexity and multi-objective (MO) nature of the problem [38]. MH algorithms emerged as effective tools for tackling such intricate problems. These algorithms, applicable to both continuous and discrete scenarios, offer multiple optimal solutions.

In this chapter, the MO Passing Vehicle Search Algorithm (MOPVS) is employed to optimize multi-modal routing in Tehran's transportation network, considering various objectives aligned with Covid-19 health protocols. The approach addresses critical objectives such as route length, interpersonal distance, traffic, surface contact, and ventilation within transportation modalities. The ongoing effort to mitigate the pandemic's impact on urban transportation systems is contributed to by integrating diverse objectives into the routing process. The results are obtained and compared using other algorithms.

2.2 RELATED WORK ON MULTI-MODAL ROUTING

The origins of the routing problem can be traced back to the inception of the traveling salesman problem (TSP) [39]. As cities grew, transportation networks became increasingly intricate, leading to the emergence of multi-modal transportation networks where paths involve a combination of different modes of transport [40]. This prompted extensive research into finding integrated multi-modal routes within urban transportation networks. Initially, the focus was on analyzing the multi-modal routing problem as a single-objective challenge, typically centered on path length. However, individuals have diverse and non-coextensive preferences when choosing a route, prompting a shift toward viewing the problem of multi-modal routing as a MO optimization issue [41].

Stochastic approaches are well-suited to addressing this problem due to its intrinsic nature as an optimization challenge. Gharib et al. successfully employed the Genetic Algorithm (GA) to solve the TSP problem [42]. Deng et al. [43] tackled the multi-modal routing problem using the GA, considering a variety of transport modes. Pahlavani et al. addressed the multi-modal routing problem with a focus on objectives such as travel time, cost, and modal transitions, utilizing the NSGA-II algorithm. Their study involved a multi-modal network integrated into Tehran's transportation system, encompassing walking, buses, metro, and taxis paths. Pahlavani et al. prioritized the resulting set of solutions through the TOPSIS method, ultimately demonstrating the suitability of NSGA-II for this problem [44].

Following the outbreak of the Covid-19 pandemic, researchers embarked on investigations and studies related to this issue. Patlins et al. detailed government actions aimed at reducing the impacts of Covid-19, categorizing them into three groups, that is, travelers, organizations, and carriers, while also identifying common mistakes in public transportation usage and proposing recommendations for mitigating Covid-19 effects [45]. Gutiérrez et al. delved into the consequences of the Covid-19 pandemic on public transportation [46], whereas Przybylowski et al. collected data on factors contributing to citizens' sense of safety when using public transportation during the pandemic through questionnaires in Gdańsk [23]. Additionally, Charoennapharat et al. investigated the effective factors of multi-modal transportation during the Covid-19 pandemic in Thailand [47]. Abreu et al. presented a qualitative analysis of changes in the New York transportation system during the Covid-19 pandemic, along with suggestions for future public transportation management [48]. While previous studies focused on the effects of Covid-19 on public transportation and recommendations for future pandemics, no practical approach to enhancing public transportation utilization by travelers was presented in the literature.

This research aims to examine precautionary measures and Covid-19 control efforts by governments and individuals, as well as the impact of Covid-19 on public transportation, suggesting public actions. In this study, we adopt a scenario where the situation remains immutable, necessitating an individual practical solution. Our proposed solution entails commuters choosing a multi-modal path with minimal Covid-19 infection risk. Amid the absence of a declaration from WHO indicating the end of Covid-19, and the possibility of future epidemics, preparations for managing this pandemic and future virus infections are crucial. The primary innovation of this study lies in presenting the optimal path for commuters with minimal risk of coronavirus infection.

2.3 PASSING VEHICLE SEARCH (PVS) ALGORITHM

In this section, the methodology of the Passing Vehicle Search (PVS) algorithm is addressed, specifically by focusing on the details of both the single-objective PVS and the MO PVS, highlighting their distinct approaches and use cases.

2.3.1 Single-Objective PVS

PVS represents an innovative single-objective population-based MH [49]. It functions as a global optimization algorithm, drawing inspiration from a specific natural phenomenon. Similar to other algorithms of its kind, PVS commences its search process with a randomly generated set of solutions, referred to conventionally as a population. It iteratively refines this population based on other explored solutions, ultimately converging to an optimized set of solutions. The algorithm emulates the overtaking behavior observed on two-lane highways, translating it into a simplified mathematical model. Noteworthy for its simplicity, efficiency, and lack of parameters, PVS adjusts its population according to three straightforward mathematical conditions related to the overtaking behavior of vehicles. The key factor in this overtaking process is ensuring a secure opportunity, influenced by various interconnected and intricate criteria such as the gap between vehicles, individual vehicle speed, road conditions, driver proficiency, traffic patterns, and weather.

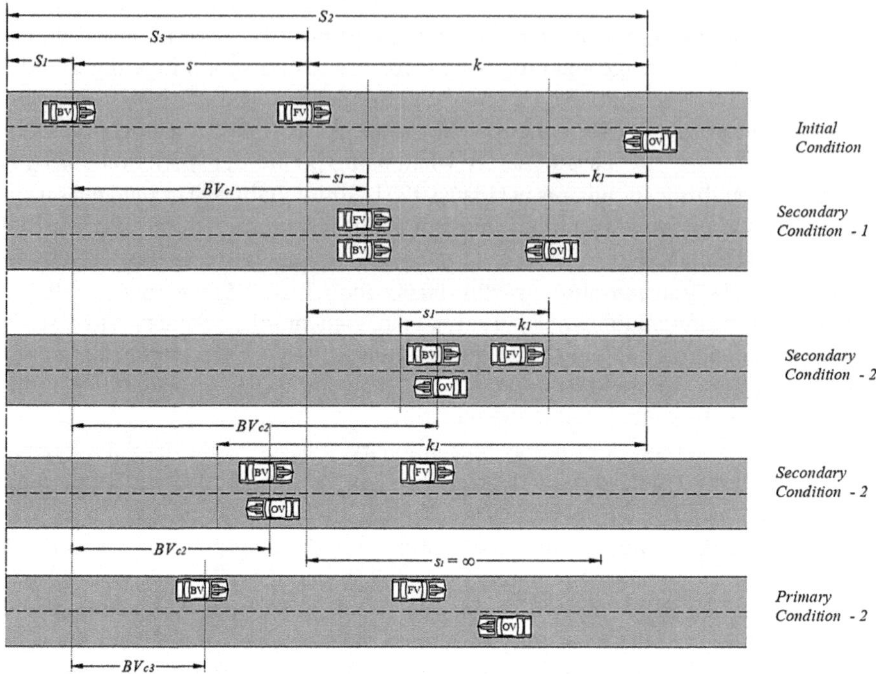

FIGURE 2.1 Flowchart of the fundamental PVS algorithm [49].

Three scenarios may unfold when a faster vehicle approaches a slower one: (a) the fast vehicle overtakes the slower one; (b) the vehicle follows the slow one until a suitable opportunity for overtaking arises; or (c) the vehicle continues to trail the slower one without the intention of overtaking.

In cases where a vehicle cannot overtake slower vehicles, a platoon emerges, impacting the vehicle's intended speed once more. This complicates the mathematical modeling of the overtaking process in two-lane traffic. Nevertheless, Savsani and Savsani [49] proposed a simplified model, depicted in Figure 2.1, to facilitate easier comprehension.

The PVS algorithm involves three distinct vehicles: the front vehicle (FV), back vehicle (BV), and oncoming vehicles (OV), all engaged in the overtaking mechanism on the highway. The overtaking process is initiated only when the velocity of BV surpasses that of FV; otherwise, overtaking is not feasible. This highlights the overtaking process's dependence on the speeds, positions of OV, and the distances between each vehicle, along with their respective velocities. This dependence gives rise to various conditions in the overtaking process on the two-lane highway, outlined as follows:

Let's consider the following variables:

s : Distance between BV and FV

k : Distance between FV and OV

S_1, S_2, S_3: Distances from the reference line
v_1, v_2, v_3: Velocities of BV, OV, and FV, respectively

At any given moment on the two-lane highway, the velocities of BV, OV, and FV are denoted by v_1, v_2, and v_3, respectively. Consequently, considering the velocity of FV, two distinct conditions arise: either FV is slower than BV, or vice versa. In the former scenario, the overtaking possibility arises, implying that BV can pass FV. In this scenario, passing is only feasible if the distance from FV at which overtaking occurs is shorter than the distance covered by OV. Conversely, if BV's speed is slower than that of FV, overtaking becomes impractical, and BV continues to move at its current speed. Thus, considering these three vehicles, the following conditions will be encountered:

Case 1. When BV is faster than FV ($v_3 < v_1$) (referred to as primary condition-1)

(a) $(k - k_1) > s_1$ (secondary condition-1)
(b) $(k - k_1) < s_1$ (secondary condition-2)

Case 2. BV is slower than FV ($v_3 > v_1$) (referred to as primary condition-2)

Primary condition-1: In this scenario, BV has a faster velocity than FV, leading to two sub-cases represented as secondary conditions 1 and 2. The detailed mathematical formulations are as follows.

Secondary condition-1: Assume that after FV covers a distance s_1, BV can catch up with and pass FV simultaneously, with the corresponding time denoted by 't' (illustrated in Figure 2.1). Hence, within the time interval 't', the distances traveled by FV and BV are given by, respectively:

$$s_1 = v_3 t, \ \ s_1 + s_1 = v_1 t \tag{1}$$

On substituting Eq. (1), the following result is obtained:

$$s_1 = \frac{v_3 s}{v_1 - v_3}, \quad t = \frac{s}{v_1 - v_3} \tag{2}$$

Within the same time interval 't', OV covered a distance of

$$k_1 = v_2 t \tag{3}$$

After substituting the value of 't' from Eq. (2) into Eq. (3), we obtain

$$y_1 = \frac{v_2 x}{v_1 - v_3} \tag{4}$$

The position alteration in the BV is

$$BV_{c1} = s + s_1 = s\left(\frac{v_1}{v_1 - v_3}\right) = \left(S_3 - S_1\right)\left(\frac{v_1}{v_1 - v_3}\right) \tag{5}$$

Thus, considering the reference line, the following equation is obtained:

$$s_1 + BV_{c1} = S_1 + \left(S_3 - S_1\right)\left(\frac{v_1}{v_1 - v_3}\right) \tag{6}$$

Secondary condition-2: This situation involves two distinct scenarios, one with a positive outcome and the other with a negative outcome, as depicted in Figure 2.1. If either of these possibilities exists, BV cannot overtake FV before OV has successfully passed BV. To prevent a potential motor vehicle accident, it is crucial that BV does not change lanes before OV completes its overtaking maneuver. The interaction between BV and OV occurs at a distance situated between the initial positions of BV and OV. The change in BV's position is then expressed as

$$BV_{c2} = rand\left(s + k\right) \tag{7}$$

where "rand" is a uniformly distributed random number, denoted by $rand \in [0,1]$. The adjustment in BV's position from the reference line is given by

$$s_1 + BV_{c2} = S_1 + rand\left(s + k\right) = S_1 + rand\left(S_2 - S_1\right) \tag{8}$$

Primary condition-2: According to Figure 2.1, BV cannot overtake FV when FV is faster than BV. Therefore, the alteration in BV's position can be expressed as

$$s_1 + BV_{c1} = S_1 + \left(S_3 - S_1\right)\left(\frac{v_1}{v_1 - v_3}\right) \tag{9}$$

In the PVS algorithm, various vehicles on a two-lane highway are analogously associated with distinct sets of solutions. The velocities of these vehicles are reflective of fitness values or an objective function, with the vehicle having the highest velocity considered the optimal fitness value. The position of a vehicle on the highway is akin to design variables. Consequently, PVS initiates its search process with a population of vehicles, representing a set of solutions.

During the reproduction phase, three solutions (vehicles) are randomly chosen. BV corresponds to the current solution among the three chosen vehicles, while FV and OV represent the other two solutions. The relative distances between vehicles and their corresponding velocities are assigned based on the population size and

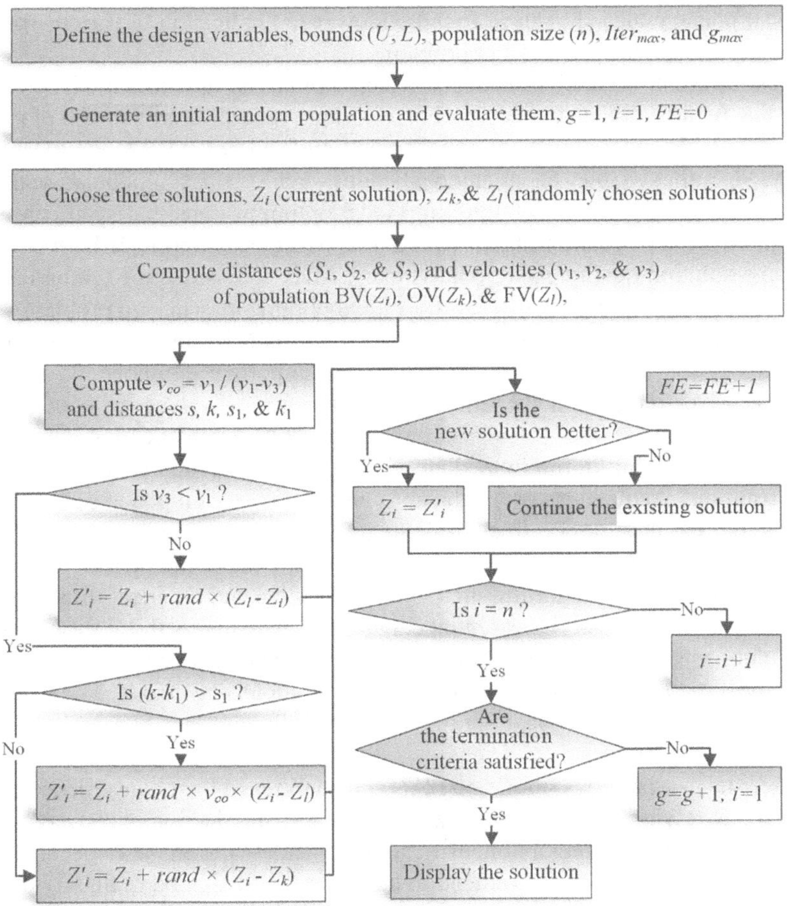

FIGURE 2.2 Flowchart of the fundamental PVS algorithm [49].

fitness values. Overtaking conditions are then examined following the allocation of distance and velocity. Consequently, vehicles adjust their positions on the highway in accordance with the applied conditions. A comprehensive explanation of the PVS algorithm is illustrated in Figure 2.2.

2.3.2 MO PVS

MO engineering design problems are more reflective of practical conditions, making them extensively applied and researched. However, these problems come with challenges such as nonlinearity, diverse objectives, multi-modal search domains, and strict bounds making it challenging for designers to discover optimal solutions. MHs have emerged as a promising alternative for tackling these complex MO design problems, proving to be robust and efficient algorithms that mimic biological and physical phenomena. They have found widespread application in various industrial, scientific, and engineering design problems.

MHs present several advantages when compared to traditional algorithms, including being gradient-free, adaptable to various variable types, devoid of problem-specific knowledge, requiring minimal setup time, and not mandating objective convexity. Nevertheless, they encounter challenges, such as a lack of population diversity, premature convergence, and falling into local optimum traps that need careful consideration. In large-scale problems, MHs may struggle in terms of both computing time and solution quality. Additionally, many MHs necessitate the tuning of control parameters, making them inefficient for global optimization design problems.

A critical factor influencing the effectiveness of MHs in discovering optimum solutions is the delicate balance between two important components: intensification and diversification. Diversification (exploration) involves generating diverse solutions to explore the global search space, while intensification (exploitation) focuses on searching in the immediate vicinity by exploiting current knowledge of good solutions. Achieving a proper balance between these components is essential for finding the best feasible solution within an acceptable computational time. However, there is no predefined set of conditions for balancing these parameters, necessitating the development of an algorithm that is both effective and practical. In general, efficient MHs must possess two essential capabilities. The first is the ability to generate new solutions with a higher likelihood of replacing existing solutions and the capacity to explore crucial areas where global optimum solutions may reside. The second capability is to escape from local optima solutions, allowing the MH to avoid getting trapped in local modes.

Nevertheless, previous studies on MHs and analyses of their convergence behavior suggest that relying solely on diversification leads to a reduction in convergence rate, while exclusively focusing on intensification enhances convergence speed. However, an excessive emphasis on diversification increases the likelihood of finding the global optimal solution, but decreases algorithm efficiency. Conversely, excessive intensification tends to trap the algorithm in local optima. Therefore, effective MHs need to strike a balance between a suitable level of local intensification and an appropriate magnitude of global diversification.

In MO problems, a set of optimal solutions known as the Pareto-optimal set is sought, as opposed to a single solution in single-objective design problems. Without prior knowledge, it is challenging for designers to determine the best solution among them. Due to the inherent complexity of MO design issues, there is a constant demand for high-performance algorithms. Moreover, the No Free Lunch theorem posits that no MH can proficiently solve all design issues. This implies that an algorithm effective in solving one specific design problem may not perform well in another type of optimization problem. Therefore, there is a requirement for a successful MH capable of addressing MO design issues and possessing the potential to find a global or near-optimal solution with high accuracy.

To address the aforementioned challenges, we propose a population-based algorithm called MOPVS for solving MO structural optimization problems [50]. The algorithm integrates the dominance approach into PVS to create an MO version, MOPVS. Dominance is defined such that if design solutions X_1 and X_2 result in function vectors

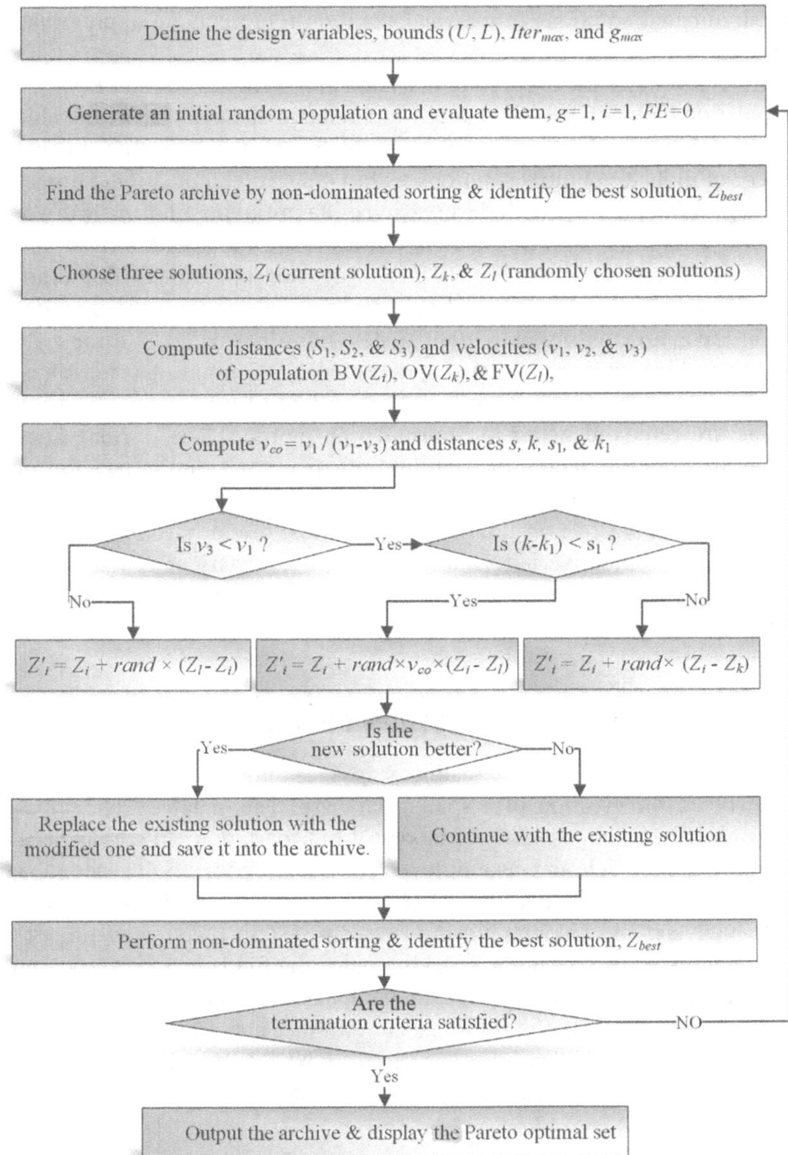

FIGURE 2.3 Flowchart of MOPVS [50].

f_1 and f_2 respectively, X_1 dominates X_2 (for minimization) if i) all elements in f_1 are less than or equal to their corresponding elements in f_2, and ii) at least one element of f_1 is strictly less than its corresponding element in f_2. In the context of this definition, non-dominated solutions within a population are those not dominated by any other solution in the set. MOMHs operate by iteratively reproducing a population, identifying non-dominated solutions, and storing them in a Pareto archive. The set of non-dominated solutions at the final iteration represents an approximate Pareto optimal set.

MOPVS leverages the overtaking mechanism of vehicles on the freeway to update the population based on vehicle velocity and relative distances between vehicles. The algorithm exhibits high efficiency, with a notable convergence rate, facilitating the discovery of globally optimal solutions within reduced computational time.

The algorithm initiates with the creation of an initial population through random function evaluations. Non-dominated solutions are identified, sorted, and stored in the initial Pareto archive. The PVS reproduction operator is then employed to generate a new set of offspring. These offspring, along with the members in the Pareto archive, are combined to identify new non-dominated solutions. The population and archive undergo iterative updates until a termination criterion is met. To manage the archive size, if it exceeds a predefined limit, certain non-dominated solutions are removed using the normal line method. The normal line method serves as an archiving technique to filter out some non-dominated solutions, optimizing computer memory usage. This step is crucial as MOMHs often explore an extensive number of non-dominated solutions, potentially leading to memory constraints. The objective is to maintain a high diversity of remaining solutions while screening out some to conserve memory. The detailed process of MOPVS is outlined in Figure 2.3.

2.4 STATE OF THE OPTIMIZATION PROBLEM

This study was executed in three distinct stages, as outlined in Figure 2.4. The initial phase involved the creation of a spatial database. This database encompasses five distinct objective functions and four modes of transportation. The urban public transportation system of Tehran was employed for this purpose. In the second stage, the multi-modal problem was investigated through the utilization of a multi-modal graph, network analysis, and Geographic Information System (GIS) modeling. The issue was then addressed using MOPVS. Additionally, the fundamental GA was employed for result evaluation. Finally, the results were assessed according to factors such as convergence during runs, algorithm execution time, and a comparison between the path achieved by MOPVS and that derived from NSGA-II and MOPSO. The refined path was also visualized to enhance comprehension.

2.4.1 Considered Area for the Study

The focal area of investigation encompasses the Tehran metropolis, which serves as both the capital and the most densely populated city of Iran. Geographically, Tehran spans 751 km^2 and is situated between 51° 6' to 51° 38' east longitude and 35° 34' to 35° 51' north latitude. The population of Tehran surpasses 8.6 million individuals.

The city has the most intricate and robust transportation network in Iran, which may result in unprecedented challenges, increased traffic levels, and a series of associated issues. Refer to Figure 2.5 for an illustration of the study area's layout.

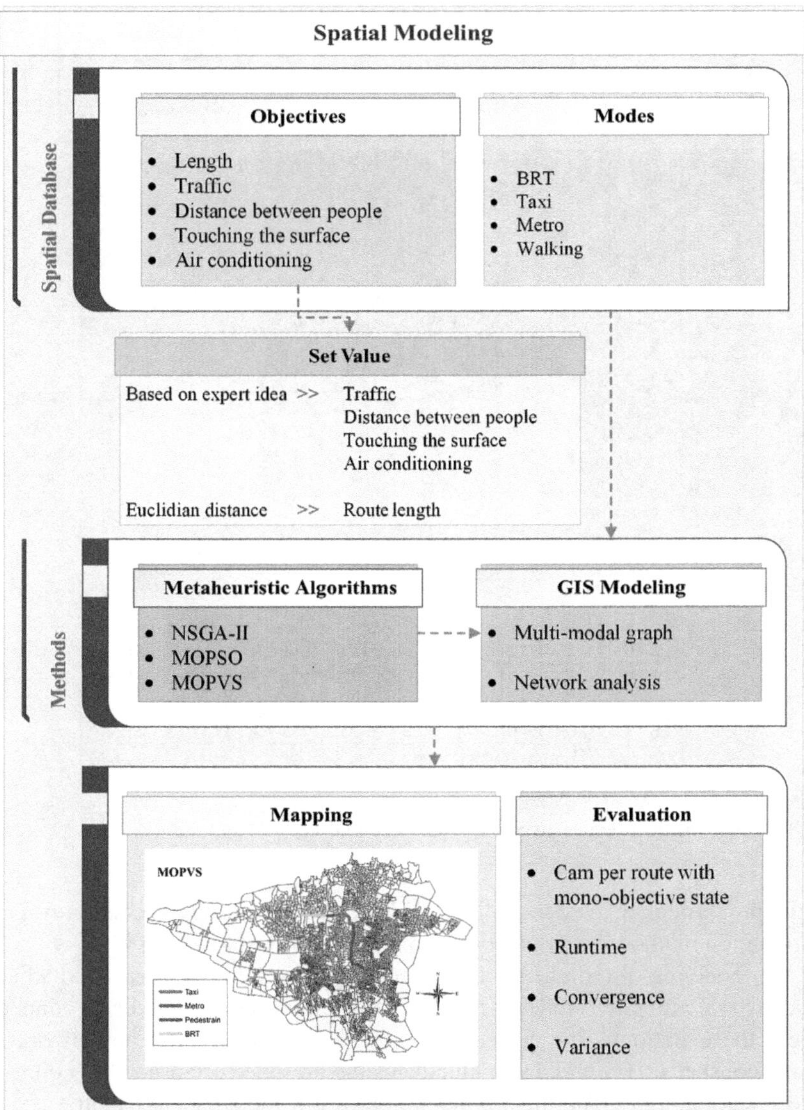

FIGURE 2.4 Framework of the routing process.

2.4.2 Multi-Modal Routing

The concept of a multi-modal network involves the integration of various travel modes, such as taxi, metro, and walking, in a combined manner for network analysis [23]. Modeling transportation systems, particularly when dealing with an increased number of travel modes, can be challenging. A Multi-modal Transportation System encompasses a blend of all possible travel states and transportation systems,

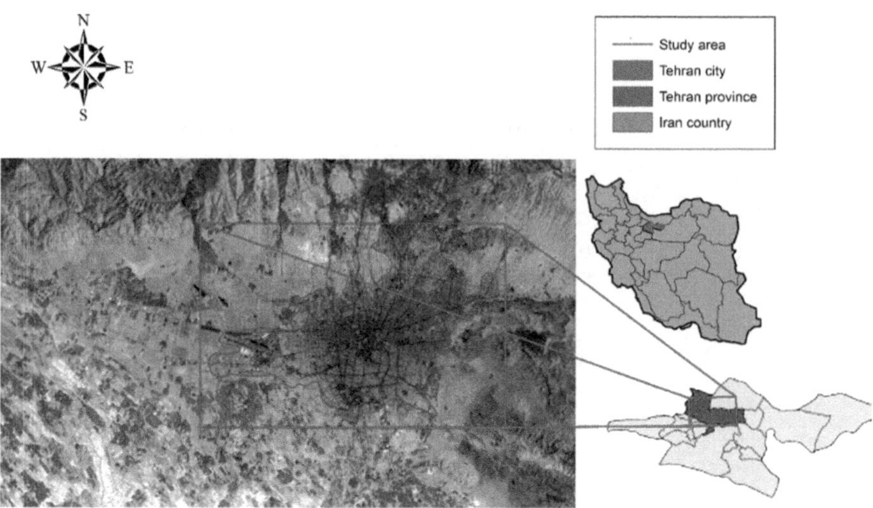

FIGURE 2.5 Adjacency matrix for multi-modal network modeling.

$$\begin{bmatrix} 0 & 1 & 1 & 0 & 1 & 1 \\ 1 & 0 & 1 & 1 & 1 & 1 \\ 1 & 1 & 0 & 0 & 1 & 1 \\ 0 & 1 & 0 & 0 & 1 & 0 \\ 1 & 1 & 1 & 1 & 0 & 1 \\ 1 & 1 & 1 & 0 & 1 & 0 \end{bmatrix} \quad \begin{bmatrix} 0 & 1 & 2 & 0 & 1 & 3 \\ 1 & 0 & 1 & 4 & 3 & 1 \\ 2 & 1 & 0 & 0 & 4 & 1 \\ 0 & 4 & 0 & 0 & 2 & 0 \\ 1 & 3 & 4 & 2 & 0 & 1 \\ 3 & 1 & 1 & 0 & 1 & 0 \end{bmatrix}$$

(a) (b)

FIGURE 2.6 Multi-modal matrix: (a) adjacency matrix, (b) label matrix.

functioning through diverse modalities. This implies that commuters employ a mix of transportation methods simultaneously to reach their destinations [35].

In the modeling of a multi-modal network, graphs, label matrices, and adjacency matrices come into play. Nodes symbolize stations, while edges indicate connections between these stations. The adjacency matrix is congruent with the network structure and consists of 0 s and 1 s. It elucidates the presence or absence of connections between stations [23]. The main diagonal entries of the adjacency matrix are set to zero, reflecting the absence of self-loops. The label matrix aligns with the network and allocates modes to edges, essentially assigning labels to connections. In Figure 2.6, a demonstration of both adjacency and label matrices is provided. In Figure 2.6(a), " 1 " signifies the existence of a connection between two nodes, whereas " 0 " denotes the absence of a connection. In Figure 2.6(b), each number designates the type of travel structure between two nodes (stations). For instance, in the second entry, the number " 1 " specifies the travel structure type between stations numbered 1 and 2 .

Figure 2.7 illustrates a representative example of a multi-modal route. Within this route, four modes of transportation are incorporated: taxis, metro, buses, and walking. Walking is utilized as a mode of transition between the other modes. The route comprises ten nodes and nine edges.

FIGURE 2.7 Example of a multi-modal route.

TABLE 2.1
Number of stations in each mode

Mode	Edge	Vertex
Metro	93	90
Taxi	82	83
Bus	237	235
Total	412	408

In the context of public transportation in Tehran, the principal components contain walking, buses, taxis, and metro. Given this emphasis, the study concentrated on these four modes. In the analyzed network, a total of 408 stations were employed, as outlined in Table 2.1. Notably, the walking mode is characterized by its absence of dedicated stations and its purpose of facilitating transfers between different modes of transportation.

GIS has proven to be highly effective in handling spatial data and is applied across a wide range of domains, including transportation systems. Specifically, GIS in Transportation (GIS-T) leverages GIS functionalities to address transportation-related challenges. Beyond the conventional GIS tools, like buffers and overlays, GIS-T is particularly adept at resolving transportation concerns. Its key strengths lie in tasks such as shortest path analysis, network flow modeling, and urban transportation analysis [39]. Notably, network analysis within GIS serves as a robust approach for modeling multi-modal route problems, which has been precisely employed in this research.

2.4.3. Modeling

The multi-modal problem explored in this research encompasses multiple objectives, including individual distance (empty space), path length, surface contact, air conditioning, and traffic. Furthermore, a constraint is imposed on the maximum allowable stay time in each mode, limited to 15 minutes. The multi-modal network encompasses metro, buses, pedestrian pathways, and taxis. The initial step involves determining the method for defining responses to model the problem using the MOPVS.

A multi-modal path is constructed based on a sequence of stations and the corresponding mode type. Each solution of the algorithm represents a potential route. Solutions within a multi-modal problem consist of station vertices and mode types. Specifically, odd-numbered variables correspond to stations, whereas even-numbered variables indicate the mode type between the adjacent variables.

2.4.3.1 Objective Function

Individuals have distinct criteria when selecting a route. For example, during virus pandemics, many prioritize paths that minimize the risk of health-related issues. Therefore, factors like individual distance (spacing), surface contact, air conditioning, and time spent in each mode are influential in virus transmission within public transportation.

The objectives of this study encompass path length, individual distance (spacing), surface contact, air conditioning, and traffic. Furthermore, the algorithm incorporates a constraint on modal durations, limiting them to 15 minutes. The optimal route, in this context, entails a shorter path length, reduced traffic, and minimized surface contact. Moreover, factors like air conditioning and spacing between individuals are taken into account. Path length is computed independently of mode type through the equation of the Euclidean distance or

$$Fitness\ Function = Min \sum D_{i,j}$$

$$D_{i,j} = \sqrt{\left(x_j, x_i\right)^2 + \left(y_j, y_i\right)^2} \tag{10}$$

Four distinct objective functions are under consideration: individual distance (spacing), surface contact, air conditioning, and traffic. These functions are ranked based on professional judgment. As the optimization problem involves minimization goals, the highest-priority objective function is assigned a value of 1, whereas the lowest-priority function is assigned a value of 4.

Considering Table 2.2, we observed that the traffic function has the lowest value within the metro mode, indicating higher priority. For the other three objective functions, the metro mode is assigned a lower priority. Conversely, the walking mode has the highest priority for three functions: individual distance, surface contact, and air conditioning. In simpler terms, the walking mode features the least contact surface and the most effective air conditioning.

TABLE 2.2
Goal prioritization according to the modal type

Objective mode	Traffic	Surface contact	Air conditioning	Individual distance
Bus	3	3	2	3
Metro	1	4	4	4
Taxi	4	2	2	2
Walking	2	1	1	1

Route 1	1	4	197	3	14	1	196	4	183	3	200
Route 2	1	3	125	3	79	3	184	2	200		
Route 3	1	3	98	2	161	3	104	2	200		

a) Stations and edge modes achieved by NSGA-II

Route 1	1	4	152	2	43	2	144	3	42	1	77	4	121	4	200
Route 2	1	4	82	2	163	2	91	3	102	1	89	4	162	4	200
Route 3	1	4	152	2	43	2	144	3	102	1	89	4	162	4	200

b) Stations and edge modes achieved by MOPSO

Route 1	1	4	15	3	198	3	15	2	200
Route 2	1	4	15	3	198	3	118	2	200
Route 3	1	4	15	3	89	3	15	2	200

c) Stations and edge modes achieved by MOPVS

FIGURE 2.8 Solutions of the algorithms.

2.5 RESULTS AND DISCUSSION OF THE CASE STUDIES

This study focused on multi-modal routing guided by Covid-19 protocols, envisioning a scenario where existing public transportation is managed by individuals. This implies navigating a situation devoid of established policies or plans for Covid-19 management in urban public transportation. Under such circumstances, passengers are tasked with identifying routes with the lowest risk of coronavirus transmission while maintaining efficiency.

In this section, the values of the objective function, convergence trends, and execution times of the algorithms are compared. The runtimes of NSGA-II, MOPSO, and MOPVS, were 1.8590, 1.7080, and 1.6916 s in 100 iterations, respectively. The solutions of the algorithms are shown in Figure 2.8. Odd cells represent the stations and even cells indicate the edge mode type ("1" = bus mode, "2" = metro mode, "3" = pedestrian mode, and "4" = taxi mode). Figure 2.9 shows the route found by algorithms; the green line is related to the bus mode, the blue line to the pedestrian mode, and the red/pink lines to the metro and taxi modes on the track, respectively. According to Figure 2.9, the route found by MOPVS is better because the route length is shorter and also more efficient.

The value of the fitness function considering each objective is shown in Figure 2.10 based on the best response of each iteration. The value of the fitness function for different methods is presented in Table 2.3. According to the achieved results, the convergence of the MOPVS is better than those of MOPSO and NSGA-II. This means that MOPVS is able to achieve optimal values faster.

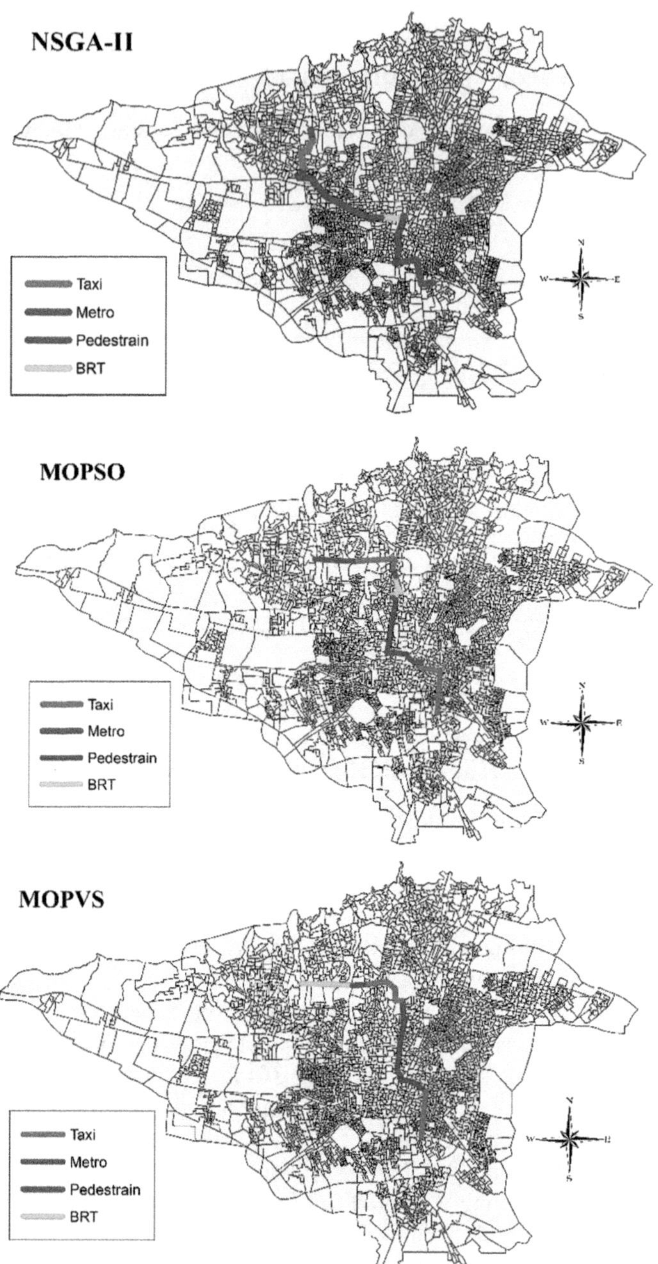

FIGURE 2.9 Multi-modal route found from the MOPVS and NSGA-II algorithms.

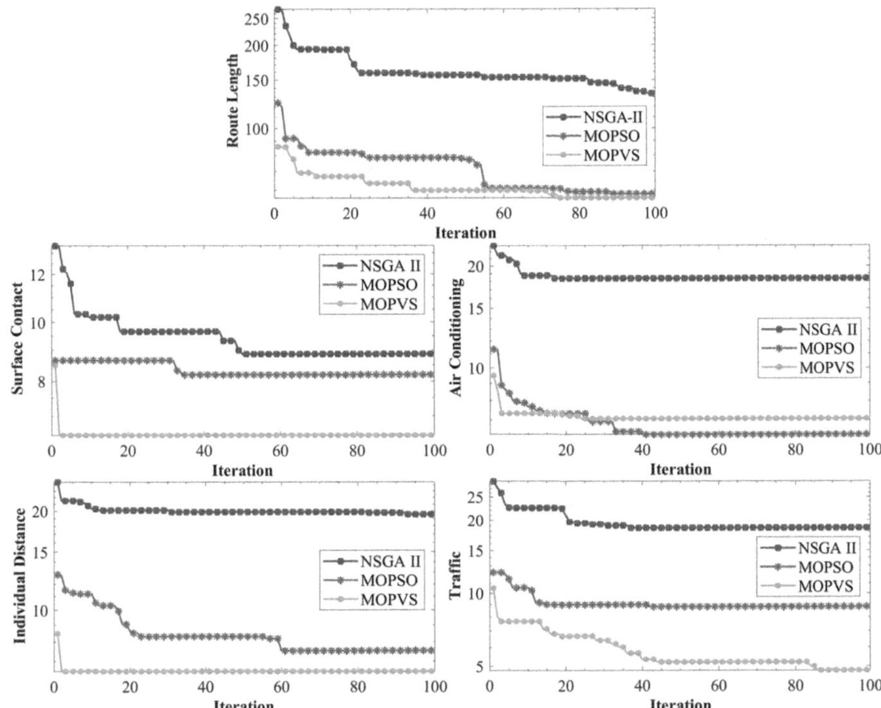

FIGURE 2.10 Convergence of different objectives considering various methods.

TABLE 2.3
Values of different objectives

Methods/ Objective	Route	Length	Traffic	Surface contact	Air conditioning	Individual distance	Runtime
NSGA-II	1	93.10	9	5	16	9	1.8590
	2	85.24	10	7	6	10	
	3	49.32	8	7	7	5	
MOPSO	1	108.50	11	11.5	12	11	1.7080
	2	184.23	12	12	12	12	
	3	200	13	18	14	9	
MOPVS	1	63.5	6	8	7	4	1.6916
	2	91.83	8	6	5	5	
	3	**70.41**	**3**	**5**	**8**	**3**	

Efficiency entails selecting routes with minimal traffic and distance, enabling passengers to reach their destinations swiftly and safely from the virus. Existing studies on Covid-19 and urban public transportation have primarily focused on examining the virus's effects, transmission modes, and operational dynamics. While some studies have proposed solutions for public transportation management during the pandemic, these efforts typically hinge on governmental resource allocation and policy adjustments. By contrast, this study offers a solution tailored to immutable existing conditions. Results indicate the merit of MOPVS in solving this multi-modal problem in comparison with other methods.

2.6 CONCLUSIONS

Due to the sudden outbreak of the Covid-19 pandemic, societies were unprepared, leading to rapid transmission of the virus and negative consequences. Urban planning was challenged in controlling the spread of the virus, and public transportation played a key role in facilitating its transmission. As a response, this chapter suggested a route selection strategy that adheres to health protocols to mitigate the spread of the virus.

Routing in the transport network is a complicated problem, where complexity can be due to, for example, the existence of different combinations of modes and the expansion of the solution space. In this chapter, the multi-modal network was based on the assumption of the bus, metro, and walking modes. A maximum of five mode changes were added, meaning the route that took more than five-mode changes was eliminated. In this study, the routing problem was implemented in the multi-modal transportation network using MOPVS. The objective function is based on the minimization of trajectory length. The simulated transport network consisted of 200 stations in an area of 1600 m^2. To evaluate the proposed algorithm, the convergence criteria, value of the objective function, optimal routes, and algorithms' execution time were compared.

The proposed algorithm found more efficient routes than NSGA-II and MOPSO. The findings indicate that MO routes conforming to health protocols, as determined by MOPVS, are rational and acceptable. These solutions hold promise for aiding citizens in utilizing urban public transportation efficiently amid future pandemics that may emerge globally. In such scenarios, objective functions should be tailored to adhere to the health protocols dictated by the virus.

In future research, other methods can also be utilized to solve the multi-modal routing problem. Other modes such as bicycles and scooters can also be considered in the multi-modal transportation network.

REFERENCES

1. Farahmand-Tabar S., Abdollahi F., Fatemi M. (2023). Robust Conjugate Gradient Methods for Non-Smooth Convex Optimization and Image Pro-cessing Problems. In: Kulkarni, A.J., Gandomi, A.H. (eds) *Handbook of For-mal Optimization*. Springer, Singapore.
2. Farahmand-Tabar S., Ashtari P. (2020). Simultaneous size and topology optimization of 3D outrigger-braced tall buildings with inclined belt truss us-ing genetic algorithm.

The Structural Design of Tall and Special Buildings, 29(13), e1776. https://doi.org/10.1002/tal.1776
3. Farahmand-Tabar S., Rashid T.A. (2023). Steel Plate Fault Detection using the Fitness Dependent Optimizer and Neural Networks. In: Kulkarni, A.J., Gandomi, A.H. (eds) *Handbook of Formal Optimization*. Springer, Singapore.
4. Farahmand-Tabar S., Ashtari P. (2023). Bilinear Fuzzy Genetic algorithm and its Application on the Optimum Design of Steel Structures with Semi-rigid Connections. In: Kulkarni, A.J., Gandomi, A.H. (eds) *Handbook of Formal Optimization*. Springer, Singapore.
5. Farahmand-Tabar S., Ashtari P. (2024). Intelligent cross-entropy optimizer: A novel machine learning-based meta-heuristic for global optimization. *Swarm and Evolutionary Computation*, 91, 101739. https://doi.org/10.1016/j.swevo.2024.101739
6. Farahmand-Tabar S., Ashtari P., Babaei M. (2024). Gaussian cross-entropy and organizing intelligence for design optimization of the outrigger system with inclined belt truss in real-size tall buildings. *Probabilistic Engineering Mechanics*, 76, 103616. https://doi.org/10.1016/j.probengmech.2024.103616
7. Farahmand-Tabar S. (2023). Memory-Driven Metaheuristics: Improving Optimization Performance. In: Kulkarni, A.J., Gandomi, A.H. (eds) *Handbook of Formal Optimization*. Springer, Singapore.
8. Farahmand-Tabar S., Ashtari P., Babaei M. (2023). Dynamic Intelligence of Self-Organized Map in the Frequency-Based Optimum Design of Structures. In: Kulkarni, A.J., Gandomi, A.H. (eds) *Handbook of Formal Optimization*. Springer, Singapore.
9. Farahmand-Tabar S., Shirgir S. (2023). Synergistic Collaboration of Motion-Based Metaheuristics for the Strength Prediction of Cement-Based Mortar Materials using TSK Model. In: Kulkarni, A.J., Gandomi, A.H. (eds) *Hand-book of Formal Optimization*. Springer, Singapore.
10. Farahmand-Tabar S., Sadrekarimi N. (2023). Overcoming Constraints: The Critical Role of Penalty Functions as Constraint Handling Methods in Struc-tural Optimization. In: Kulkarni, A.J., Gandomi, A.H. (eds) *Handbook of Formal Optimization*. Springer, Singapore.
11. Farahmand-Tabar S. (2023). Genetic Algorithm and Accelerating Fuzzifica-tion for Optimum Sizing and Topology Design of Real-Size Tall Building Systems. In: Dey, N. (eds) *Applied Genetic Algorithm and Its Variants. Springer Tracts in Nature-Inspired Computing*. Springer, Singapore. https://doi.org/10.1007/978-981-99-3428-7_9
12. Ashtari P., Karami R., Farahmand-Tabar S. (2021). Optimum geometrical pattern and design of real-size diagrid structures using accelerated fuzzy-genetic algorithm with bilinear membership function. *Applied Soft Computing*, 110, 107646. https://doi.org/10.1016/j.asoc.2021.107646
13. Farahmand-Tabar S., Shirgir S. (2023). Boosting the Efficiency of Metaheu-ristics through Opposition-Based Learning in Optimum Locating of Control Systems in Tall Buildings. In: Kulkarni, A.J., Gandomi, A.H. (eds) *Handbook of Formal Optimization*. Springer, Singapore.
14. Farahmand-Tabar S., Shirgir S. (2023). Positron-Enabled Atomic Orbital Search Algorithm for Improved Reliability-Based Design Optimization. In: Kulkarni, A.J., Gandomi, A.H. (eds) *Handbook of Formal Optimization*. Springer, Singapore.
15. Farahmand-Tabar S., Babaei M. (2023). Memory-assisted adaptive multiverse optimizer and its application in structural shape and size optimization. *Soft Computing*, 27, 11505–11527. https://doi.org/10.1007/s00500-023-08349-9

16. Farahmand-Tabar S., Shirgir S. (2023). Incorporating Nelder Mead Simplex as an Accelerating Operator to Improve the Performance of Metaheuristics in Nonlinear System Identification. In: Kulkarni, A.J., Gandomi, A.H. (eds) *Handbook of Formal Optimization*. Springer, Singapore.
17. Farahmand-Tabar S., Shirgir S. (2023). Opposed Pheromone Ant Colony Optimization for Property Identification of Nonlinear Structures. In: Dey N. (ed) *Applications of Ant Colony Optimization and its Variants*. Springer.
18. Farahmand-Tabar S., Shirgir S. (2023). Antlion-Facing Ant Colony Optimization in Parameter Identification of the MR Damper as a Semi-Active Control Device. In: Dey N. (ed) *Applications of Ant Colony Optimization and its Variants*. Springer.
19. Farahmand-Tabar S., Shirgir S. (2023). Frequency-Based Optimization of Truss Dome Structures using Ant Colony Optimization (ACOR) with Multi-Trail Pheromone Memory. In: Dey N. (ed) *Applications of Ant Colony Optimization and its Variants*. Springer.
20. Shirgir S., Farahmand-Tabar S., Aghabeigi P. (2023). Optimum design of real-size reinforced concrete bridge via charged system search algorithm trained by nelder-mead simplex. *Expert Systems with Applications*, 238, 121815. https://doi.org/10.1016/j.eswa.2023.121815
21. Shelat S., Cats O., van Cranenburgh S. (2022). Traveller behaviour in public transport in the early stages of the COVID-19 pandemic in the Netherlands. *Transportation Research Part A: Policy and Practice*, 159, 357–371.
22. Simić V., Ivanović I., Đorić V., Torkayesh, A.E. (2022). Adapting urban transport planning to the COVID-19 pandemic: An integrated fermatean fuzzy model. *Sustainable Cities and Society*, 79, 103669.
23. Przybylowski A., Stelmak S., Suchanek M. (2021). Mobility behaviour in view of the impact of the COVID-19 pandemic-public transport users in gdansk case study. *Sustainability (Switzerland)*, 13(1), 1–12. https://doi.org/10.3390/su13010364
24. Kim K.E. (2022). Ten takeaways from the COVID-19 pandemic for transportation planners. *Transportation Research Record*, 03611981221090515.
25. Kwok K.O., Li K.K., Chan H.H.H., Yi Y.Y., Tang A., Wei W.I., Wong S.Y.S. (2020). Community responses during early phase of COVID-19 epidemic, Hong Kong. *Emerging Infectious Diseases*, 26(7), 1575.
26. Abdullah M., Ali N., Hussain S.A., Aslam A.B., Javid, MA. (2021). Measuring changes in travel behavior pattern due to COVID-19 in a developing country: A case study of Pakistan. *Transport Policy*, 108, 21–33.
27. Mogaji E., Adekunle I., Aririguzoh S., Oginni A. (2022). Dealing with impact of COVID-19 on transportation in a developing country: Insights and policy recommendations. *Transport Policy*, 116, 304–314.
28. Faroqi H., Saadi Mesgari M. (2016). Performance comparison between the multi-colony and multi-pheromone ACO algorithms for solving the multi-objective routing problem in a public transportation network. *The Journal of Navigation*, 69(1), 197–210. https://doi.org/10.1017/S0373463315000594
29. Litman T. (2017). *Evaluating Transportation Equity*. Victoria Transport Policy Institute.
30. Brinchi S., Carrese S., Cipriani E., Colombaroni C., Crisalli U., Fusco G., Gemma A., Isaenko N., Mannini L., Patella S.M. (2020). On transport monitoring and forecasting during COVID-19 pandemic in Rome. *Transport and Telecommunication*, 21(4), 275–284.

31. Fazayeli S., Eydi A., Kamalabadi I.N. (2018a). Location-routing problem in multi-modal transportation network with time windows and fuzzy demands: Presenting a two-part genetic algorithm. *Computers and Industrial Engineering*, 119, 233–246. https://doi.org/10.1016/j.cie.2018.03.041
32. Fazayeli S., Eydi A., Kamalabadi I.N. (2018b). Location-routing problem in multimodal transportation network with time windows and fuzzy demands: Presenting a two-part genetic algorithm. *Computers & Industrial Engineering*, 119, 233–246.
33. Huang X., Chen Y. (2015). An optimization model for multimodal transportation decisions based on congestion information. *Proceedings of the World Congress on Intelligent Control and Automation (WCICA), 2015-March (March)*, 610–615. https://doi.org/10.1109/WCICA.2014.7052784
34. Idri A., Oukarfi M., Boulmakoul A., Zeitouni K., Masri A. (2017). A new time-dependent shortest path algorithm for multimodal transportation network. *Procedia Computer Science*, 109, 692–697.
35. Kai K., Haijiao N., Yuejie Z., Weicun, Z. (2009). Improved integrated optimization Model research of mode and route in multimodal transportation. *2009 International Conference on Information Management, Innovation Management and Industrial Engineering*, vol. 1, 621–624.
36. Liu Y., Chen J., Wu W., Ye J. (2019). Typical combined travel mode choice utility model in multimodal transportation network. *Sustainability*, 11(2), 549.
37. Yang X., Ban X.J., Mitchell J. (2018). Modeling multimodal transportation network emergency evacuation considering evacuees' cooperative behavior. *Transportation Research Part A: Policy and Practice*, 114, 380–397.
38. Pahlavani P., Ghaderi, F. (2017). Multimodal multi-objective route planning using non-dominated sorting genetic algorithm-II and TOPSIS method. *Engineering Journal of Geospatial Information Technology*, 4(4), 123–142.
39. Gutin G., Yeo A. (2001). TSP tour domination and Hamilton cycle decompositions of regular digraphs. *Operations Research Letters*, 28(3), 107–111.
40. Pajor T. (2009). Multi-Modal Route Planning. 171.
41. Caramia M., Dell'Olmo, P. (2020). Multi-Objective Optimization. In *Multi-Objective Management in Freight Logistics* (pp. 21–51). Springer.
42. Gharib A., Benhra J., Chaouqi, M. (2015). A performance comparison of PSO and GA applied to TSP. *International Journal of Computer Applications*, 975, 8887.
43. Deng Y., Hu S. (2011). Route optimization of multi-modal travel based on improved genetic algorithm. *Proceedings 2011 International Conference on Transportation, Mechanical, and Electrical Engineering (TMEE)*, 1701–1704.
44. Chen S., Tan J., Claramunt C., Ray C. (2011). Multi-scale and multi-modal GIS-T data model. *Journal of Transport Geography*, 19(1), 147–161. https://doi.org/10.1016/j.jtrangeo.2009.09.006
45. Patlins A. (2021). Adapting the Public Transport System to the COVID-19 challenge, ensuring its sustainability. *Transportation Research Procedia*, 55, 1398–1406. https://doi.org/10.1016/j.trpro.2021.07.125
46. Gutiérrez A., Miravet D., Domènech, A. (2020). COVID-19 and urban public transport services: Emerging challenges and research agenda. *Cities & Health*, 1–4. https://doi.org/10.1080/23748834.2020.1804291
47. Charoennapharat T., Chaopaisarn P. (2022). Factors affecting multimodal transport during CoViD-19: a Thai service provider perspective. *Sustainability*, 14(8), 4838.

48. Abreu L., Conway A. (2021). A qualitative assessment of the multimodal passenger transportation system response to COVID-19 in New York City. *Transportation Research Record*, 03611981211027149.
49. Savsani P., Savsani V. (2016). Passing Vehicle Search (PVS): A novel metaheuristic algorithm. *Applied Mathematical Modelling*, 40(5–6), 3951–3978. https://doi.org/10.1016/j.apm.2015.10.040
50. Kumar S, Tejani G.G., Pholdee N., Bureerat S. (2021). Multi-Objective Passing Vehicle Search algorithm for structure optimization. *Expert Systems with Applications*, 169, 114511. https://doi.org/10.1016/j.eswa.2020.114511

3 Hybrid Equilibrium Optimizer and Slime Mold Algorithm for the Multi-objective Optimization of Truss Structures

Salar Farahmand-Tabar

3.1 INTRODUCTION

Addressing real-world engineering design challenges requires solutions that are both intricate and multi-faceted, as indicated by numerous studies [1, 2]. In these complex domains, designers often face a multitude of objectives, collectively referred to as multi-objective (MO) goals, which are interconnected and frequently conflict with each other [3, 4]. Consequently, rather than offering a single optimal solution, these situations typically yield a set of two or more potential solutions [5, 6]. Optimization serves as a critical component in system design, aiming to maximize or minimize system's outcomes while conserving resources [7, 8]. Structural optimization, for instance, involves minimizing costs or weight while simultaneously maximizing the strength or performance of the structure, posing significant design challenges [9, 10]. Despite the development of various deterministic optimization methods over the past three decades, these methods often face difficulties in finding optimum solutions due to complexities such as premature convergence and inefficient handling of nonlinear problems [11].

To address these challenges, researchers have increasingly turned to gradient-free and robust algorithms like metaheuristics (MHs) [12–14]. MHs, inspired by natural phenomena, offer advantages such as simplicity and adaptability, finding applications across diverse industrial and engineering sectors [15, 16]. Moreover, in recent decades, MHs have gained prominence in addressing structural optimization problems with multiple conflicting objectives [17, 18]. Unlike single-objective design problems, MO design challenges typically result in a set of optimum solutions referred to as the Pareto optimum set [19, 20]. Managing a design problem with

DOI: 10.1201/9781003601555-3

multiple objective functions generally involves exploring potential Pareto optimum solutions and selecting one or more solutions from the achieved set [21, 22].

Although initially designed for single-objective optimization, MHs have been adapted for MO optimization by leveraging their inherent characteristics, which include relying on populations or sets of design solutions and incorporating randomization [23]. The non-dominated sorting operator, crucial in this transition, plays a pivotal role. Initially known as MO Evolutionary Algorithms (MOEAs) and later broadened to encompass diverse philosophical approaches to MH-based search, first-generation MOMHs included MO Genetic Algorithms [24], Non-Dominated Sorting Genetic Algorithms (NSGA-II) [25], and Strength Pareto Evolutionary Algorithms [26]. Numerous MOMHs have since been developed alongside their single-objective counterparts, incorporating various strategies to enhance their MO capabilities that involve the incorporation of non-dominance strategies [27], decomposition-based methods [28], elitism strategies [29–32], grid-based approaches [33], preference-based approaches [34], population-based guided methods [35], and opposition-based learning strategies [36–38]. Despite their capacity to navigate through a single optimization process, improving MOMHs remains an ongoing challenge, particularly in handling many-objective optimization problems.

This chapter introduces the MO Hybrid Equilibrium Optimizer and Slime Mold Algorithm (MOEOSMA) for structural optimization, targeting two conflicting objectives: minimizing the structural weight while keeping structural responses within an acceptable limit. To evaluate the performance of the proposed MOEOSMA method, two challenging constrained problems of truss structures, 37-bar and 120-bar trusses, are investigated in this study.

3.2 RELATED STUDIES ON MULTI-OBJECTIVE TRUSS OPTIMIZATION

Table 3.1, updated from [39] for the context of this chapter, provides a summary of key information related to structural optimization problems. This table encompasses details such as the dimensionality of the problem (planar or space structure), the structural type (truss, tower, bridge, etc.), the specific objective functions targeted, constraints involved, and the algorithms referenced in the reviewed literature. The holistic consideration of objectives, constraints, and automatic member grouping, highlighted in this chapter, is similar to the approaches outlined in Table 3.1.

In Table 3.1, W represents the overall structural weight or mass, f_1 denotes the initial natural vibration frequency, u denotes the maximum displacement, λ^m denotes the constraint for member m within the structure regarding buckling, σ denotes the permissible stress, and $NCST$ denotes the number of distinct cross-sections. FRF represents the frequency response function, whereas $\overline{FRF}$ represents the average value of the parameters of FRF concerning the natural vibration frequencies f_1, f_2, and f_3. FT denotes the parameter for the force transmissibility concerning f_i, whereas $\overline{FT}$ denotes the mean value of FT at f_1, f_2, and f_3. CMA is the average mass under constraints, $SDCV$ represents the variability in constraint violations, RI denotes the index of reliability incorporating reliability constraints

TABLE 3.1
Literature review of multi-objective truss optimization

Structure type	Method	Objective Func.	Constraints	Ref.
Planar truss	SLDM-NSGA-II	W, u	σ	[40]
	MOPSO	W, u	σ	[41]
	CoSAR	W, RI	RC	[42]
		$W, f_1 + f_2 + f_3$	λ^m	
		$W,\ 1/\sum_{i=1}^{3} FRF(f_i)$	Ad hoc	
		$W, 1/\sum_{i=1}^{3} FT(f_i)$		
	AMISS-MOP	W, u	σ	[43]
Planar truss bridge	PAES, NSGA-II, SPEA2, PBIL, PSO	$W, 1/f_1$	σ	[44]
Truss tower	SPEA2, PBIL, AMOSA	$M, 1/\sum_{i=1}^{3}(F_i\ u_i)$	σ	[45]
Space truss	KSR	W, u	σ	[46]
	gM-PAES	W, u	σ	[47]
	MO-CLFBA, MO-FA	W, u	σ	[48]
	CSS	W, u	σ	[49]
	MOAS, MOACS	W, u	σ	[50]
	AICSAMO	W, u	σ	[51]
	MOASOS	W, u	σ	[52]
	GDE3, GDE3+APM	W, u	σ	[53]
	MOCBO	W, u	σ	[54]
	MOVPS, MOCBO	W, u	σ	[55]
	MOMASOS	W, u	σ	[56]
	MSOGP	W, u	σ	[57]
	MOMHTS	W, u	σ	[21]
	MOHTS	W, u	σ	[13]
	MOWCA, MOEA/D NSGA-II, MODA MOSWCA	Mass, compliance	σ, u	[58]
	User preference-oriented R-GDE3, R-DE3+APM R-NSGA-II	W, u	σ	[59]
	GDE3	W, f_1	σ, λ_1, u	[39]
		W, λ_1	σ, f_1, u	
		W, u	σ, λ_1, f_1	
	GDE3, SHAMODE, SHAMODE-WO, MM-IPDE	W, f_1, u	σ, f_1	[60]
		W, λ_1, u	σ, u	
		W, f_1, λ_1	σ	
		W, f_1, u, λ_1	σ	
	FC-MOPSO	W, u, f_1, TPE	σ	[61]
		W, u, f_1	σ, u	
		W, u	σ	

(RC), TPE denotes the total potential energy, LCC denotes the life-cycle costs, GC denotes the geometric constraints, and λ_1 is the initial load factor concerning global stability and the elastic critical load.

3.3 HYBRID EQUILIBRIUM OPTIMIZER AND SLIME MOLD ALGORITHM

Recently, Yin et al. [62] introduced a highly efficient algorithm called the hybrid Equilibrium Optimizer and Slime Mold Algorithm (EOSMA) and assessed its performance across CEC2019, CEC2021 functions, and various engineering problems. EOSMA incorporates the Equilibrium Optimizer (EO) search operator, proposed by Faramarzi et al. [63], to guide the slime mold's oscillatory aging behavior. This integration provides a specific direction to the anisotropic search of slime mold, achieving a more balanced exploration-exploitation trade-off. Furthermore, the algorithm integrates a random differential mutation operator and a greedy selection strategy into its search process to simultaneously enhance the capabilities of exploration and exploitation.

3.3.1 Single-objective EOSMA

Initializing population and equilibrium pool: Population initialization is the initial stage of the swarm intelligence optimization algorithm. In the case of EOSMA, the initial solutions for the population are generated randomly and uniformly distributed in the search space. The initialization formula is represented as follows:

$$X_1 = LB + r.(UB - LB) \tag{1}$$

where $LB = [lb_1, lb_2, \cdots, lb_d]$ and $UB = [ub_1, ub_2, \cdots, ub_d]$ represent the lower and upper bounds of variables, respectively. X_i represents the i th initial solution and r is a vector of random numbers in the range $[0,1]$. In order to enhance both exploration and exploitation capabilities, EOSMA introduces an equilibrium pool. This pool consists of five positions, where the first four are locally optimum positions aiding exploration, and the last one is the average position of the first four, contributing to exploitation. The equilibrium pool is expressed as

$$X_{eq} = [X_{eq,1}, X_{eq,2}, X_{eq,3}, X_{eq,4}, \cdots, X_{eq,avg}] \tag{2}$$

where $X_{eq,1}, X_{eq,2}, X_{eq,3}, X_{eq,4}$ represent the four individuals with the best fitness in the current population and $X_{eq,avg} = (X_{eq,1} + X_{eq,2} + X_{eq,3} + X_{eq,4})/4$. During the iterations, a position is randomly chosen from the equilibrium pool X_{eq} as the best food source X_b, guiding the updating direction of the search agent. This approach enables slime mold to use multiple food sources simultaneously.

Optimizing procedure of EOSMA: Drawing inspiration from the foraging behavior of slime mold, the EOSMA optimization process is divided into two stages: anisotropic search and vein-like tube formation. The anisotropic search stage is replaced with the search operator of EO, whereas the vein-like tube formation stage is updated using the search operator of the Slime Mold Algorithm (SMA). The process can be represented by

$$X_i^* = \begin{cases} X_b + (X_i - X_b).F + (G./\lambda).(1-F) & r_1 < z \\ X_i + vb.(W_i.X_{R1} - X_{R2}) & r_2 < P \\ X_b + (W_i.X_{R1} - X_{R2}) & others \end{cases} \tag{3}$$

where X_i^* represents the updated i th solution, $i = 1,2,\cdots,N$, where N is the population size. X_i represents the current solution, X_b is a solution randomly chosen from the equilibrium pool, F is the exponential term coefficient, $F = a_1 sign(r-0.5)$ $\bullet(e^{-\lambda t_1} - 1)$, $t_1 = (1 - t/max\_t)^{(a_2 t/max\_t)}$, and $a_1 = 2$ and $a_2 = 1$ are adaptive parameters adjusting exploration and exploitation. max_t and t represent the maximum and current iterations, respectively, G is the mass generation rate, λ is a vector of random numbers in the range $[0,1]$, i is the fitness weight of the i th slime mold individual, vb is a vector of random numbers in the range $[-a, a]$, $a = atanh(1 - t/max\_t)$, vc decreases linearly from 1 to 0, r_1 and r_2 are random numbers in the range $[0,1]$, X_{R1} and X_{R2} denote two randomly chosen solutions from the current population, $z = 0.6$ is a hybrid parameter, $p = tanh|Fit_i - BF|$, Fit_i represents the fitness of the ith individual, and BF is the best fitness. The mass generation rate is calculated as follows:

$$G = \begin{cases} 0.5 r_1.(X_b - \lambda.X_i).F & r_2 \geq 0.5 \\ 0 & others \end{cases} \tag{4}$$

where r_1 and r_2 are random numbers in the range $[0,1]$. The adaptive weight is generated as follows:

$$W(FitIdx_i) = \begin{cases} 1 + r.log\left(\dfrac{bF - Fit_i}{bF - wF} + 1\right) & i < N/2 \\ 1 - r.log\left(\dfrac{bF - Fit_i}{bF - wF} + 1\right) & others \end{cases}$$

$$FitIdx = sort(Fit) \tag{5}$$

where $FitIdx$ represents the fitness sorted in ascending order, r is a vector of random numbers in the range $[0,1]$, wF and bF are the worst and best fitness values in the current population, respectively, and Fit_i denotes the fitness of the ith individual. Population size is represented by N.

After updating the search agent according to Eq. (3), a greedy selection strategy and random differential mutation mechanism are utilized to improve the exploration and exploitation capabilities, assisting in escaping local optima. The mathematical expression for the random differential mutation operator is

$$X_i^* = \begin{cases} X_i + CF\left(LB + r_3\left(UB - LB\right)\right).L & r_4 < q \\ X_i + SF\left(X_{R1} - X_{R2}\right) & others \end{cases} \tag{6}$$

where $CF = \left(1 - t/max\_t\right)^{\left(a_1 t/max\_t\right)}$ represents an adaptive compression factor, L is a vector with elements 0 or 1, SF is a random number in the range $\left[0.2, 1\right]$, r_3 and r_4 are random numbers in the range $\left[0, 1\right]$, and $q = 0.2$ is a tunable parameter. To prevent invalid searches, following the update of the search agent's location, solutions outside the boundary are adjusted utilizing the dichotomy approach as follows:

$$X_{ij}^* = \begin{cases} \left(X_{ij} + ub_j\right).2 & X_{ij}^* > ub_j \\ \left(X_{ij} + lb_j\right).2 & X_{ij}^* < lb_j \\ X_{ij}^* & others \end{cases} \tag{7}$$

where $i = 1, 2, \cdots, N$, $j = 1, 2, \cdots, d$, where N is the population size, d is the dimension of the decision variable, and ub_j and lb_j are the upper and lower bounds of the jth decision variable, respectively. After conducting boundary checking, the fitness of each search agent is evaluated, and a greedy selection strategy is implemented to retain the superior individuals as follows:

$$X_i^* = \begin{cases} X_i^* & F\left(X_i^*\right) < F\left(X_i\right) \\ X & others \end{cases} \tag{8}$$

where $F(\cdot)$ represents the evaluation of an individual's fitness, X_i^* denotes the current individual, and X_i denotes the individual from the previous iteration. In EOSMA, a greedy selection method and equilibrium pool are integrated. The greedy selection method manages an archive with a capacity equal to the population size, storing the best solutions each search individual has discovered. Following each iteration, the archive is refreshed using Eq. (8). This strategy enhances memory by providing search agents with a powerful tool to recall successful foraging areas from the past. The equilibrium pool consists of five solutions, including four local optimum solutions and the central position of these four solutions. This allows slime mold to simultaneously explore multiple food sources, increasing the likelihood of discovering the global optimal solution. For a detailed understanding, the flowchart and pseudocode of EOSMA are available in [62].

3.3.2 MOEOSMA

Similar to other MO algorithms [64–66], EOSMA has its scenario for discovering Pareto solutions. Aligned with the foraging behavior of slime mold, the EOSMA optimization process is divided into two stages: the anisotropic search stage and the vein-like tube formation stage. The anisotropic search stage is altered using the EO search operator, with the objective of guiding the search agents' trajectory, expanding the exploration range, and mitigating premature convergence. Meanwhile, the vein-like tube formation stage is updated using the primary search operator of SMA. Notably, slime molds in these two stages coexist within the population simultaneously, as given by

$$X_i^* = \begin{cases} X_b + (X_i - X_b).F + (G./\lambda).(1-F) & r < z \\ X_b + vb(W_i.X_{R1} - X_{R2}) & others \end{cases} \tag{9}$$

where X_i^* and X_i represent the updated ith solution and current solution, respectively. Additionally, X_b is a solution randomly chosen from the equilibrium pool, F is the exponential term coefficient, $F = a_1 sign(r-0.5) \bullet (e^{-\lambda t_1} - 1), t_1 = (1 - t/max\_t)^{(a_2 t/max\_t)}$, and a_1 and a_2 are generated by

$$a_1 = \left(1 + (1 - t/max\_t)^{(2t/max\_t)}\right) \times r$$

$$a_2 = \left(2 - (1 - t/max\_t)^{(2t/max\_t)}\right) \times r \tag{10}$$

where G is calculated using Eq. (4), λ is a vector of random numbers in the range [0, 1], W_i is the fitness weight of the ith slime mold individual, vb is a vector of random numbers in the range $[-a, a]$, $a = atanh(1 - t/max\_t)$, r is a random number in the range [0, 1], X_{R1} and X_{R2} denote two randomly chosen solutions from the current population, and $z = 0.6$ is an adjustable parameter controlling the balance of exploration and exploitation. The adaptive weight is computed according to

$$W(FitIdx_i) = \begin{cases} 1 + r \,.\, log\left(\dfrac{bF - Fit_{ik}}{bF - wF} + 1\right) & i < N/2 \\ 1 - r \,.\, log\left(\dfrac{bF - Fit_{ik}}{bF - wF} + 1\right) & others \end{cases}$$

$$FitIdx = sort(Fit_k), k = rem(t, M) + 1 \tag{11}$$

where $rem(\cdot)$ is the remainder function, t is the current iteration number, M is the number of objective functions, r is a vector of random numbers in the range $[0,1]$, wF

and bF are the worst and best fitness values in the current population, respectively, Fit_{ik} denotes the fitness of the kth objective function of the ith individual, and N is the population size. Following the update of the slime mold's location by Eq. (9), the greedy selection method and random difference mutation operator are introduced to enhance the exploration and exploitation capabilities. This helps the search agent escape local optima and attain solutions with higher accuracy. The mathematical formula for the mutation operator is

$$X_i^* = X_i + SF\left(X_{R1} - X_{R2}\right) \tag{12}$$

where X_i^* represents the updated ith solution, X_i is the current solution, SF is a random number in the range [0.2, 1], X_{R1} and X_{R2} denote two randomly chosen solutions from the current population. To ensure valid searches, after updating the location of slime mold, a check is performed to determine whether X^* is within the search range. If not, the location is updated using Eq. (7). Following this, fitness is assessed, and the greedy selection method, as described in Eq. (8), is employed to preserve the superior slime mold individuals. The pseudocode for MOEOSMA is outlined in Algorithm 1.

Algorithm 1 Pseudocode for MOEOSMA [67]

Input the parameters *z*, *N*, *d*, *max_t*, *As*
Output the *PS* and its fitness *PF*
1. Initialize the locations of the search agent X_i (1,2,…,N);
2. **while** $t \leq max\_t$
3. Check the boundary using Eq. (7) and calculate the fitness;
4. Save the Pareto optimal solutions to the archive;
5. Rank the *PS* based on the crowding distance;
6. **if** *Archive overflow*
7. Remove crowded solutions using the roulette method;
8. Rank the *PS* based on the crowding distance;
9. **end**
10. Put the solution with the smallest crowding distance into the equilibrium pool;
11. Retain better solutions using Eq. (8);
12. Calculate the fitness weight of the slime mold using Eq. (11);
13. Calculate the values of adaptive variables a_1, a_2, t_1, a,;
14. Update the random variables r, r_1, r_2, r_n, v_b;
15. **for** $i = 1$ to N
16. Select a solution as the best food source X_b from the equilibrium pool;
17. **if** *rand* < *z*
18. Update X_i^* using the EO operator;
19. **else**
20. Update X_i^* using the SMA operator;
21. **end if**
22. **end for**
23. Check the boundary using Eq. (7) and calculate the fitness;
24. Save the Pareto optimal solutions to the archive;
25. Rank the *PS* based on the crowding distance;

26. **if** *Archive overflow*
27. Remove more crowded solutions using the roulette method;
28. **end if**
29. Retain better solutions using Eq. (8);
30. Update X using Eq. (12);
31. $t = t+1;$
32. **end while**

3.4 MATHEMATICAL FORMULATION OF THE MULTI-OBJECTIVE TRUSS OPTIMIZATION PROBLEM

In essence, practical engineering design problems [68] can be effectively represented as MO problems characterized by inherent conflicts among objectives. For instance, consider the trade-off between enhancing structural strength and structural weight, which impacts the cost. Moreover, these design challenges typically exhibit multimodality and nonlinearity, and often entail vast search spaces, rendering them inherently complex. The complexity of such design problems arises from the multitude of potential solutions, in contrast to the single optimum solution typically found in single-objective design problems. Designers are faced with the daunting task of selecting a single solution from the optimum solution set, known as the Pareto optimum set, based on predefined criteria—a task that is highly challenging.

Furthermore, in many instances, these Pareto optimal solutions remain largely unexplored for a given design problem. Hence, there is a crucial need to seek out multiple optimal solutions rather than confining the design process to a single solution. The typical mathematical representation for MO truss optimization problems is as follows:

$$Find\ Z = \{Z_1, Z_2, \ldots, Z_m\}$$

to optimize the structural mass and max. nodal displacement, miňimine

$$f_1(Z) = \sum_{i=0}^{m} Z_i \rho_i L_i\ a\check{n}d\ maximine\ f_2(Z) = max\left(\left|\delta_j\right|\right)$$

Subject to:

$$Behavior\ constraints(\text{Stress}):\ g(Z): \left|\sigma_i\right| - \sigma_i^{max} \leq 0$$

$$Construction\ constraints(\text{Cross-sectional area}):\ Z_i^{min} \leq Z_i \leq Z_i^{max} \tag{13}$$

In this context, Z_i represents a design variable vector, ρ_i denotes the density of the elements' mass, and L_i denotes their length. Additionally, Ei and σi denote the elasticity modulus and stress of the ith element, respectively. It is also important to note that the permissible lower and upper limits are denoted by the superscripts *max* and *min*, respectively.

An exceptionally effective strategy for addressing constrained problems involves converting these design challenges into dynamic unconstrained optimization problems utilizing a dynamic penalty method. Particularly in design scenarios, the multiplication-based penalty function emerges as notably effective:

$$f(X)*(1+\varepsilon_1*C)^{\varepsilon_2},\ C=\sum_{i=1}^{m}C_i,\ C_i=\left|1-\frac{\left|\sigma_i^{max}-\sigma_i\right|}{\sigma_i^{max}}\right| \tag{14}$$

Users have the flexibility to define the values of the parameters ε_1 and ε_2, with both of them being set to 3 in this case.

3.5 RESULTS AND DISCUSSION OF NUMERICAL EXAMPLES

The performance assessment of the algorithm under consideration involved their application to two complex structural examples [13, 21, 56, 57] with the properties according to Table 3.2. Additionally, the evaluation involved the utilization of standard MO structural functions to gauge the search efficiency of the algorithm. For this study, a population size of 100 was employed, in line with previous research practices. To obtain results, each algorithm underwent 10 independent runs.

Four performance indicators, Hypervolume (HV), Generational Distance (GD), Inverted GD (IGD), and Space-to-Extend (STE), were evaluated to measure the search performance of the algorithms. The lower the values of the indicators, the better the results, except for HV, for which the higher the values, the better the results.

3.5.1 The 37-Bar Truss Structure

The investigation of the simply supported Pratt-Type 37-bar truss (depicted in Figure 3.1) has been carried out by researchers [58] using different methods, such as the NHGA algorithm and evolutionary method. This truss presents an optimization problem with 14 size variables, five shape variables, and three frequency constraints. The additional load of -10 kg is applied on the nodes of the lower chord with a rectangular cross-sectional area of 0.4 cm^2.

TABLE 3.2
Data for the design of truss structures

	37-bar truss	120-bar truss
Parameters	**Value**	**Value**
Design variables	$A_i, i = 1, 2, \ldots, 14$	$A_i, i = 1, 2, \ldots, 7$
Constraints (N/m^2)	$\sigma_{max} = 400$	$\sigma_{max} = 400$
Elasticity modulus $E\,(\text{N/m}^2)$	2.0×10^{11}	2.0×10^{11}
Density of material $\rho\,(\text{kg/m}^3)$	7850	7850
Cross-sectional area (m^2)	$0.0001 < A < 0.001$	$0.0001 < A < 0.01293$

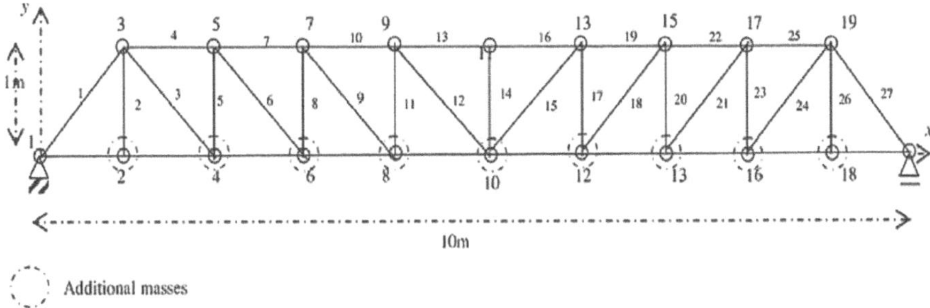

FIGURE 3.1 The 37-bar truss.

The figures within this study play a key role in depicting the efficacy of MOEOSMA in comparison to NSGA-II for resolving intricate MO truss optimization challenges. In Figure 3.2, the best Pareto fronts are presented, assessed based on two main performance metrics: HV and IGD.

Initially, let's focus on the Pareto fronts illustrated in Figure 3.2. These fronts offer a visual depiction of the trade-offs between conflicting objectives for both MOEOSMA and NSGA-II. Notably, the Pareto front resulting from the consideration of HV and IGD for MOEOSMA appears superior when compared with the one obtained with NSGA-II. This inference suggests that MOEOSMA excels in optimizing the 37-bar truss structure when multiple, potentially conflicting, objectives are taken into account. On closer scrutiny of the performance indicators illustrated in Figure 3.2, several results emerge. The HV values, serving as a gauge of the volume encompassed by the Pareto front, exhibit a notably higher magnitude for MOEOSMA in comparison to NSGA-II. This signifies that MOEOSMA achieves a broader coverage of the Pareto front, indicative of a richer diversity of solutions, a highly desirable attribute in MO optimization.

In addition to HV, GD and STE values are lower for MOEOSMA compared to NSGA-II. A lower GD indicates that the solutions found by MOEOSMA are closer to the true Pareto front, showcasing the algorithm's capability to generate solutions that are not only diverse but also closer to the optimal trade-off between objectives. The lower STE indicates that the solutions produced by MOEOSMA are more consistent, exhibiting less variation, which is crucial for reliable optimization. Collectively, these findings emphasize the exceptional performance and reliability of MOEOSMA in tackling the MO optimization problem of the 37-bar truss structure. The ability of MOEOSMA to yield a superior Pareto front, alongside higher HV, and lower GD and STE values, illustrates its effectiveness in providing well-balanced solutions that effectively reconcile multiple objectives.

3.5.2 The 120-bar Truss Structure

Figure 3.3 displays an example of size optimization, namely a 120-bar truss dome [58]. In this configuration, the non-structural masses are applied as follows: 3000 kg at node 1, 1500 kg at nodes 2–13, and 500 kg at the remaining nodes. Table 3.2

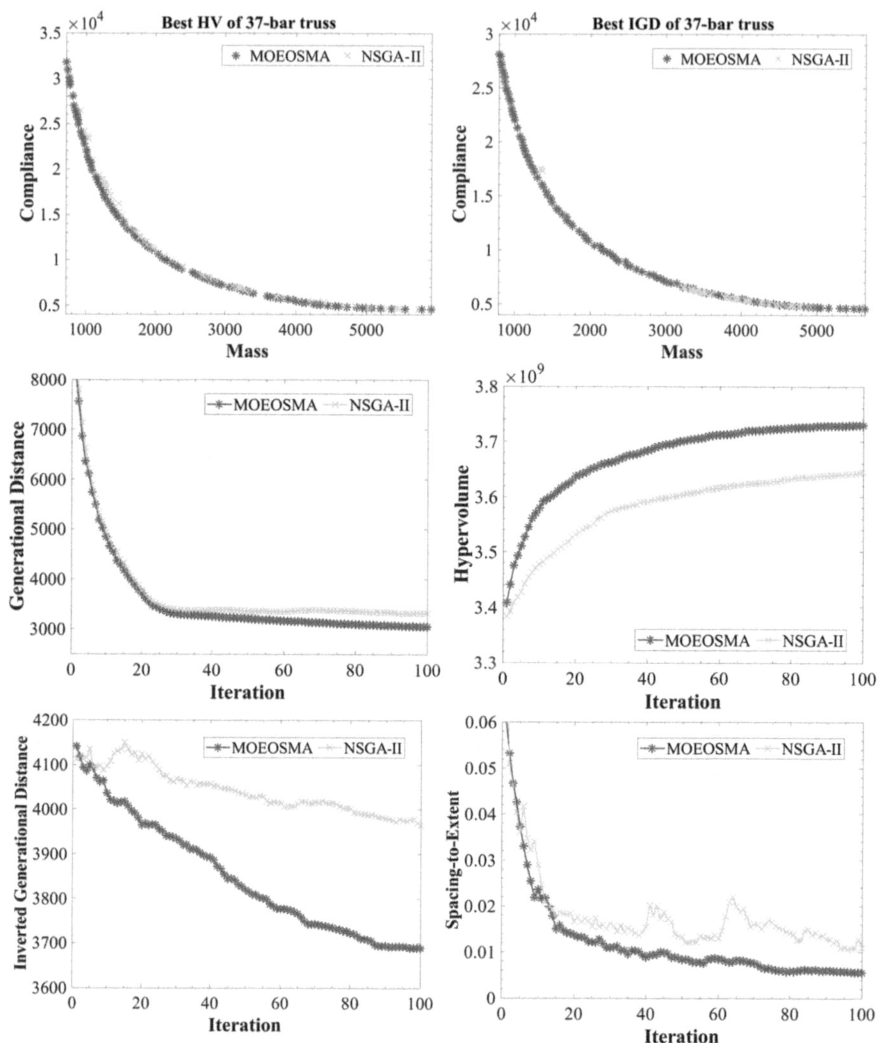

FIGURE 3.2 Optimum result of the 37-bar truss: Pareto fronts and performance indicators.

provides comprehensive details regarding the problem, such as material properties and design variable constraints.

A comprehensive analysis of the performance indicators unveils compelling insights according to Figure 3.4. Particularly, MOEOSMA exhibits significantly higher HV values compared to NSGA-II. This suggests that MOEOSMA excels in offering a broader coverage of the Pareto front, signaling a diverse array of solutions—an essential trait in MO optimization pursuits.

Moreover, examining the GD and STE values, it becomes evident that MOEOSMA surpasses NSGA-II in both aspects. The lower GD values for MOEOSMA indicate that its solutions are closer to the true Pareto front, highlighting its ability to generate

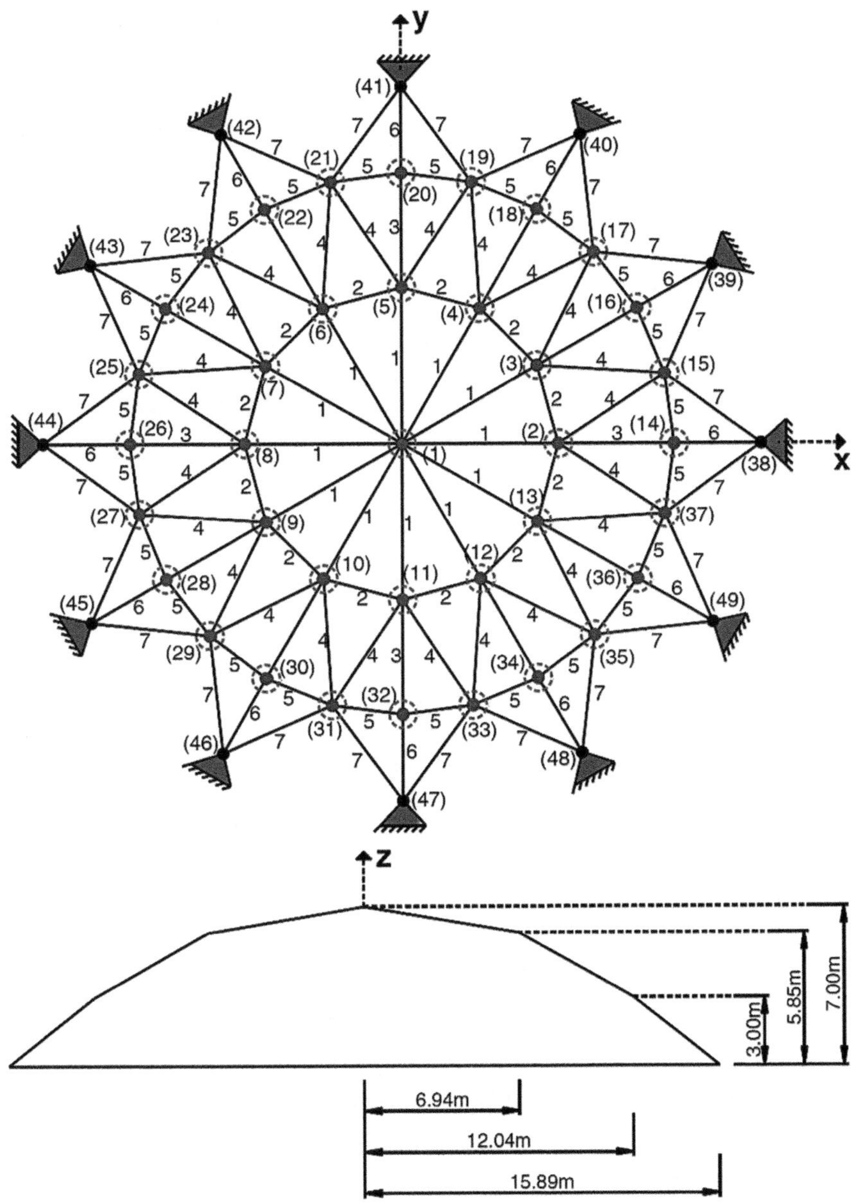

FIGURE 3.3 The 120-bar truss.

solutions that offer a superior trade-off between objectives. Additionally, the reduced STE values signify that solutions of MOEOSMA are more consistent and exhibit less variability, emphasizing its reliability. These results underscore the remarkable performance and reliability of MOEOSMA when applied to the intricate MO optimization challenges associated with the 120-bar truss structure. MOEOSMA not only

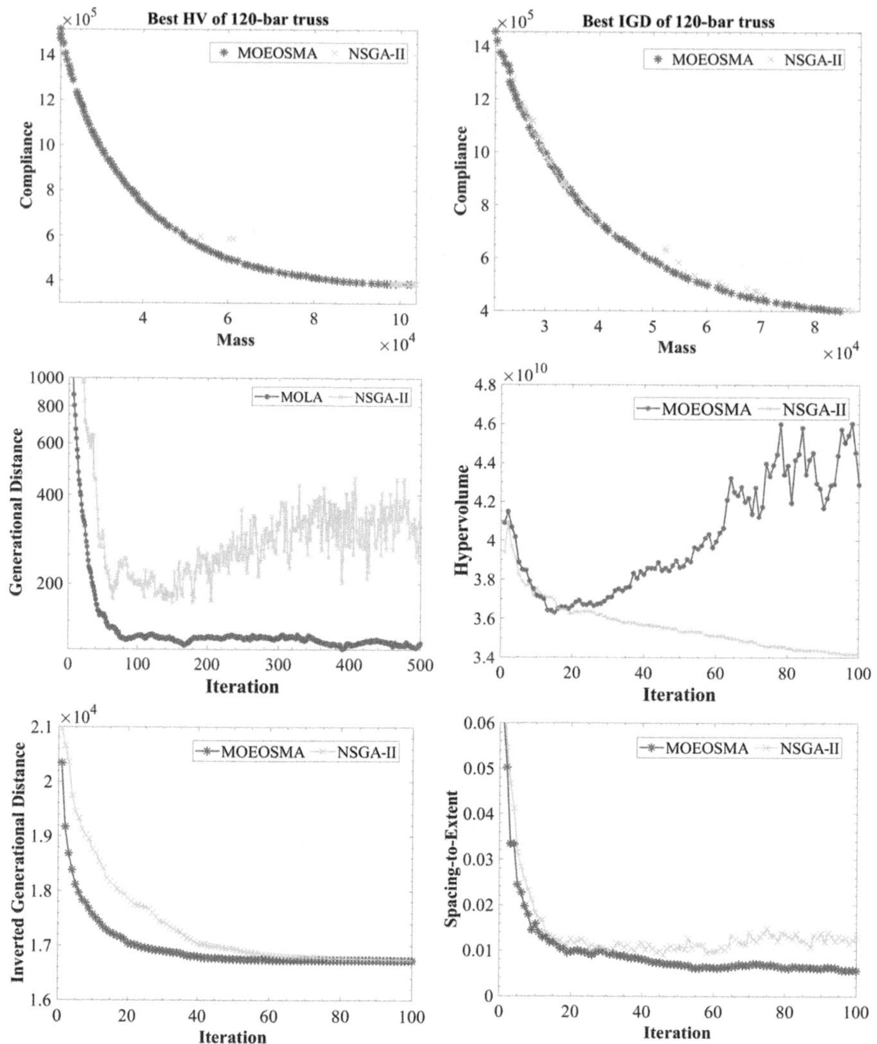

FIGURE 3.4 Optimum result of the 120-bar truss: Pareto fronts and performance indicators.

excels in producing an enhanced Pareto front but also achieves higher HV values while maintaining lower GD and STE values. These findings establish MOEOSMA as a robust and effective tool for addressing the complexities of MO optimization in the realm of structural engineering.

3.6 CONCLUSIONS

The hybrid EOSMA was applied and explored in the study for the MO optimization of truss structures. Testing of the algorithm was conducted on two complex MO structural optimization problems (37-bar and 120-bar truss), characterized by design variables that adhere to practical engineering constraints. Comparative analysis was performed against NSGA-II using the same input conditions. Evaluation was based on several performance metrics, such as the Pareto Front HV and Front STE, to gauge the quality and diversity of the non-dominated solution set. The results indicated the superior performance of MOEOSMA, showing significant enhancements in performance metrics. These findings underscore the efficiency of MOEOSMAin tackling real-world optimization challenges within the realm of structural engineering.

REFERENCES

1. Dey N, Ashour A, Bhattacharyya S. (2020). *Applied nature-inspired computing: Algorithms and case studies*. Singapore: Springer.
2. Farahmand-Tabar S, Ashtari P, Babaei M. (2024). Gaussian cross-entropy and organizing intelligence for design optimization of the outrigger system with inclined belt truss in real-size tall buildings. *Probabilistic Eng Mech* 76, 103616. https://doi.org/10.1016/j.probengmech.2024.103616
3. Blondet G, Duigou JL, Boudaoud N. (2019). A knowledge-based system for numerical design of experiments processes in mechanical engineering. *Expert Syst Appl* 122, 289–302. https://doi.org/10.1016/j.eswa.2019.01.013
4. Wu J, Wang YG, Burrage K, Tian YC, Lawson B, Ding Z. (2020). An improved firefly algorithm for global continuous optimization problems. *Expert Syst Appl* 149, 113340. https://doi.org/10.1016/j.eswa.2020.113340
5. Shirgir S, Farahmand-Tabar S, Aghabeigi P. (2023). Optimum design of real-size reinforced concrete bridge via charged system search algorithm trained by nelder-mead simplex. *Expert Syst Appl* 121815. https://doi.org/10.1016/j.eswa.2023.121815
6. Farahmand-Tabar S, Abdollahi F, Fatemi M. (2023). Robust Conjugate Gradient Methods for Non-Smooth Convex Optimization and Image Processing Problems. In: Kulkarni AJ, Gandomi AH (eds) *Handbook of Formal Optimization*. Springer, Singapore.
7. Farahmand-Tabar S, Ashtari P. (2023). Bilinear Fuzzy Genetic algorithm and its application on the optimum design of steel structures with semi-rigid connections. In: Kulkarni AJ, Gandomi AH (eds) *Handbook of Formal Optimization*. Springer, Singapore.
8. Farahmand-Tabar S, Ashtari P. (2020). Simultaneous size and topology optimization of 3D outrigger-braced tall buildings with inclined belt truss using genetic algorithm. *Struct Des Tall Spec Build* 29(13): e1776. https://doi.org/10.1002/tal.1776
9. Ashtari P, Karami R, Farahmand-Tabar S. (2021). Optimum geometrical pattern and design of real-size diagrid structures using accelerated fuzzy-genetic algorithm with bilinear membership function. *Appl Soft Comput* 110: 107646. https://doi.org/10.1016/j.asoc.2021.107646
10. Farahmand-Tabar S. (2023). Genetic Algorithm and Accelerating Fuzzification for Optimum Sizing and Topology Design of Real-Size Tall Building Systems. In: Dey N (eds) *Applied Genetic Algorithm and Its Variants. Springer Tracts in Nature-Inspired Computing*. Springer, Singapore. https://doi.org/10.1007/978-981-99-3428-7_9
11. Farahmand-Tabar S, Rashid TA. (2023). Steel Plate Fault Detection using the Fitness Dependent Optimizer and Neural Networks. In: Kulkarni AJ, Gandomi AH (eds) *Handbook of Formal Optimization*. Springer, Singapore.

12. Farahmand-Tabar S, Ashtari P. (2024). Intelligent cross-entropy optimizer: A novel machine learning-based meta-heuristic for global optimization. *Swarm Evol Comput* 91, 101739. https://doi.org/10.1016/j.swevo.2024.101739
13. Tejani GG, Kumar S, Gandomi AH. (2021). Multi-objective heat transfer search algorithm for truss optimization. *Engi Comput* 37: 641–662. https://doi.org/10.1007/s00366-019-00846-6
14. Farahmand-Tabar S, Shirgir S. (2023). Incorporating Nelder Mead Simplex as an Accelerating Operator to Improve the Performance of Metaheuristics in Nonlinear System Identification. In: Kulkarni AJ, Gandomi AH (eds) *Handbook of Formal Optimization*. Springer, Singapore.
15. Farahmand-Tabar S, Sadrekarimi N. (2023). Overcoming Constraints: The Critical Role of Penalty Functions as Constraint Handling Methods in Structural Optimization. In: Kulkarni, AJ, Gandomi AH (eds) *Handbook of Formal Optimization*. Springer, Singapore.
16. Farahmand-Tabar S, Shirgir S. (2023). Synergistic Collaboration of Motion-Based Metaheuristics for the Strength Prediction of Cement-Based Mortar Materials using TSK Model. In: Kulkarni AJ, Gandomi AH (eds) *Handbook of Formal Optimization*. Springer, Singapore.
17. Farahmand-Tabar S, Shirgir S. (2023). Positron-Enabled Atomic Orbital Search Algorithm for Improved Reliability-Based Design Optimization. In: Kulkarni AJ, Gandomi AH (eds) *Handbook of Formal Optimization*. Springer, Singapore.
18. Farahmand-Tabar S, Ashtari P, Babaei M. (2023). Dynamic Intelligence of Self-Organized Map in the Frequency-Based Optimum Design of Structures. In: Kulkarni AJ, Gandomi AH (eds) *Handbook of Formal Optimization*. Springer, Singapore.
19. Farahmand-Tabar S, Shirgir S. (2023). Antlion-Facing Ant Colony Optimization in Parameter Identification of the MR Damper as a Semi-Active Control Device. In: Dey, N. (eds) *Applications of Ant Colony Optimization and its Variants*. Springer, Singapore.
20. Ho-Huu V, Hartjes S, Visser HG, Curran R. (2018). An improved MOEA/D algorithm for bi-objective optimization problems with complex Pareto fronts and its application to structural optimization. *Expert Syst Appl* 92, 430–446. https://doi.org/10.1016/j.eswa.2017.09.051
21. Kumar S, Tejani GG, Pholdee N, Bureerat S. (2021). Multi-objective modified heat transfer search for truss optimization. *Eng Comput* 37, 3439–3454. https://doi.org/10.1007/s00366-020-01010-1
22. Kaveh A, Laknejadi K. (2011). A novel hybrid charge system search and particle swarm optimization method for multi-objective optimization. *Expert Syst Appl* 38(12), 15475–15488. https://doi.org/10.1016/j.eswa.2011.06.012
23. Pardalos PM, Romeijn HE. (2013). *Handbook of global optimization*. Springer Science & Business Media.
24. Fonseca CM, Fleming PJ. (1993). Genetic Algorithms for Multiobjective Optimization: Formulation Discussion and Generalization. In: *Proceedings of the 5th International Conference on Genetic Algorithms*: 416–423. San Francisco, CA, USA: Morgan Kaufmann Publishers Inc.
25. Deb K, Pratap A, Agarwal, S, Meyarivan T. (2002). A fast and elitist multiobjective genetic algorithm: NSGA-II. *IEEE Transac Evol Comput* 6(2), 182–197. https://doi.org/10.1109/4235.996017
26. Zitzler E, Laumanns M, Thiele L. (2001). SPEA2: Improving the Strength Pareto Evolutionary Algorithm. In *Evolutionary Methods for Design Optimization and Control with Applications to Industrial Problems*: 95–100. https://doi.org/10.1.1.28.7571

27. Tejani GG, Pholdee N, Bureerat S, Prayogo D. (2018). Multiobjective adaptive symbiotic organisms search for truss optimization problems. *KnowlBased Syst* 161, 398–414. https://doi.org/10.1016/j.knosys.2018.08.005
28. Zhang Q, Li H. (2007). MOEA/D: A multiobjective evolutionary algorithm based on decomposition. *IEEE Transac Evol Comput* 11(6), 712–731. https://doi.org/10.1109/TEVC.2007.892759
29. Deb K, Agrawal S, Pratap A, Meyarivan T. (2000). A Fast Elitist Non-dominated Sorting Genetic Algorithm for Multi-objective Optimization: NSGA-II. In *International Conference on Parallel Problem Solving from Nature*: 849–858. Berlin, Heidelberg: Springer Berlin Heidelberg.
30. Farahmand-Tabar S, Babaei M. (2023). Memory-assisted adaptive multiverse optimizer and its application in structural shape and size optimization. *Soft Comput* 27(16): 11505–10527. https://doi.org/10.1007/s00500-023-08349-9
31. Farahmand-Tabar S. (2023). Memory-Driven Metaheuristics: Improving Optimization Performance. In: Kulkarni AJ, Gandomi AH (eds) *Handbook of Formal Optimization*. Springer, Singapore.
32. Farahmand-Tabar S. (2023). Frequency-Based Optimization of Truss Dome Structures using Ant Colony Optimization (ACOR) with Multi-Trail Pheromone Memory. In: Dey N (eds) *Applications of Ant Colony Optimization and its Variants*. Springer, Singapore.
33. Knowles J, Corne D. (1999). The Pareto Archived Evolution Strategy: A New Baseline Algorithm for Pareto Multiobjective Optimisation. In: *Proceedings of the 1999 Congress on Evolutionary Computation-CEC99* (Cat. No. 99TH8406) (pp. 98–105). https://doi.org/10.1109/CEC.1999.781913
34. Bureerat S, Srisomporn S. (2010). Optimum plate-fin heat sinks by using a multiobjective evolutionary algorithm. *Eng Optim* 42(4), 305–323. https://doi.org/10.1080/03052150903143935
35. Got A, Moussaoui A, Zouache D. (2020). A guided population archive whale optimization algorithm for solving multiobjective optimization problems. *Expert Syst Appl* 141, 112972. https://doi.org/10.1016/j.eswa.2019.112972
36. Gupta S, Deep K, Heidari AA, Moayedi H, Wang M. (2020). Opposition-based learning Harris Hawks optimization with advanced transition rules: Principles and analysis. *Expert Syst Appl* 158, 113510. https://doi.org/10.1016/j.eswa.2020.113510
37. Farahmand-Tabar S, Shirgir S. (2023). Boosting the Efficiency of Metaheuristics through Opposition-Based Learning in Optimum Locating of Control Systems in Tall Buildings. In: Kulkarni AJ, Gandomi, A.H. (eds) *Handbook of Formal Optimization*. Springer, Singapore.
38. Farahmand-Tabar S, Shirgir S. (2023). Opposed Pheromone Ant Colony Optimization for Property Identification of Nonlinear Structures In: Dey N (eds) *Applications of Ant Colony Optimization and its Variants*. Springer, Singapore.
39. Lemonge AC, Carvalho JP, Hallak PH, Vargas DE. (2021). Multi-objective truss structural optimization considering natural frequencies of vibration and global stability. *Expert Syst Appl* 165: 113777.
40. Vo-Duy T, Duong-Gia D, Ho-Huu V, Nguyen-Thoi T. (2020). An effective couple method for reliability-based multi-objective optimization of truss structures with static and dynamic constraints. *Int J Comput Methods* 17(06): 1950016.
41. Hosseini SS, Hamidi SA, Mansuri M, Ghoddosian A. (2015). Multiobjective particle swarm optimization (MOPSO) for size andshape optimization of 2D truss structures. *Periodica Polytechnica Civil Eng* 59(1): 9.

42. Greiner D, Hajela P. (2012). Truss topology optimization for massand reliability considerations – Co-evolutionary multiobjective formulations. *Struct Multidiscip Optim* 45(4): 589–613.
43. Su R, Wang X, Gui L, Fan Z. (2011). Multi-objective topology and sizing optimization of truss structures based on adaptive multiislandsearch strategy. *Struct Multidiscip Optim* 43(2): 275–286.
44. Noilublao C, Bureerat S. (2009). Simultaneous Topology, Shape and Sizing Optimisation of Skeletal Structures using Multiobjective Evolutionary Algorithms. In: *Evolutionary Computation.* Intech Open.
45. Noilublao N, Bureerat S. (2011). Simultaneous topology, shapeand sizing optimisation of a three-dimensional slender truss tower using multiobjective evolutionary algorithms. *Comput Struct* 89(23–24): 2531–2538.
46. Richardson JN, Adriaenssens S, Bouillard P, Coelho RF. (2012). Multiobjective topology optimization of truss structures with kinematic stability repair. *Struct Multidiscip Optim* 46(4): 513–532.
47. Kaveh A, Laknejadi K. (2013). A hybrid evolutionary graph-based multi-objective algorithm for layout optimization of truss structures. *Acta Mech* 224(2): 343–364.
48. Gholizadeh S, Asadi H, Baghchevan A. (2014). Optimal design of truss structures by improved multi-objective firefly and bat algorithms. *Iran Univ Sci Technol* 4(3): 415–431.
49. Kaveh A, Massoudi M. (2014). Multi-objective optimization of structures using charged system search. *Sci Iran Trans A Civil Eng* 6: 1845–60.
50. Angelo JS, Bernardino HS, Barbosa HJ. (2015). Ant colony approaches for multiobjective structural optimization problems with a cardinality constraint. *Adv Eng Softw* 80: 101–115.
51. Xie L, Tang H, Hu C, Xue S. (2016). An adaptive multi-objective immune algorithm for optimal design of truss structures. *J Asian Archit Build Eng* 15(3): 557–564.
52. Tejani GG, Pholdee N, Bureerat S, Prayogo D. (2018). Multiobjective adaptive symbiotic organisms search for truss optimization problems. *Knowl-Based Syst* 161: 398–414.
53. Vargas DE, Lemonge AC, Barbosa HJ, Bernardino HS. (2019). Differential evolution with the adaptive penalty method for structural multi-objective optimization. *Optim Eng* 20(1): 65–88
54. Kaveh A, Mahdavi VR. (2019). Multi-objective colliding bodies optimization algorithm for design of trusses. *J Comput Des Eng* 6(1): 49–59.
55. Kaveh A, Ghazaan MI. (2019). A new VPS-based algorithm for multi-objective optimization problems. *Eng Comput* 36: 1029–1040.
56. Tejani GG, Pholdee N, Bureerat S, Prayogo D, Gandomi AH. (2019). Structural optimization using multi-objective modified adaptive symbiotic organisms search. *Expert Syst Appl* 25: 425–441.
57. Assimi H, Jamali A, Nariman-Zadeh N. (2019). Multi-objective sizing and topology optimization of truss structures using genetic programming based on a new adaptive mutant operator. *Neural Comput Appl* 31(10): 5729–5749.
58. Kumar S, Tejani G.G., Pholdee N., Bureerat S. (2021). Multiobjecitve structural optimization using improved heat transfer search. *Knowl-Based Syst* 219: 106811. https://doi.org/10.1016/j.knosys.2021.106811.
59. Vargas DE, Lemonge AC, Barbosa HJ, Bernardino HS. (2021). Solving multi-objective structural optimization problems usingGDE3 and NSGA-II with reference points. *Eng Struct* 239: 112187.

60. Carvalho JPG, Carvalho ÉCR, Vargas DEC, Hallak PH, Lima BSLP, Lemonge ACC. (2021). Multi-objective optimum design of truss structures using differential evolution algorithms. *Comput Struct* 252: 106544.
61. Mokarram V, Banan MR. (2018). A new PSO-based algorithm formulti-objective optimization with continuous and discrete design variables. *Struct Multidiscip Optim* 57(2): 509–533.
62. Yin S, Luo Q, Zhou Y. (2022b). EOSMA: An equilibrium optimizer slime mould algorithm for engineering design problems. *Arab J Sci Eng* 47: 10115–10146. https://doi.org/10.1007/s13369-021-06513-7
63. Faramarzi A, Heidarinejad M, Stephens B, Mirjalili S. (2020). Equilibrium optimizer: A novel optimization algorithm. *Knowl-Based Syst* 191: 105190. https://doi.org/10.1016/j.knosys.2019.105190
64. Farahmand-Tabar S, Shirgir S. (2024). Multi-objective Adaptive Guided Differential Evolution for Passively Controlled Structures Equipped with a Tunned Mass Damper. In: Dey N (eds) *Applied Multi-objective Optimization. Springer Tracts in Nature-Inspired Computing*. Springer, Singapore. https://doi.org/10.1007/978-981-97-0353-1_3
65. Farahmand-Tabar S. (2024). Multi-objective Lichtenberg Algorithm for the Optimum Design of Truss Structures. In: Dey N (eds) *Applied Multi-objective Optimization. Springer Tracts in Nature-Inspired Computing*. Springer, Singapore. https://doi.org/10.1007/978-981-97-0353-1_5
66. Farahmand-Tabar S, Afrasyabi P. (2024). Multi-modal Routing in Urban Transportation Network Using Multi-objective Quantum Particle Swarm Optimization. In: Dey N (eds) *Applied Multi-objective Optimization. Springer Tracts in Nature-Inspired Computing*. Springer, Singapore. https://doi.org/10.1007/978-981-97-0353-1_7
67. Luo Q, Yin S, Zhou G, et al. (2023). Multi-objective equilibrium optimizer slime mould algorithm and its application in solving engineering problems. *Struct Multidisc Optim* 66, 114. https://doi.org/10.1007/s00158-023-03568-y
68. Farahmand-Tabar S, Aghani K. (2023). *Practical programming of finite element procedures for solids and structures with MATLAB: From elasticity to plasticity*. Elsevier.

4 GMOCSO

Multi-objective Cat Swarm Optimization Algorithm based on a Grid System

Aram M. Ahmed, Bryar A. Hassan, Tarik A. Rashid, Kaniaw A. Noori, Soran Ab. M. Saeed, Omed H. Ahmed and Shahla U. Umar

4.1 INTRODUCTION

The term optimization means trying to select the best solution for a specific problem among many alternative solutions. The objective can be minimization, such as minimizing cost or time, or it can be maximization, such as maximizing profit or production [1]. In terms of Mathematics, optimization problems can be formulated in a generic form as follows:

$$\min_{x \in R^n} f_i(x), (i = 1, 2, \ldots, I)$$

$$\text{Subject to} \begin{cases} c_j(x) = 0, (j = 1, 2, \ldots, J) \\ c_k(x) \geq 0, (k = 1, 2, \ldots, K) \end{cases} \quad (1)$$

If we suppose that the parameters of the search space are denoted in a vector like "x", then $f_i(x)$ represents the objective functions, sometimes called fitness functions. $f(x)$ can be a minimizing or maximizing function depending on the design at hand. Finally, $c_j(x)$ and $c_k(x)$ are the equality and inequality constraints of the design problem, respectively [2].

Optimization problems can be classified according to the objectives, constraints, landscape, form functions, and variables. When the classification is based on the number of objectives, there are two kinds of optimization problems: single and multi-objective. If Eq. (1) is taken, the problem has a single objective, when $I = 1$. This means that there is only one goal in the search space and all the factors are used to reach that goal. However, when $I > 1$, it is a multi-objective problem, which contains multiple conflicting objectives or goals. Many real-world problems are multi-objective types, for example, maximizing profit and meanwhile minimizing time or cost. When these objectives clash with each other, algorithms require a trade-off

 DOI: 10.1201/9781003601555-4

between them, which leads to the generation of a set of compromised solutions commonly known as Pareto-optimal solutions [3]. In addition, constraints in optimization problems are those conditions that a solution must fulfill. Optimization problems can also be classified according to the number of these constraints. For example, in the previous equation: if $j = k = 0$, this means there are no constraints and it is known as an "unconstrained optimization problem". As for Eq. (1), if $j > 0$ or $k > 0$, the problem is known as an "equality-constrained optimization problem" or "inequality-constrained optimization problem", respectively. Furthermore, depending on the nature of the problems, some models may require a discrete set of values to be used to represent the solution sets [4]. These types of optimization are called discrete optimization. However, some other models, known as continuous optimization, are only comprehended when actual real numbers are employed to represent the solution sets. It is worth mentioning that continuous optimization algorithms can also be used to solve discrete problems. Moreover, problems can be called linear if the change in the output is proportional to the change in the input. On the other hand, nonlinear problems are exactly the opposite, thus their outputs are unpredictable, and this makes them harder to solve. Finally, unimodal problems are problems that have one global optimum in the search space, whereas multimodal problems have multiple global optima in the search space [4].

When dealing with multi-objective problems, decision and objective spaces need to be taken into consideration. The decision space is a space that contains all the possible solutions, whereas the evaluations of these solutions are presented in a space called the objective space. It is worth mentioning that, in real-world scenarios, some of these solutions, which are known as infeasible regions, are not applicable due to the design limitations of the problem [5]. In single-objective optimization problems, one can easily compare two solutions and decide which one is better or more optimum. However, in multi-objective optimization problems, the concept of dominance is required to compare the solutions.

Let x and y be two solutions for a minimization multi-objective optimization problem. Thus, the objectives are as follows:

$$\min_{x \in X}\left(f1(x), f2(x), f3(x), \ldots, fk(x)\right)$$

$$\min_{y \in X}\left(f1(y), f2(y), f3(y), \ldots, fk(y)\right) \tag{2}$$

where $k \geq 2$ and the set X includes all the feasible solutions.
Then, x dominates y if

1. $fi(x) \leq fi(y)$ for all indices $i \in \{1, 2, \ldots, k\}$
2. $fj(x) < fj(y)$ for at least one index $j \in \{1, 2, \ldots, k\}$.

Therefore, x dominates y, that is, $x \prec y$ if

1. solution y is not better than solution x in any of the objectives
2. solution x is better than solution y in at least one objective.

In the process of multi-objective optimization, solutions are often obtained, where neither of them dominates the others, that is, one solution dominates the other for an objective function, meanwhile, the other solution dominates this solution for a different objective function. These types of solutions, which cannot be dominated by the other solutions, are known as a non-dominated solution set [6].

In addition, the non-dominated solution set belonging to the whole feasible decision space is known as the Pareto front. Pareto front solutions that cannot be further improved are called the true Pareto front. If an evolutionary multi-objective algorithm is not able to find the true Pareto front, it attempts to obtain an approximate set of solutions that is close to it. This set of solutions is called the approximate Pareto front.

Several methods have been proposed for multi-objective problems. Schafer initially proposed the novel notion of using evolutionary optimization methods for multi-objective optimization in 1984 [2]. Since then, many scholars have tried to create various multi-objective algorithms. For example, the Non-dominated Sorting Genetic Algorithm 2 (NSGA-II) [7], Multi-objective Particle Swarm Optimization (MOPSO) [8], Multi-objective Evolutionary Algorithm based on Decomposition (MOEA/D) [9], Pareto Archived Evolution Strategy (PAES) [10], and Strength Pareto Evolutionary Algorithm (SPEA/SPEA2) [11] are among the most well-known algorithms in the literature. The Cat Swarm Optimization (CSO) Algorithm is a robust metaheuristic swarm-based algorithm originally proposed by Chu et al. in 2006 [12]. Pradhan and Panda proposed multi-objective CSO (MOCSO) by extending CSO to deal with multi-objective problems [13]. MOCSO is combined with the concept of an external archive and Pareto dominance to handle non-dominated solutions. Orouskhani et al. proposed a multi-objective algorithm by combining the CSO algorithm with Borda ranking [14]. They also proposed an enhanced version by combining the CSO algorithm with an opposition-based learning technique and the elitism idea [15]. Dwivedia et al. extended the quantum-inspired CSO as a multi-objective algorithm, which uses the idea of qubits [16]. Murtza et al. extended the integer CSO as a multi-objective algorithm, which can be used for binary multi-objective problems [17].

Multi-objective algorithms have two main goals to pursue while searching for optimum solutions: convergence and diversity preservation [18]. In the first one, the approximate Pareto front has to move toward the true Pareto front as much as possible, hence, the algorithm obtains better and more optimum solutions. However, in diversity preservation, the set of solutions in the approximate Pareto front has to be well distributed in the area. The term "well distributed" means that they have to cover the whole area of the Pareto front and have uniform distances in between as much as possible. The reason behind this goal is to provide decision-makers with solutions from all areas and let them choose their region of interest. Therefore, this paper presents a multi-objective version of the CSO algorithm, which is called the Grid-based MOCSO Algorithm (GMOCSO). To accomplish these objectives, we start by substituting a greedy method for the original CSO algorithm's roulette wheel mechanism. The grid structure and double archive method are then incorporated as essential ideas from PAES. The pressure vessel design challenge is a real-world case study that is utilized along with several test functions to gauge how well the suggested method performs. Using various measures, including the Reversed Generational

Distance (rGD), Spacing metric, and Spread metric, the experiment compares the proposed method against other well-known algorithms such as the Multi-Objective Mayfly Optimization Algorithm (MMA) [19], Multi-objective Dragonfly Algorithm (MODA) [20], and SPEA2 [11]. The rest of the paper is organized as follows: Section 4.2 presents the proposed GMOCSO algorithm. Section 4.3 discusses the performance evaluation framework for the GMOCSO algorithm. Section 4.4 shows the obtained results and finally, Section 4.5 provides the conclusion and future research directions.

4.2 GRID-BASED MULTI-OBJECTIVE CAT SWARM OPTIMIZATION ALGORITHM

The proposed algorithm adopts two key concepts from the PAES algorithm [10] to extend the CSO algorithm into a multi-objective scheme. First, the proposed algorithm has two populations called internal and external populations. The internal population is used to store the initial population and the dominated individuals, whereas the external archive (repository) is used to store the non-dominated individuals in each iteration. This external archive is empty at the beginning and has a predefined size that cannot be exceeded. Second, the proposed algorithm uses a hyper-grid system to determine the least crowded area in the Pareto front of the objective space. This hyper-grid scheme divides the objective space into several non-overlapping hyper-boxes, where each box might contain several agents. Figure 4.1 explains the hyper-grid system.

The fitness values of these agents denote their coordinates in the objective space. Consequently, this can be used to specify which agent belongs to which hyper-box. Agents inside the least crowded box are called leaders and one of them is used as a global attraction mechanism, where the rest of the agents tend to move toward it. In addition, the roulette wheel method of the CSO algorithm is replaced with the greedy

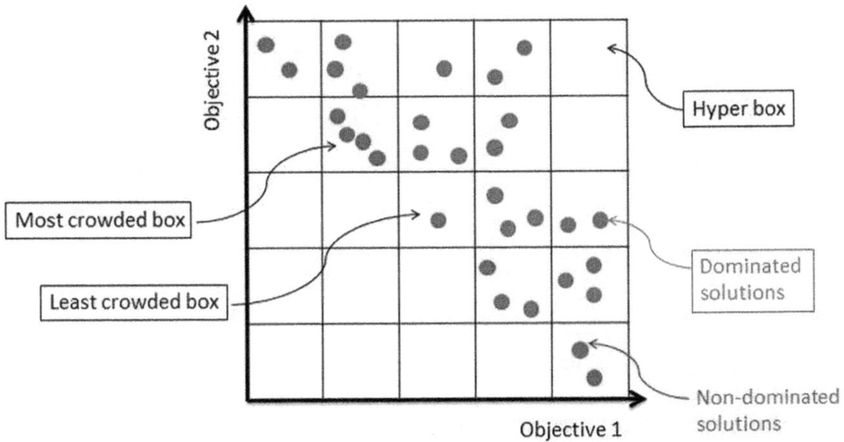

FIGURE 4.1 Explains the hyper-grid system.

method to drive the agents to move toward the global optimum and hence increase the exploitation ratio of the algorithm.

Therefore, the algorithm takes the following steps in the optimization process:

1. Randomly generate the initial population of cats.
2. Calculate the fitness cost for all cats.
3. Determine the non-dominated cats and store them in an external archive.
4. Use the hyper-grid system to create the hyper-boxes.
5. The archive has a predefined size (e.g., archive = 100). If the archive was overfilled, select the most crowded box and delete the extra members.
6. Select the least crowded box and randomly select a member from the box to be the leader.
7. Send the leader to the tracing mode.
8. Assign all the cats to the tracing mode.
   - Update the velocities for all cats using

$$V_{k,d} = V_{k,d} + c_1 * \text{rand} * \left(X_{leader,d} - X_{k,d}\right) \tag{3}$$

   where $V_{k,d}$ is the velocity for the k^{th}cat in the d^{th}dimension; $X_{leader,d}$ is the position of the leader cat, which was chosen from the least crowded box; and $X_{k,d}$ is the current position of the k^{th}cat in the dth dimension. c_1 is a constant and *rand* is a single uniformly distributed random number in the range [0, 1].
   - Update the positions of all cats using

$$X_{k,d} = X_{k,d} + \text{V}_{\text{k,d}} \tag{4}$$

   where $X_{k,d}$ is the position of acat in the dth dimension.
9. Assign the cats to the seeking mode:
   - Use the CDC and SMP parameters to generate the candidate positions, where the original position is one of the candidates.
   - Determinate the non-dominated candidate position to be a new position; if there was more than one non-dominated candidate, randomly select one of them.
10. If any cat dimension is out of the boundary, it is equal to the limits, as in

$$X_{k,d} = \begin{cases} L_d & if \quad X_{k,d} < L_d \\ U_d & if \quad X_{k,d} > U_d \end{cases} \tag{5}$$

   where L_d and U_d are the lower and upper bounds for the search space, respectively.
11. Check if the termination condition is satisfied, then terminate the program. Otherwise, repeat Steps 2–10; see Figure 4.2.

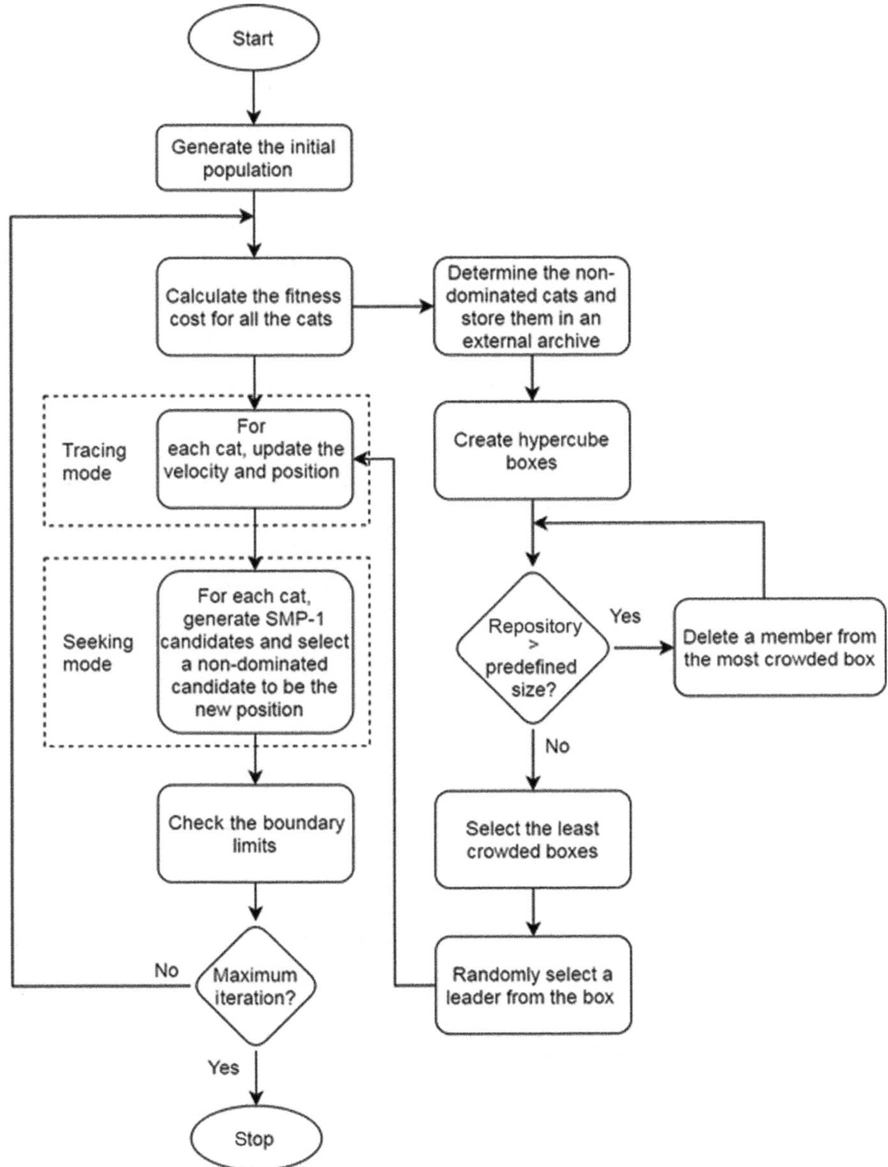

FIGURE 4.2 Flowchart of the GMOCSO algorithm.

4.3 PERFORMANCE EVALUATION FRAMEWORK FOR THE GMOCSO ALGORITHM

In this section, the performance of the proposed GMOCSO algorithm is evaluated using a general performance evaluation framework, as shown in Figure 4.3. The framework consists of two different criteria as follows:

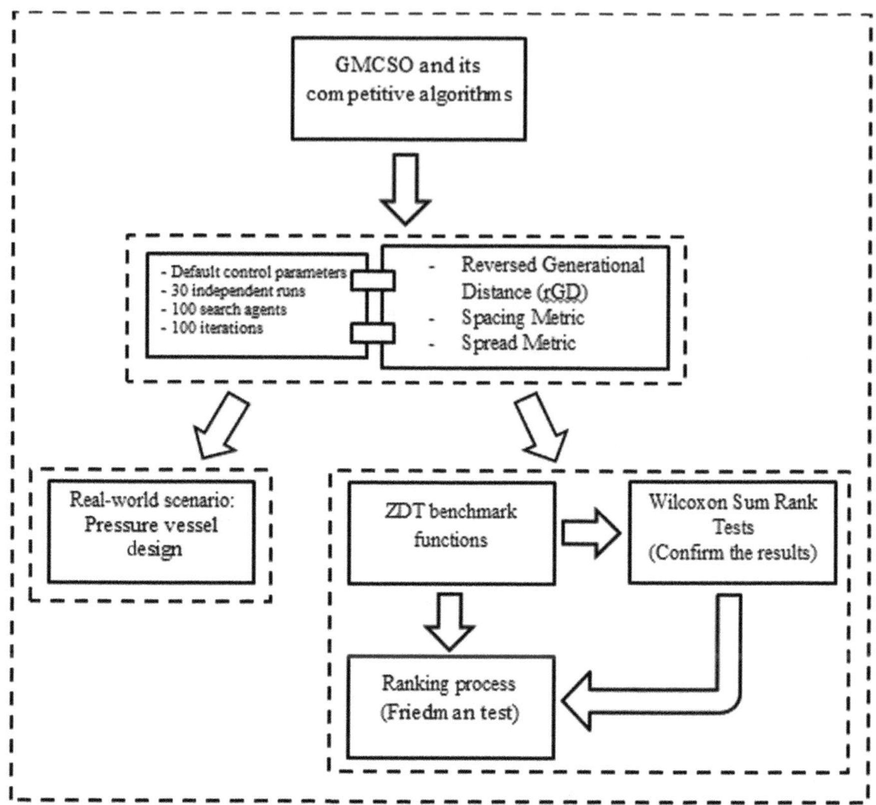

FIGURE 4.3 General performance evaluation framework for the GMOCSO algorithm.

i. Applying the algorithm on ZDT benchmark functions.
ii. Applying the algorithm to a real-world scenario.

4.3.1 Benchmark Functions

To assess the performance of the proposed GMOCSO algorithm, it was first applied to five common benchmark functions: ZDT1, ZDT2, ZDT3, ZDT4, and ZDT6 [21]. Table 4.1 presents their mathematical details

Experiment Setting:
For each benchmark function, the algorithm was executed for 30 independent runs. For each run, the number of both search agents and iterations was equal to 100. Furthermore, the results were compared with three well-known algorithms: MMA [19], MODA [20], and SPEA2 [11]. Parameter settings for these algorithms are presented in Table 4.2.

Then, three essential metrics were used to assess the obtained results, which are listed below.

TABLE 4.1
Mathematical descriptions of the ZDT functions

Function name	Mathematical definition
ZDT1	$f_1(x) = x_1$ $f_2(x) = g(x)\left[1 - \sqrt{x_1/g(x)}\right]$ where $g(x)$ is defined as $g(x) = 1 + 9\left(\sum_{i=2}^{n} x_i\right)/(n-1)$
ZDT2	$f_1(x) = \mathrm{x}_1$ $f_2(x) = g(x)\left[1 - \left(\frac{x_1}{g(x)}\right)^2\right]$ where $g(x)$ is defined as $g(x) = 1 + 9\left(\sum_{i=2}^{n} x_i\right)/(n-1)$
ZDT3	$f_1(x) = x_1$ $f_2(x) = g(x)\left[1 - \sqrt{x_1/g(x)} - \frac{x_1}{g(x)}\sin(10\pi x_1)\right]$ where $g(x)$ is defined as $g(x) = 1 + 9\left(\sum_{i=2}^{n} x_i\right)/(n-1)$
ZDT4	$f_1(x) = x_1$ $f_2(x) = g(x)\left[1 - \sqrt{x_1/g(x)}\right]$ where $g(x)$ is defined as $g(x) = 1 + 10(n-1) + \sum_{i=2}^{n}\left[x_i^2 - 10\cos(4\pi x_i)\right]$
ZDT6	$f_1(x) = 1 - \mathrm{e}^{(-4x_1)}\sin^6(6\pi x_1)$ $f_2(x) = g(x)\left[1 - \left(\frac{f_1(x)}{g(x)}\right)^2\right]$ where $g(x)$ is defined as $g(x) = 1 + 9\left[\frac{\sum_{i=2}^{n} x_i}{(n-1)}\right]^{\frac{1}{4}}$

1. Reversed Generational Distance

The metric rGD is calculated by determining the Euclidean distance between the points of the true Pareto front and the closest point on the approximate Pareto front [22]. The main purpose of this criterion is to show the ability of the different algorithms to determine a set of non-dominated individuals that have the shortest

TABLE 4.2
Parameter setting details for the selected multi-objective algorithms

Algorithms	Parameters	Values
GMOCSO	Population size	100
	Max. iteration	100
	Inertia weight W(I)	1
	SMP	2
	CDC	1
	Constant (C)	1
	No. of grids	10
	SRD	1
	Archive size	100
MMA	Population size	100
	Max. iteration	100
	Inertia weight	0.8
	Inertia weight damping ratio	1
	Personal learning coefficient	1.0
	Global learning coefficient	1.5
	Beta	2
	Mutation coefficient	dance = 0.77, dance_damp=0.99
	Random flight	f l = 0.77, fl_damp = 0.99
	No. of crossover	20
	Mutation rate	0.02
	Archive size	100
MODA	Population size	100
	Max. iteration	100
	Inertia weight W(I)	[0.2 – 0.9]
	Random values	[0,1]
	Archive size	100
SPEA2	Population size	100
	Max. iteration	100
	Archive size	100

distance from the true Pareto optimal front. Therefore, it can be seen that algorithms with the smallest generation distance (rGD) produce the best convergence to the true Pareto front, that is, the approximate Pareto front is the closest to the true Pareto front. Furthermore, the best rGD value that can be obtained by an algorithm is zero, where the approximate Pareto front and true Pareto front are similar. The mathematical details of this metric are as follows:

$$rGD(t) = \frac{\sum_{i=1}^{|\mathcal{P}^*(t)|} d_i}{|\mathcal{P}^*(t)|} \quad (6)$$

where d_i is the Euclidean distance between the ith member of $P^*(t)$ and its closest point from $Q(t)$; $P^*(t)$ is the true Pareto front and $Q(t)$ is the obtained Pareto front.

2. Spacing Metric

The main purpose of the spacing metric (S) is to demonstrate how well the non-dominated individuals are distributed along the approximate Pareto front. The best distribution is where the individuals are equally distributed and the distance between them is equal. The mathematical details of the space metric [23] are as follows:

$$S = \sqrt{\frac{1}{n_{pf}} \sum_{i=1}^{n_{pf}} \left(d_i - \bar{d}\right)^2} \tag{7}$$

where n_{pf} is the number of members in the approximate Pareto front and d_i is the Euclidean distance between the member within the approximate Pareto front and the nearest member in the approximate Pareto front. Therefore, the algorithm with the minimum S value has the best distribution of its individuals. In the best case, the S value is equal to zero, where the solutions are completely uniform and the distance between the non-dominated solutions is equal.

3. Spread Metric or Delta Metric

This metric is also called the Delta metric, which is used to show how far the non-dominated solutions have extended and how well they are spread along the approximate Pareto front. The mathematical details of this metric are as follows:

$$\Delta = \frac{d_f + d_l + \sum_{i=1}^{n_{pf}} \left|d_i - \bar{d}\right|}{d_f + d_l + \left(n_{pf} - 1\right)\bar{d}} \tag{8}$$

where d_f and d_l are the Euclidean distances between the extreme solutions in the true Pareto front and the approximate Pareto front, respectively. Further, d_i is the Euclidean distance between each point in the approximate Pareto front and the closest point in the true Pareto front. n_{pf} and $\bar{d}$ are defined as the total number of members in the approximate Pareto front and the average of all distances, respectively [23].

Subsequently, based on the results obtained for the three metrics, the Friedman test was used to rank the algorithms and determine their overall performances. For this, the average values obtained from the 30 independent runs were used. Meanwhile, the Wilcoxon sum rank test was used to calculate the P-values and confirm the result.

In addition to these well-known metrics, another metric called the elapsed time measurement was used. The elapsed time calculates the amount of time each algorithm requires to finish one independent run.

4.3.2 Applying the GMOCSO Algorithm to a Real-World Scenario

The proposed GMOCSO algorithm was also applied to a real-world scenario called the pressure vessel design problem [24]. A cylindrical pressure vessel is topped at both ends by curved heads, as presented in Figure 4.4.

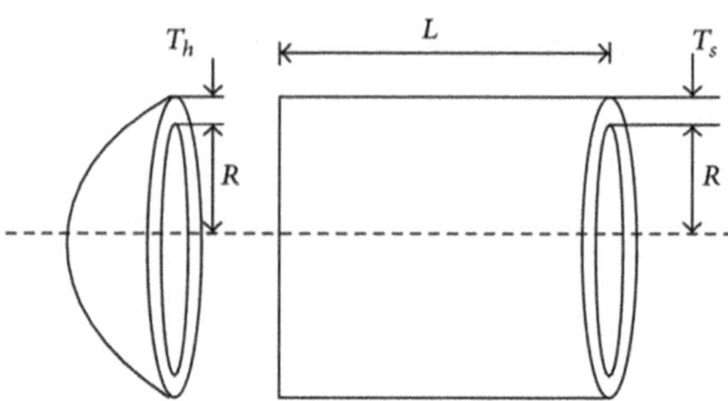

FIGURE 4.4 Pressure vessel design problem, adopted from [24].

The problem considers the design of an air storage tank with a working pressure of 1000 psi and minimum volume of 750 ft³.

The first objective of the problem is to reduce the total cost of a cylindrical pressure vessel. This cost involves the cost of the material and the cost of forming and welding):

$$f_1(x) = 0.6224x_1x_3x_4 + 1.7781x_2x_3^2 + 3.1661x_1^2x_4 + 19.84x_1^2x_3 \quad (9)$$

Subject to

$$g_1(x) = x_1 - 0.0193x_3 \geq 0$$

$$g_2(x) = x_2 - 0.00954x_3 \geq 0$$

$$g_3(x) = \pi x_3^2 x_4 + \frac{4}{3}\pi x_3^3 - 1296000 \geq 0$$

where$x_1, x_2 \in \{1,\ldots..,100\}$, $x_3 \in [10,200]$, and $x_4 \in [10,240]$. x_1 and x_2 are integer multiples of 0.0625.

$x_1 = T_s$, which represents the thickness of the shell.
$x_2 = T_b$, which represents the head of the pressure vessel.
$x_3 = R$, which represents the inner radius.
$x_4 = L$, which represents the length of the cylindrical section.

The second objective of the problem is the sum of the three constraint violations:

$$f_2(x) = \sum_{i=1}^{3} max\{g_i(x), 0\} \quad (10)$$

The experiment settings and parameter settings are all the same as those in the previous section.

4.4 RESULTS AND DISCUSSION

Based on the evaluation criteria mentioned in the previous section, the following results were obtained.

4.4.1 Results for the benchmark functions

The proposed GMOCSO algorithm was first benchmarked on six test functions called the ZDT functions. In the experiment, three more algorithms were chosen for comparison: MMA [19], MODA [20], and SPEA2 [21].

The results of the benchmark functions are presented in Table 4.3.

Based on the results in Table 4.3, the Freidman test was conducted to rank the algorithms and determine the overall performance of the algorithms. According to the results of the test, which are presented in Table 4.4 and Figure 4.5, GMOCSO ranked second. Then, the Wilcoxon sum rank test was also conducted to find the *P*-values and further confirm the results. The results of the Wilcoxon sum rank test are shown in Table 4.5. As can be seen, the *P*-values for all the results were less than 0.05, except for four cases, which are highlighted in bold. Thus, it can be concluded that the results are significant.

Finally, the algorithms were compared using the elapsed time measurement. The results are presented in Table 4.6. As shown, the proposed GMOCSO algorithm was the most efficient and took the least time, whereas the MMA algorithm took the most time. Therefore, it can be concluded that, even though the GMOCSO algorithm performed worse than the MMA algorithm in terms of the rGD, *S*, and Delta metrics, it recompensed the loss by its efficiency.

4.4.2 Results and Discussion of the Pressure Vessel Design Problem

The GMOCSO algorithm was also applied to a real-world scenario called the pressure vessel design problem. Similar to the previous section, the proposed algorithm was compared with three more algorithms: MMA [19], MODA [20], and SPEA2 [21]. The algorithms were run 30 independent times, and each time, the number of search agents and iterations was 100. Then, the average value and standard deviation of these 30 independent runs were taken. Furthermore, the algorithms were also compared using the elapsed time measurement. The results are all presented in Table 4.7. It can be noticed that the proposed GMOCSO algorithm yielded very competitive results.

4.5 CONCLUSION

This paper presented a multi-objective version of the CSO algorithm, which is called GMOCSO. For this, two main modifications were made. First, the seeking mode of the algorithm was modified and the roulette wheel approach was replaced with a greedy method. This technique increased the convergence rate of the algorithm. Second, to preserve the diversity of the non-dominated solutions in the objective space, two key concepts from the PAES algorithm were adopted: the external population and hyper-grid box techniques. In addition, several benchmark functions and a

TABLE 4.3
Comparison results between GMOCSO and the selected algorithms for the ZDT test functions in terms of the average and standard deviation

		GMOCSO		MMA		MODA		SPEA2	
F	Metrics	Average	STD	Average	STD	Average	STD	Average	STD
ZDT1	RGD	0.008251645	0.000645212	**0.004779893**	0.000222114	0.100799368	0.09082163	0.053992	0.017841
	S	0.077806532	0.011078946	0.075503344	0.006606084	**0.027136347**	0.044158279	0.878917	0.238826
	Delta	0.853871219	0.032721397	**0.399518442**	0.039293129	1.364857245	0.160130313	0.853213	0.099897
ZDT2	RGD	0.028268889	0.109917947	**0.004856085**	2.07E-04	0.074477018	0.062369162	34254.15	8481.733
	S	0.075916272	0.016613567	0.076431697	0.006518955	**0.017571498**	0.013397005	463969.7	94282.35
	Delta	0.839039408	0.052525346	**0.395613038**	0.029195821	1.428002616	0.127234485	0.827969	0.139153
ZDT3	RGD	0.009046667	0.001527072	**0.005300991**	0.000211342	0.086483057	0.062537721	0.116075	0.027901
	S	0.217135677	0.025091761	0.269998037	0.02158768	**0.080144325**	0.084676525	0.895159	0.260512
	Delta	0.939763609	0.080065515	**0.559479212**	0.016761967	1.345371429	0.127584933	0.784542	0.100351
ZDT4	RGD	8.22E-03	0.000785028	**0.004776174**	0.000244893	1.83E-01	0.096584815	5.01E-02	0.014521
	S	**8.22E-03**	0.000785028	0.075192952	0.006955307	2.11E-01	0.491719343	6.30E+00	1.252434
	Delta	8.62E-01	0.040205206	**0.39837872**	0.033276381	1.28E+00	0.186181632	9.18E-01	0.174281
ZDT6	RGD	0.007520491	0.001834684	**0.003779973**	0.000120089	0.030610889	0.031934169	0.134059	0.08455
	S	0.256361178	0.310232383	0.233597523	0.339582389	**0.158387045**	0.258385158	3.093395	0.51416
	Delta	1.016366296	0.201172776	**0.589973634**	0.382808442	1.518800475	0.158049857	0.809812	0.144761

TABLE 4.4
Ranking of the selected algorithms for the ZDT test functions (Friedman test)

TF	Metrics	Ranking GMOCSO	Ranking MMA	Ranking MODA	Ranking SPEA2
ZDT1	RGD	2	1	4	3
	S	3	2	1	4
	Delta	3	1	4	2
ZDT2	RGD	2	1	3	4
	S	2	3	1	4
	Delta	3	1	4	2
ZDT3	RGD	2	1	3	4
	S	2	3	1	4
	Delta	3	1	4	2
ZDT4	RGD	2	1	4	3
	S	1	2	3	4
	Delta	2	1	4	3
ZDT6	RGD	2	1	3	4
	S	3	2	1	4
	Delta	3	1	4	2
RGD subtotal		10	5	17	18
RGD ranking		2	**1**	3.4	3.6
S subtotal		11	12	7	20
S ranking		2.2	2.4	**1.4**	4
Delta subtotal		14	5	20	11
Delta ranking		2.8	**1**	4	2.2
Total		35	22	44	49
Overall ranking		2.333333	**1.466667**	2.933333	3.266667

real-world scenario called the pressure vessel design problem were used to evaluate the performance of the proposed algorithm. The results were then compared with three other well-known algorithms. Next, the Friedman test and Wilcoxon sum rank were used to rank the algorithms and confirm the results. Furthermore, the elapsed time measurement was used to test the efficiency of the algorithm. Finally, it can be concluded that the proposed algorithm efficiently converged toward the Pareto front solutions and still preserved its diversity along the area. In the future, we can replace the hyper-grid box with the indicator-based method to obtain better diversity perseverance. We noticed that, in recent years, optimization algorithms have still been considered one of the most promising techniques in terms of being integrated with other technologies. This means that the changes and developments in them are continuing to come up with new methods that can be more effective and capable of yielding more satisfying results. For further study, the reader can optionally read the following research works [25–41].

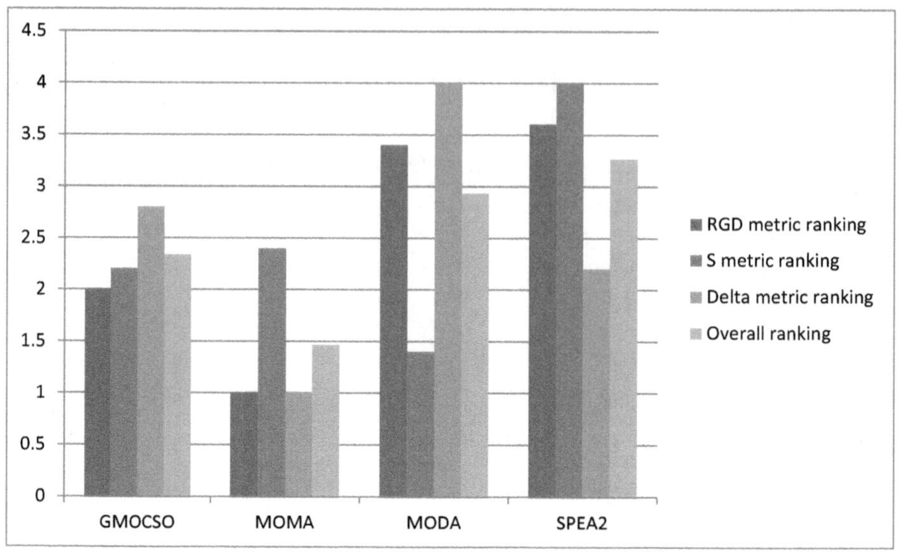

FIGURE 4.5 Ranking of the selected algorithms in the experiment for the ZDT test functions using the Friedman test.

TABLE 4.5
Wilcoxon sum rank tests between the GMOCSO algorithm and the selected algorithms for the ZDT test functions

Functions	Metric	GMOCSO vs. MMA (*P*-values)	GMOCSO vs. MODA (*P*-values)	GMOCSO vs. SPEA2 (*P*-values)
ZDT1	RGD	1.691123e-17	1.691123e-17	1.691123e-17
	S	**0.708193**	5.998246e-13	1.691123e-17
	Delta	1.691123e-17	1.691123e-17	**0.820070**
ZDT2	RGD	1.691123e-17	4.841517e-13	4.841517e-13
	S	**0.320778**	4.841517e-13	3.017967e-11
	Delta	1.691123e-17	1.691123e-17	**0.431747**
ZDT3	RGD	1.691123e-17	1.691123e-17	1.691123e-17
	S	5.798703e-11	4.035171e-09	1.691123e-17
	Delta	1.691123e-17	2.029348e-16	1.727306e-08
ZDT4	RGD	1.691123e-17	1.691123e-17	1.691123e-17
	S	1.691123e-17	1.085811e-10	1.691123e-17
	Delta	1.691123e-17	1.120285e-12	0.035767
ZDT6	RGD	1.691123e-17	7.610055e-16	1.691123e-17
	S	0.004530	0.000452	1.691123e-17
	Delta	0.000011	9.745944e-14	0.000054

TABLE 4.6
Comparison results between the GMOCSO and the selected algorithms for the ZDT test function in terms of the elapsed time measurement

Functions	Elapsed time for GMOCSO (seconds)	Elapsed time for MMA (seconds)	Elapsed time for MODA (seconds)	Elapsed time for SPEA2 (seconds)
ZDT1	**23.18582**	141.5477	28.87808	100.2301
ZDT2	**24.16883**	140.7262	29.17383	101.101
ZDT3	**22.78487**	143.8483	23.47593	100.472
ZDT4	**22.40294**	141.8431	29.39307	100.5177
ZDT6	**19.91391**	150.2155	27.35512	100.095

TABLE 4.7
Comparison results between the GMOCSO and the selected algorithms for the pressure vessel design problem

		Pressure vessel design problem			
Algorithms	Metrics	RGD	S	Delta	Elapsed time (Second)
GMOCSO	Average	**12450.82**	337249.75	0.912582	22.47755
	STD	13194.535	58915.063	0.168256	
MMA	Average	20813.508	309937.61	1.0772376	116.2052
	STD	13987.115	52101.146	0.131381	
MODA	Average	19041.145	**114826.8**	1.5380991	24.88461
	STD	4497.0885	51727.737	0.0962503	
SPEA2	Average	34254.15	463969.7	**0.827969**	100.3218
	STD	8481.733	94282.35	0.139153	

REFERENCES

1. Ahmed, A. M., Rashid, T. A., & Saeed, S. A. M. (2021). Dynamic Cat Swarm Optimization algorithm for backboard wiring problem. *Neural Computing and Applications*, 33(20), 13981–13997.
2. Baskan, O. (Ed.). (2016). *Optimization Algorithms: Methods and Applications*. BoD–Books on Demand.
3. Sharifi, M. R., Akbarifard, S., Qaderi, K., & Madadi, M. R. (2021). A new optimization algorithm to solve multi-objective problems. *Scientific Reports*, 11(1), 1–19.
4. Xin-She, Y. (2010). *An Introduction with Metaheuristic Applications. Engineering Optimization*. John Wiley.
5. Fausto, F., Reyna-Orta, A., Cuevas, E., Andrade, Á. G., & Perez-Cisneros, M. (2020). From ants to whales: Metaheuristics for all tastes. *Artificial Intelligence Review*, 53(1), 753–810.

6. Emmerich, M. T., & Deutz, A. H. (2018). A tutorial on multiobjective optimization: Fundamentals and evolutionary methods. *Natural Computing*, 17(3), 585–609.
7. Deb, K., Agrawal, S., Pratap, A., & Meyarivan, T. (2000, September). A fast elitist non-dominated sorting genetic algorithm for multi-objective optimization: NSGA-II. In *International Conference on Parallel Problem Solving from Nature* (pp. 849–858). Springer.
8. Coello, C. C., & Lechuga, M. S. (2002, May). MOPSO: A proposal for multiple objective particle swarm optimization. In *Proceedings of the 2002 Congress on Evolutionary Computation. CEC'02 (Cat. No. 02TH8600)* (Vol. 2, pp. 1051–1056). IEEE.
9. Zhang, Q., & Li, H. (2007). MOEA/D: A multiobjective evolutionary algorithm based on decomposition. *IEEE Transactions on Evolutionary Computation*, 11(6), 712–731.
10. Knowles, J. D., & Corne, D. W. (2000). Approximating the nondominated front using the Pareto archived evolution strategy. *Evolutionary Computation*, 8(2), 149–172.
11. Zitzler, E., Laumanns, M., & Thiele, L. (2001). SPEA2: Improving the strength Pareto evolutionary algorithm. TIK-report, 103.
12. Ahmed, A. M., Rashid, T. A., & Saeed, S. A. M. (2020). Cat swarm optimization algorithm: A survey and performance evaluation. *Computational Intelligence and Neuroscience*, 2020, 4854895.
13. Sharafi, Y., Khanesar, M. A., & Teshnehlab, M. (2013). Discrete binary cat swarm optimization algorithm. In *Computer, Control & Communication (IC4), 2013 3rd International Conference on 2013 Sep 25* (pp. 1–6). IEEE.
14. Pradhan, P. M., Panda, G. (2012, February 15). Solving multiobjective problems using cat swarm optimization. *Expert Systems with Applications,* 39(3), 2956–64.
15. Orouskhani, M., Teshnehlab, M., & Nekoui, M. A. (2016). Integration of cat swarm optimization and Borda ranking method for solving dynamic multi-objective problems. *International Journal of Computational Intelligence and Applications*, 15(03), 1650014.
16. Orouskhani, M., Teshnehlab, M., & Nekoui, M. A. (2018). EMCSO: An Elitist Multi-Objective Cat Swarm Optimization. *Journal of Optimization in Industrial Engineering*, 11(2), 107–117.
17. Dwivedi, A. K., & Patel, R. N. (2017). Digital filter design using quantum-inspired multiobjective cat swarm optimization algorithm. In *Quantum Inspired Computational Intelligence* (pp. 327–359). Morgan Kaufmann.
18. Zhou, A., Qu, B. Y., Li, H., Zhao, S. Z., Suganthan, P. N., & Zhang, Q. (2011). Multiobjective evolutionary algorithms: A survey of the state of the art. *Swarm and Evolutionary Computation*, 1(1), 32–49.
19. Zervoudakis, K., & Tsafarakis, S. (2020). A mayfly optimization algorithm. *Computers & Industrial Engineering*, 145, 106559.
20. Mirjalili, S. (2016). Dragonfly algorithm: A new meta-heuristic optimization technique for solving single-objective, discrete, and multi-objective problems. *Neural Computing and Applications*, 27(4), 1053–1073.
21. Hassannejad, R., Tasoujian, S., & Alipour, M. R. (2016). Breathing crack identification in beam-type structures using cat swarm optimization algorithm. *Modares Mechanical Engineering*, 15(12), 17–24.
22. Tosun, U. (2015). On the performance of parallel hybrid algorithms for the solution of the quadratic assignment problem. *Engineering Applications of Artificial Intelligence*, 39, 267–278.
23. Applegate, D. L., Bixby, R. E., Chvátal, V., Cook, W., Espinoza, D. G., Goycoolea, M., & Helsgaun, K. (2009). Certification of an optimal TSP tour through 85,900 cities. *Operations Research Letters*, 37(1), 11–15.

24. Yang, X. S., Huyck, C., Karamanoglu, M., & Khan, N. (2013). True global optimality of the pressure vessel design problem: A benchmark for bio-inspired optimisation algorithms. *International Journal of Bio-Inspired Computation*, 5(6), 329–335.
25. Hassan, B. A., Tayfor, N. B., Hassan, A. A., Ahmed, A. M., Rashid, T. A., & Abdalla, N. N. (2024, July). From A-to-Z Review of Clustering Validation Indices. *Neurocomputing*, 601, 128198. doi: 10.1016/J.NEUCOM.2024.128198.
26. Umar, S. U., Rashid, T. A., Ahmed, A. M., Hassan, B. A., & Baker, M. R. (2024, July). Modified Bat Algorithm: A newly proposed approach for solving complex and real-world problems. *Soft Computing*, 1–16. doi: 10.1007/S00500-024-09761-5/METRICS.
27. Hamarashid, H. K., Hassan, B. A., & Rashid, T. A. (2024, September). Modified-improved fitness dependent optimizer for complex and engineering problems. *Knowledge Based Systems*, 300, 112098. doi: 10.1016/J.KNOSYS.2024.112098.
28. Muhammed, R. K., et al. (2024, May). Comparative analysis of AES, Blowfish, Twofish, Salsa20, and ChaCha20 for Image Encryption. *Kurdistan Journal of Applied Research*, 9(1), 52–65. doi: 10.24017/SCIENCE.2024.1.5.
29. Rashid, T. A., et al. (2024). NSGA-II-DL: Metaheuristic optimal feature selection with deep learning framework for HER2 classification in breast cancer. *IEEE Access*, 12, 38885–38898. doi: 10.1109/ACCESS.2024.3374890.
30. Ahmed, A. M., et al. (2024, January). Balancing exploration and exploitation phases in whale optimization algorithm: An insightful and empirical analysis. In *Handbook of Whale Optimization Algorithm: Variants, Hybrids, Improvements, and Applications* (pp. 149–156).doi: 10.1016/B978-0-32-395365-8.00017-8.
31. Hassan, B. A., et al. (2024, January). Equitable and fair performance evaluation of whale optimization algorithm. In *Handbook of Whale Optimization Algorithm: Variants, Hybrids, Improvements, and Applications* (pp. 157–168). doi: 10.1016/B978-0-32-395365-8.00018-X.
32. Abdalla, M. H., et al. (2023). Sentiment analysis based on hybrid neural network techniques using binary coordinate ascent algorithm. *IEEE Access*, 11, 134087–134099. doi: 10.1109/ACCESS.2023.3334980.
33. Hassan, B. A. (2023, November). Ontology learning using formal concept analysis and WordNet., Accessed: July 24, 2024. [Online]. Available: https://arxiv.org/abs/2311.14699v1
34. Rashid, T. A., et al. (2023, June). Awareness requirement and performance management for adaptive systems: A survey. *Journal of Supercomputing*, 79(9), 9692–9714. doi: 10.1007/S11227-022-05021-1/METRICS.
35. Abdulkhaleq, M. T., et al. (2023, January). Fitness dependent optimizer with neural networks for COVID-19 patients. *Computer Methods and Programs in Biomedicine Update*, 3, 100090. doi: 10.1016/J.CMPBUP.2022.100090.
36. Hassan, B. A., & Rashid, T. A. (2021). Artificial Intelligence algorithms for natural language processing and the semantic web ontology learning.
37. Hassan, B. A. (2021). CSCF: A chaotic sine cosine firefly algorithm for practical application problems. *Neural Computing Application*, 33(12). doi: 10.1007/s00521-020-05474-6.
38. Bryar, T. A. R., Hassan, A. (2020). Formal context reduction in deriving concept hierarchies from Corpora using adaptive evolutionary clustering algorithm. *Complex & Intelligent Systems.*
39. Hassan, B. A., & Rashid, T. A. (2021). A multidisciplinary ensemble algorithm for clustering heterogeneous datasets. *Neural Computing Application.* doi: 10.1007/s00521-020-05649-1.

40. Hassan, B. A., & Rashid, T. A. (2020). Operational framework for recent advances in backtracking search optimisation algorithm: A systematic review and performance evaluation. *Applied Math Computing*, 370, 124919.
41. Hassan, B. A., Rashid, T. A., & Mirjalili, S. (2021). Formal context reduction in deriving concept hierarchies from corpora using adaptive evolutionary clustering algorithm star. *Complex & Intelligent Systems*, 1–16.

5 A Task Scheduling Algorithm for Cloud Computing Based on Bio-inspired Multi-objective Manta Rays

Abubakr S. Issa, Sajjad Shamkhi Jaber, Yusra H. Ali and Tarik A. Rashid

5.1 INTRODUCTION

All the expansion of cloud computing's scalability, adaptability, and cost-efficiency has propelled it into a widely embraced platform for executing extensive applications. By provisioning resources such as processing capabilities, storage, and network bandwidth on demand, cloud computing empowers users to distribute their programs across numerous computing nodes (Houssein et al., 2021). With the continuous surge in cloud computing's popularity and adoption, the necessity for efficient task scheduling mechanisms has escalated. These mechanisms play a pivotal role in optimally allocating resources and enhancing overall performance. Task scheduling is a critical element within cloud computing, facilitating the streamlined execution of diverse computational tasks and processes on distributed resources (Pradeep and Jacob, 2018, Abdulredha et al., 2021).

In response to this challenge, the research community has concentrated on devising proficient task scheduling algorithms that cater to multiple service level agreements, augment load balancing, and minimize wait and execution times (Bhalaji, 2019). A spectrum of techniques is employed by cloud task scheduling algorithms, encompassing cost and performance assessment, resource allocation, workflow organization, energy conservation, load balancing, and workload prediction. The core objectives of these techniques are to curtail task execution latency, curb cloud resource consumption, and enhance computational parallelism. To realize these objectives, researchers have introduced an array of task scheduling and resource management algorithms, which have been deployed across diverse cloud environments (Atlam et al., 2018, Chen et al., 2020).

Task scheduling in cloud computing can be categorized into two main types: static and dynamic. Static scheduling involves arranging all tasks before execution based

DOI: 10.1201/9781003601555-5

on available information, whereas dynamic scheduling doesn't have all the task information in advance, making it challenging for the cloud provider to plan effectively. Tasks in cloud computing can take on various configurations depending on their interactions. A "Bag of Tasks" (BoT) is a specific type where multiple tasks can be executed concurrently at the same processing level without interfering with each other (Jaber et al., 2022, Chhabra et al., 2022). The problem of cloud computing task scheduling is complex and difficult, often considered NP-hard (Bacanin et al., 2019). Traditional methods struggle to effectively solve this problem. To address this, recent research has turned to heuristic and metaheuristic algorithms. These algorithms are well suited for optimizing task scheduling in cloud computing due to their effectiveness in solving complex, large-scale problems (Ben Alla et al., 2018).

Workflows in cloud computing involve intricate tasks with interdependencies, making scheduling even more complex. Multi-objective metaheuristic methods have been developed to find near-optimal scheduling solutions considering conflicting optimization goals like cost minimization and makespan reduction (Ibrahim et al., 2020, Chamkoori and Katebi, 2021, Li et al., 2017). Since this problem is NP-complete, finding an optimal solution within a reasonable timeframe is challenging. Swarm intelligence methods, such as those used in optimizing cloud computing work scheduling, have been applied to tackle this difficulty (Alkayal et al., 2016). The research presented in this study focuses on addressing work scheduling challenges within cloud computing, a sophisticated and distributed computer architecture commonly utilized for managing applications like BoT.

To tackle these challenges, a method named the Multi-objective Manta Ray Foraging Optimization Algorithm (MOMRFOA) is introduced. This algorithm draws inspiration from the foraging behavior of manta rays and is employed to optimize task scheduling in cloud environments. The unique aspect of this approach is its application of an optimization algorithm that has not been previously used in cloud task scheduling. This contribution is significant as it introduces a fresh perspective to resolving the task scheduling predicament in cloud computing. The MOMRFOA algorithm aims to strike a balanced and optimal compromise between the time required to complete a set of tasks within the cloud system and the associated costs. Notably, it is designed to be adaptable and flexible, catering to the diverse requirements of various users who may hold distinct objectives concerning execution time and costs. Through comparative testing using diverse datasets of varying sizes, the proposed technique demonstrates a substantial advantage over rival algorithms such as Multi-objective Cuckoo Search (MOCS), Modified Particle Swarm Optimization (MPSO), the Bee Life Algorithm (BLA), Time-cost Aware Scheduling (TCaS), and Round Robin (RR). This superiority is manifested in terms of enhanced Quality of Service (QoS), offering swifter and more cost-effective solutions.

The rest of the chapter is organized as follows. Research is presented to explain recent studies as related work in Section 5.2. The challenge of task scheduling within the realm of cloud computing is elucidated, accompanied by a proposed remedy, in Section 5.3. The proposed algorithm is fully discussed in detail in Section 5.4 to show the ability to optimize the performance of tasks. The discussion of the results with a comparison table is explained in Section 5.5. Eventually, the conclusion is presented in Section 5.6.

5.2 RELATED WORK

With the increasing complexity of task scheduling in cloud computing, researchers have explored diverse strategies to optimize resource allocation and enhance system performance. Among these approaches, the utilization of metaheuristic algorithms for task scheduling has garnered significant attention (Abdulkhaleq, 2022, Hassan, 2021). Metaheuristic algorithms, inspired by biological evolution processes like genetic algorithms and ant colony optimization, have been employed to address task scheduling challenges in cloud computing, yielding promising outcomes. A substantial portion of research endeavors in this domain concentrate on enhancing the task scheduling process by maximizing a multitude of influencing factors, including cost, energy consumption, wait times, deadlines, makespan, and other aspects, as examined in prior investigations (Sujan and Kanniga Devi, 2016, Zhang and Zhou, 2017).

Nguyen et al. (2019) used evolutionary algorithms to improve task scheduling for Internet of Things (IoT)-based BoT applications in cloud-fog computing environments. The goal is to find a balance between the duration of execution and costs, also complying with the demands of the IoT network. Both the lion optimization algorithm and gravitational search algorithm have been hybridized through extensive research to address multi-objective scheduling problems, as demonstrated by Manikandan and Pravin (2019). Huang et al. (2020) proposed a cloud computing task scheduler employing various discrete variants of the particle swarm optimization (PSO) algorithm, the approach has only been tested on small datasets. Karaja et al. (2020) explored a heterogeneous multi-cloud budget-constrained dynamic BoT scheduling technique. An improved Cuckoo Search (CS)-based Scheduling Algorithm, CSDEO, combines Opposition-based Learning, CS, and Differential Evolution (DE) to optimize workload makespan and cloud resource energy usage, as presented by Chhabra et al. (2021). Dubey and Sharma (2021) introduced a multi-hybrid approach employing standard PSO and Ant Colony Optimization for cloud task scheduling. The LATOC approach, proposed in this study, considers crucial task aspects like processing unit requirements, data length, and completion time (Moori et al., 2022). Tekawade and Banerjee (2023) devised a scheduling approach distributing task graph nodes across diverse clouds. Fu et al. (2023) presented a Deep Reinforcement Learning-Enhanced Two-Stage Scheduling Model to tackle the NP-hard problem of operation complexity in IoT systems. This model optimizes computing resource allocation within an edge-enabled infrastructure. Cao and Deng (2023) addressed the Dependent Task Offloading (DTO) problem in a multi-user, multi-edge scenario where interdependent smaller tasks within user tasks are managed.

As stated earlier, most existing task scheduling algorithms are only effective for small datasets and do not consider the trade-offs between multiple objectives. It is important to acknowledge the limitations of such approaches in accurately reflecting real-world scenarios. To address the trade-offs between multiple competing goals, such as cost, makespan, and other relevant factors in task scheduling, the utilization of Pareto optimization techniques or multi-objective algorithms becomes crucial. These techniques enable the exploration of potential solutions that optimize multiple objectives simultaneously, providing a more comprehensive approach to solving multi-objective task scheduling problems. Task scheduling techniques today are mostly multi-objective problems, which are more practical compared to single-objective issues (Chen et al., 2020, Prasanna Kumar and Kousalya, 2020, Abd Elaziz and Attiya, 2021). In multi-objective issues, there is a need to balance multiple goals to arrive at the most appropriate decision. Therefore, this study proposes using a multi-objective approach for task scheduling in cloud environments. To accomplish this, a modified MOMRFOA algorithm is developed that is efficient in solving multi-objective optimization issues.

5.3 BACKGROUND

In this section, the theoretical background of the study is presented to introduce recent work. Many studies are introduced to highlight the problems in task scheduling and their proposed solution.

5.3.1 Problem Formulation

Allocating or scheduling many tasks on available computing resources to maximize one or more objectives presents a challenge in cloud computing systems (energy, makespan, cost, etc.). The mathematical formulation of the task scheduling problem can be formulated by presuming that the cloud system contains a set of (*Npm*) physical machines (*PMs*), and that each *PM* has a set of (*NVM*) virtual machines (*VMs*). According to the data presented by Attiya and Zhang (2017), each individual virtual machine is constructed from a number of computational entities, each of which possesses memory, storage space, processing power, and network bandwidth. Let us assume that there are a number of *NT* jobs that need to be spread among a variety of virtual machines, and that each lth task ($l = 1, 2, 3..., NT$) possesses the following qualities:

$$Tl = [SID_l,\ T\_len_l,\ ECT_l,\ PI_l] \tag{1}$$

where *SIDl* represents the unique identifier of a task, T_len_l (unit: million instructions) represents the length of time required to complete the task, ECT_l represents the earliest completion time of the task, and PI_l represents the priority index of the task. The following matrix equation shows the relationship between *NT* tasks and *NVM* virtual machines in the context of cloud computing task scheduling, where *ECT* denotes the execution completion time, which is the time it takes for a task to be completed on a virtual machine, *NT* denotes the number of tasks that need to be scheduled, and *NVM* denotes the number of available virtual machines:

$$ETC = \begin{bmatrix} \text{ETC1,1} & \text{ETC1,2} & \cdots & \text{ETC 1,Nvm} \\ \text{ETC2,1} & \text{ETC2,2} & \cdots & \text{ETC 2,Nvm} \\ \vdots & & \ddots & \vdots \\ \text{ETC NT,1} & \text{ETC NT,2} & \cdots & \text{ETC NT,Nvm} \end{bmatrix} \tag{2}$$

where *ETCl,j* represents the *ECT* for the lth task on the jth virtual machine and is defined as

$$ETCl, j = \frac{T_{lenl}}{MIPSj}, j = 1, 2, \ldots, NVM \tag{3}$$

where *MIPSj* denotes the jth virtual machine's processing power. As a result, the formulation of the makespan (*MKS*) is

$$MKS = \max_{j \in 1,2,\ldots,Nvm} \sum_{I=1}^{NT} ETCl, j \tag{4}$$

This means that the objective is to find a schedule solution that reduces *MKS* as follows:

$$M = min\ MKS \tag{5}$$

The task's cost function in a cloud environment is indicated by the symbol T_k^i. According to

$$Cost\ (T_k^i) = Cm(T_k^i) + Cp(T_k^i) + Cb(T_k^i) \tag{6}$$

the cost function is made up of the three terms $Cm\ (T_k^i)$, $Cp\ (T_k^i)$, and $Cb\ (T_k^i)$. The first term, $Cm(T_k^i)$, denotes the task's communication cost. The task's processing cost, or the number of computing resources needed to complete the work, is represented by the second term, $Cp(T_k^i)$. The task's storage cost, or the amount of memory needed to store the job, is represented by the third term, $Cb(T_k^i)$, as seen in the following three equations:

$$Cp(T_k^i) = c1 * ETC(T_k^i) \tag{7}$$

$$Cm\ (T_k^i) = c2 * Mem(T_k^i) \tag{8}$$

$$Cb(T_k^i) = c3 * Bw(T_k^i) \tag{9}$$

where *c1* represents the cost of a node's CPU consumption per time unit, whereas *c2* represents the cost of a node's memory usage per data unit. The memory required by a task is denoted by $Mem(T_k^i)$. $ETC(T_k^i)$ is defined in Eq. (3) and represents the execution time of task T_k at node *Pi*. The bandwidth consumption cost per data unit, written as *c3*, and bandwidth quantity, denoted by $Bw(T_k^i)$, are equal to the total of the sizes of the input and output files:

$$Manimize: \sum_{i=1}^{M} \left(\sum\nolimits_{Tj \in Pi} \left(Cost\left(T_k^i\right) \right) \right) \tag{10}$$

5.3.2 Manta Ray Foraging Optimization (MRFO) Algorithm

Nature has always been an inspiration for scientists and engineers. The MRFO algorithm is a bio-inspired method that mimics manta rays' natural foraging conduct. It is an algorithm based on swarm intelligence that mimics the three foraging techniques used by manta rays: chain foraging, tumbling foraging, and spiral foraging. In terms of convergence accuracy and effectiveness in addressing cloud job scheduling issues, the method has been demonstrated to surpass other intelligent algorithms including Gravitational Search Algorithm (Rashedi et al., 2009), PSO, and Artificial Bee Swarm Optimization. Additionally, Zhao et al. (2020), proposed the MRFO algorithm in 2020, which makes it very recent. It is shown in Algorithm 1. This makes it a fascinating and attractive field of research for enhancing cloud work scheduling due to its novelty. The MRFO algorithm is simpler to develop than other existing algorithms and has fewer operators and parameters.

Algorithm 1. Manta Ray Foraging Optimization Algorithm

Initialize population size *P*, maximal number of iterations *T* and each manta ray

$$Y_i(t) = Y_l + rand(Y_u - Y_l) \text{ for } i = 1,.., P \text{ and } t = 1$$

Calculate every single individual's fitness $f_i = f(Y_i)$ and achieve the greatest solution discovered thus far Y_{best}, where Y_u and Y_l are the top and lower problem space bounds, correspondingly.

While (stopped condition fails) do

FOR $i = 1$ to P Do

IF *rand* < 0.5 then // cyclone foraging

IF $\frac{t}{T_{max}} < rand$ Then

$$Y_{rand} = Y_l + rand(Y_u - Y_l)$$

$$Y_i(t+1) = \begin{cases} Y_{rand} + r(Y_{rand} - Y_i(t)) + \beta(Y_{rand} - Y_i(t)) i = 1 \\ Y_{rand} + r(Y_{i-1} - Y_i(t)) + \beta(Y_{rand} - Y_i(t)) i = 2,.,P \end{cases}$$

ELSE

$$Y_i(t+1) = \begin{cases} Y_{best} + r(Y_{best} - Y_i(t)) + \beta(Y_{best} - Y_i(t)) i = 1 \\ Y_{best} + r(Y_{i-1} - Y_i(t)) + \beta(Y_{best} - Y_i(t)) i = 2,..,P \end{cases}$$

END IF

ELSE // Chain foraging

$$Y_i(t+1) = \begin{cases} Y_i(t) + r(Y_{best} - Y_i(t)) + \alpha(Y_{best} - Y_i(t)) i = 1 \\ Y_i(t) + r(Y_{i-1} - Y_i(t)) + \alpha(Y_{best} - Y_i(t)) i = 2,\cdots,P \end{cases}$$

END IF

Fit every individual $f(Y_i(t+1))$ iff $(Y_i(t+1)) < f(Y_{best})$

THEN $Y_{best} = Y_i(t+1)$

// somersault foraging

FOR $i = 1$ TO P DO

$$Y_i(t+1) = Y_i(t) + S(r_2 Y_{best} - r_2 . Y_i(t))$$

Fit every individual $(Y_i(t+1))$. IF$f(Y_i(t+1)) < f(Y_{best})$

THEN $Y_{best} = Y_i(t+1)$

End For

End While

Return Y_{best}

The process of initialization involves the creation of a population of solutions referred to as "manta ray" within a designated search space. The algorithm's exploration and convergence are determined by the population size (*P*) and maximum number of iterations (*T*). The fitness function ($f(Y_i)$) is employed to evaluate the quality of each solution proposed by the manta ray. The aforementioned function assesses the performance of a solution in relation to the objectives of the problem. In addition, the algorithm stores Y_{best}, which represents the best answer that has been identified.

The iterative search method, known as MOMRFOA, employs an iterative process to refine population solutions. An algorithm that randomly selects between cyclone or chain foraging strategies for each individual ray determines the foraging behavior of manta rays in a group.

When the cyclone foraging strategy is chosen, the algorithm generates a new position, denoted by Y_{rand}, by randomly moving inside the search space. The position of the manta ray at time $t + 1$ ($Y_i(t + 1)$) is modified by applying random offsets and weights with respect to Y_{rand}.

The chain foraging algorithm involves the movement of the manta ray toward the optimal solution (Y_{best}) to update its position ($Y_i(t+1)$). This update employs stochastic offsets and weights akin to the foraging behavior of cyclones.

Fitness Evaluation: Reassessing the fitness value of the updated solution ($Y_i(t+1)$) following the adjustment of its position. The value of Y_{best} is changed to the new answer if it demonstrates superior performance.

The MOMRFOA framework encompasses the practice of somersault foraging, along with cyclone and chain foraging techniques. Manta rays do somersaults in a direction that is randomly determined in relation to the Y_{best} reference point. This enhances the process of exploration.

Termination: The algorithm iterates through these phases until a specified stopping condition, such as reaching a convergence threshold or completing a certain number of iterations, is satisfied.

The function MOMRFOA returns the variable Y_{best}, which represents the optimal solution obtained from the search process.

5.4 PROPOSED WORK

This section describes how the proposed task scheduling method reduces execution time and running expenses for BoT applications in the cloud. The MOMRFOA approach cannot map continuous values to computational nodes, making it unsuitable for cloud-based job scheduling. The discrete work environment has altered certain algorithm parameters.

5.4.1 Initial Population

The initial population can consist of all people who were used by the MOMRFOA to find the best answer. The population is thought to consist of *N* people. The *N*

Tasks	T1	T2	T3	T4	T5	T6	T7
Nodes	2	1	1	3	2	3	1

FIGURE 5.1 Initial solution format.

individuals are chosen at random during initialization to ensure that there is a wide range of individuals in the first generation, as well as to discover various locations in the search space. To make the next generation, certain members of the original population are changed in ways that have already been planned. The solution is shown as a list of tasks that must be done by the CPUs that are available. One potential contender for a solution is shown in Figure 5.1.

5.4.2 Fitness Function

After the initial solution is created, the fitness value of the individual is evaluated and recorded. It affects whether the best solution can be discovered and how quickly the algorithm will converge. The fitness function is usually derived from the problem's goal function. Section 5.3 Eqs. (5) and (10) are applied to define the fitness function. Considering cost and makespan, a multi-objective function is utilized.

5.4.3 MOMRFOA Algorithm

The MRFO's start is defined randomly to generate random populations for a set of individuals X, followed by getting the most desirable individual X_{best} with the best fitness value. Three strategies are used to update the individual. These foraging approaches are harnessed to produce a novel solution (vector) by utilizing the α and βparameters incorporated in the search equations. These parameters serve as initial values for generating a fresh random vector, with task limits ranging from the lowest (seed) to the highest (VM number), followed by the evaluation of the fitness of each vector. The following are the mathematical models of these strategies.

1. Chain foraging: individual X_i is modified at iteration k according to

$$X_i(t+1)=\begin{cases} X_i(t)+r(X_{best}-X_i(t))+\pm(X_{best}-X_i(t))\, i=1 \\ X_i(t)+r(X_{i-1}-X_i(t))+\pm(X_{best}-X_i(t))\, i=2,\ldots,P \end{cases} \quad (11)$$

where $X_i(t+1)$ is the new solution, $X_i(t)$ is the old solution, r is a random vector in [0, 1], α is the weight coefficient, and X_{best} is the best solution.

2. Cyclone foraging: The individual X_i is updated using

$$Y_i(t+1)=\begin{cases} X_{best}+r(X_{best}-X_i(t))+\beta(X_{best}-X_i(t))\, i=1 \\ X_{best}+r(X_{i-1}-X_i(t))+\beta(X_{best}-X_i(t))\, i=2,\ldots,P \end{cases} \quad (12)$$

The weight coefficient is denoted by the variable β. Moreover, the cyclone foraging step could modify its position inside the search space in order to enhance the exploration capabilities of the MRFO. Therefore, the present individual location is revised utilizing

$$X_i(t+1)=\begin{cases} X_{rand}+r(X_{rand}-X_i(t))+\beta(X_{rand}-X_i(t))\, i=1 \\ X_{rand}+r(X_{i-1}-X_i(t))+\beta(X_{rand}-X_i(t))\, i=2,\ldots,P \end{cases} \tag{13}$$

$$Xrand = Low + r * (Up - Low) \tag{14}$$

The variables "*Low*" and "*Up*" represent the lower and upper boundaries of the search space, respectively.

3. Somersault foraging: The act of somersault foraging refers to foraging behavior observed in certain animal species when individuals engage in somersaults. The mathematical expression employed to update the individual during this phase is represented by

$$X_i(t+1)=X_i(t)+S(r_2.X_{best}-r_3.X_i(t)) \tag{15}$$

Let S represent the somersault factor, which is assigned a value of 2. Additionally, r_2 and r_3 are random values that fall in the range 0 to 1.

The adaptation of α and βvalues is tailored to accommodate the depiction of multi-objective task scheduling in cloud computing.

5.4.4 Pareto-Optimal Techniques

The primary benefit lies in Pareto optimization's ability to offer multiple scheduling alternatives instead of a solitary solution. Pareto solutions delve into balancing diverse objectives, including minimization, cost reduction, and resource optimization, enabling adaptable and interactive decision-making. In this study, the focus is on two objectives: makespan and cost.

5.5 EXPERIMENTAL RESULTS AND COMPARISON

The forthcoming sections present the outcomes of simulations and an evaluation of the suggested approach. Both aspects rely on insights derived from preceding tests.

5.5.1 Multi-objective Cuckoo Search (MOCS) Algorithm

The inspiration for the CS algorithm is drawn from the flight patterns of birds. A phenomenon called brood parasitism, wherein a cuckoo lays eggs in the nest of a different bird species, is mirrored in this algorithm. The eggs of other birds symbolize potential solutions and their nests represent the search space. Incorporating Lévy flying behavior facilitates swift convergence toward the global optimum (Yang and Deb, 2009). The conventional CS algorithm focuses on individual optimization functions, whereas in MOCS, certain behaviors and rules from the original algorithm have been adjusted to encompass multi-objective optimization.

5.5.2 Bee Life Algorithm (BLA)

BLA is an optimization approach inspired by honeybees, introduced in 2005 (Karaboga, 2005). This algorithm emulates the foraging behavior of bees in search of food. In a related investigation, the bee swarm algorithm was utilized to enhance fog computing job scheduling, achieving equilibrium between CPU execution duration and allocated memory.

5.5.3 Modified Particle Swarm Optimization (MPSO)

MPSO is an altered version of PSO designed to conform to the specific problem being addressed (Li and Engelbrecht, 2007). In MPSO, particle velocities are updated for *pBest* and *gBest* daily, ensuring particles continuously progress toward the optimal solution attained thus far. This approach guarantees that the particles maintain a constant trajectory toward the most efficient solution achieved up to that point.

5.5.4 Experimental Settings

This research was assessed by the utilization of Jupyter notebooks on a Google Colab training server. Twelve datasets, encompassing scheduling challenges spanning from 40 to 1000 tasks, were employed to assess the MOMRFOA method. Cloud nodes exhibit distinct computational capacities and resource costs. These nodes are characterized by the million instructions per second (MIPS) processing potential, accompanied by CPU, memory, and bandwidth expenses. Consequently, cloud nodes vary in both cost and capability. Table 5.1 outlines the attributes of the 13 cloud system processing units, which are high-performance data center nodes responsible for cloud-tier tasks. These nodes execute tasks with enhanced efficiency due to their heightened computing prowess. The currency used in the simulation, grid dollars (G$), signifies the elevated energy expenses incurred by the cloud. The cloud system dissects user requests into numerous tasks, each being evaluated based on

TABLE 5.1
Components of the cloud system

Parameters	Cloud tier	Unit
CPU rate	[500, 5000]	MIPS
Memory usage cost	[0.01, 0.05]	G$/MB
CPU usage cost	[0.1, 1.0]	G$/s
Bandwidth usage cost	[0.01, 0.1]	G$/MB
Number of nodes	13	node

TABLE 5.2
Task characteristics

Property	Value	Unit
Number of instructions	[1, 100]	109 instructions
Output file size	[10, 100]	MB
Memory required	[50, 200]	MB
Input file size	[10, 100]	MB

TABLE 5.3
Proposed algorithms for conditions used to evaluate efficacy

Parameter	Acronym	Possible settings
Population Size	I	100
No. of systems	nSystem	10
No. of runs	nRun	3
No. of generations	Max_g	500
No. of nodes	nC	13
No. of tasks	m	{40,80,120,160,200,250, 300,350,400,450,500,1000}

memory, instructions, and I/O file size. The quantity of tasks and required resources for each task are determined by the demand. Experimental roles for the datasets are randomly assigned features, as depicted in Table 5.2. The experiment encompasses various scenarios, some demanding substantial computational capabilities, while others necessitate increased bandwidth or memory due to variations in task types.

5.5.5 Experimental Results

In this section, we assess the efficacy of the proposed MOMRFOA algorithm in addressing the task scheduling challenge within a cloud setting. The algorithm's performance-affecting parameter configurations are outlined in Table 5.3 for reference.

Tables 5.4 and 5.5 present the results of various task scheduling algorithms, namely MOMRFOA, MOCS, TCAS, BLA, MPSO, and RR, for three different systems. Each system was run ten times, and each run consisted of tasks with different magnitudes represented by m values ranging from 40 to 1000. The cloud system model used in the experiments had a parameter nC set to 13. Notably competitive results, indicating superior performance, are highlighted in bold in the tables. Table 5.4 specifically showcases the cost function, while Table 5.5 displays the makespan.

Table 5.4 presents a comparison of six distinct algorithms designed for various tasks, evaluating their overall costs. The cumulative cost reflects the summation of performance metrics, where lower values indicate superior algorithm performance. The highlighted values in bold correspond to the lowest total costs within each row.

TABLE 5.4
Cost comparison of the six algorithms

No. of tasks	MOMRFOA	MOCS	TCaS*	BLA*	MPSO*	RR*
40	**565.54**	666.36	733.85	730.88	740.38	755.98
80	**1111.34**	1486.42	1540.23	1540.85	1553.43	1581.82
120	**1720.65**	2142.03	2363.37	2367.59	2381.16	2418.90
160	**2381.84**	3038.37	3176.17	3183.80	3197.01	3246.69
200	**2995.74**	3692.46	3755.21	3767.37	3778.27	3835.40
250	**3805.42**	4745.66	4988.94	5017.17	5007.59	5079.06
300	**4605.87**	5776.54	5832.69	5877.41	5862.52	5935.94
350	**5356.41**	6649.92	6607.52	6653.37	6632.38	6738.19
400	**6117.86**	7645.42	7738.56	7816.04	7759.95	7875.39
450	**6941.48**	8810.02	8845.90	8926.57	8876.30	9016.59
500	**7765.10**	9693.35	9902.64	9995.97	9921.76	10097.7
1000	**19538.23**	19782.622	19898.641	19665.671	19774.671	19995.671

* Results of these algorithms from the research paper by Nguyen (2019) #1.

TABLE 5.5
Comparison of the six algorithms' makespans

No. of tasks	MOMRFOA	MOCS	TCaS*	BLA*	MPSO*	RR*
40	**171.07**	186.38	191.44	207.57	215.27	456.11
80	**303.80**	374.02	394.69	416.09	445.35	963.08
120	**483.70**	526.76	607.71	654.92	686.2	1495.6
160	**680.93**	780.35	819.97	917.67	932.33	1912.0
200	**866.94**	906.20	940.87	1067.4	1065.9	1905.7
250	**1082.4**	1187.9	1270.9	1490.6	1479.8	3054.8
300	**1293.9**	1481.4	1473.7	1765.4	1712.7	3309.1
350	**1543.7**	1681.5	1949.0	2010.7	1911.2	3638.1
400	**1731.0**	1933.3	1944.5	2421.9	2300.5	4347.6
450	**1980.3**	2298.3	2235.1	2840.7	2751.2	5350.3
500	**2229.6**	2475.3	2503.0	3174.3	3067.2	6023.7
1000	**4653.489**	4778.3239	4887.8166	4890.439	4981.439	5317.439

* Results of these algorithms from the research paper by Nguyen (2019) #1.

The table illustrates that the MOMRFOA algorithm consistently exhibits the lowest total cost across all task scenarios. The table exhibits a comparison of makespan among six distinct algorithms across varying task quantities. The term "makespan" pertains to the duration required to complete all tasks within a provided schedule. The table illustrates a consistent trend wherein the makespan increases proportionally Comparison of related work and the proposed algorithm with the task volume

TABLE 5.6
Comparison of related work and the proposed algorithm

Ref.	Used algorithm	Gap of research	Limitation	Results
Nguyen et al. (2019)	EA	Optimize the task scheduling problem	Time, transmission costs, processing resources, and energy use must be optimized to satisfy users	Greater QoS and better time-cost balance than BLA
Manikandan and Pravin (2019)	HBLBA LGSA	Cloud computing	Not suited to many situations	Outperformed other algorithms in profit, cost, and energy use.
Huang et al. (2020)	PSO	Absence of discrete task scheduling PSO algorithm performance assessments	VM homogeneity and task size uniformity	Minimized expenses and maximized resources
Karaja et al. (2020)	BoTs	Reduce makespan while considering budget	May not be ideal for online scheduling or real-world situations	Improved makespan and cost efficiency over prior methods
Chhabra et al. (2021)	CSDEO	Inadequate scheduling techniques for cloud computing systems' complexity	Requirement for broader dataset testing and premature convergence	Makespan and energy consumption were better than those for other state-of-the-art algorithms
Dubey and Sharma (2021)	Hybrid multi-faceted	Lack of a complete cloud computing job scheduling solution that can boost performance and efficiency	Assumption of homogenous VMs and static task allocation threshold	Outperformed existing methods in makespan time, task scheduling time, and cloud service provider profit
Moori et al. (2022)	List-based heuristic	Scheduling workflows on multi-cloud systems to minimize makespan, cost, or dependability	Not explicitly mentioned	State-of-the-art cost, makespan, and dependability were 12%, 11%, and 1.1%, respectively

(continued)

TABLE 5.6 (Continued)
Comparison of related work and the proposed algorithm

Ref.	Used algorithm	Gap of research	Limitation	Results
Tekawade and Banerjee (2023)	DRL	Complexity of end-edge-cloud IoT computing processes	Necessity for lots of training data and great computational complexity	Outperformed three scheduling methods in learning efficiency and scheduling performance
Fu et al. (2023)	WOA IWC	Need for more efficient and effective cloud computing task scheduling	Need more testing and optimization for larger systems	WC approach surpassed conventional methods in convergence speed, accuracy, and resource use
Cao and Deng (2023)	HGSWC	Need for more efficient and effective cloud task scheduling metaheuristic algorithms	Further testing on larger and more complicated workloads, and premature convergence are possible.	Excelled in makespan and cloud resource use
Proposed work	MOMRFOA	Cloud computing	No limitations observed	Compared to other task scheduling methods in cost, job completion time, resource consumption, and workload balance

for all algorithms. Notably, the RR algorithm displays the highest makespan, while the MOMRFOA algorithm consistently showcases the lowest makespan across all task counts. The remaining algorithms exhibit intermediate makespans. The table's numerical values are presented with varying degrees of precision based on the methodology and task quantity.

Figures 5.2 and 5.3 offer a juxtaposition of total cost and makespan for six distinct algorithms. The total cost serves as a metric encompassing the overall expenditure associated with accomplishing a task set. The objective of this comparison is to identify the algorithm with the least total cost among the six. Makespan, on the other hand, gauges the time required to conclude a task set. Meanwhile, Figure 5.4 presents a contrast between time and makespan specific to the MOMRFOA algorithm. In this instance, the MOMRFOA algorithm is evaluated against itself at

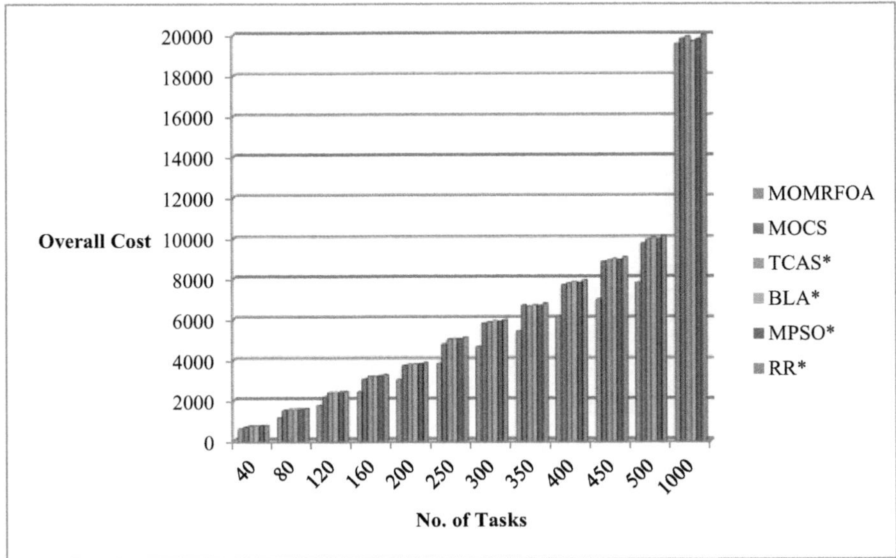

FIGURE 5.2 Comparison of the six algorithms' total costs.

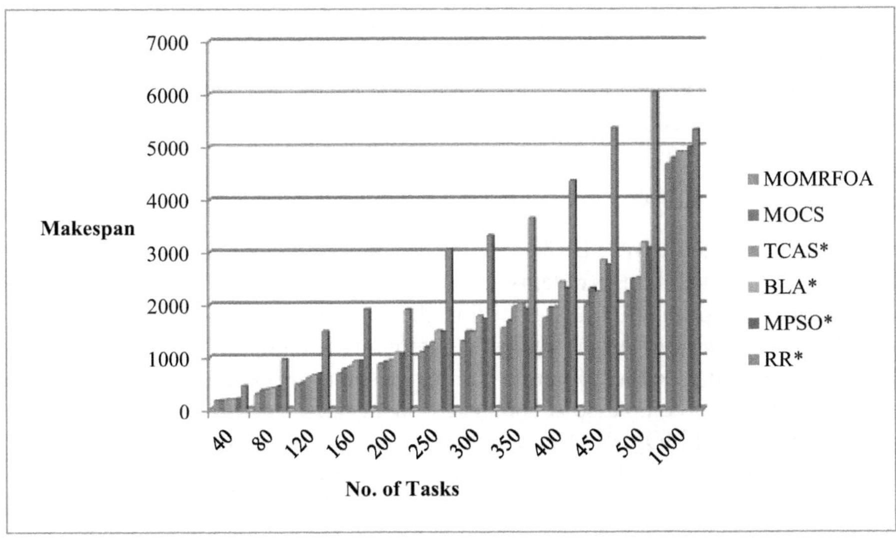

FIGURE 5.3 Comparison of the six algorithms by makespan.

varying time intervals to discern alterations in its performance over time. The time and makespan parameters are depicted on a single graph to elucidate their interrelation. Lastly, Table 5.6 shows the comparison between the proposed method and related work.

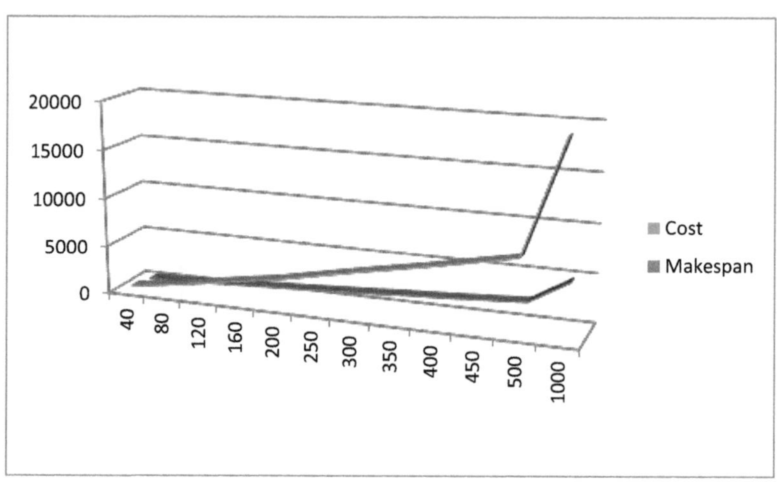

FIGURE 5.4 Comparison of the MOMRFOA algorithms' time and makespan.

In summary, it can be observed that the MOMRFOA algorithm exhibits superior optimization performance, lower cost, reduced makespan, and a higher level of load balancing when compared to alternative metaheuristic algorithms. The MOMRFOA algorithm demonstrates enhanced optimization skills through its ability to escape local optima. The findings presented in this study provide evidence of the efficacy of the suggested method in addressing job scheduling challenges in cloud computing.

5.6 CONCLUSION

This work presents the MOMRFOA algorithm, which is designed to address multi-objective task scheduling problems in the context of cloud computing for BoT applications. The primary objective of this algorithm is to improve system performance by effectively utilizing the available computing resources. The method helps to minimize cost and execution duration in comparison to other algorithms. The results of the experimental comparisons conducted on 12 data sets of tasks within a cloud system indicate that MOMRFOA outperforms other methods such as MOCS, MPSO, TCaS, BLA, and RR in effectively identifying optimal trade-off solutions between makespan and cost. Future endeavors will explore the incorporation of additional QoS parameters to enhance the framework.

REFERENCES

Abd Elaziz, M. & Attiya, I. 2021. An improved Henry gas solubility optimization algorithm for task scheduling in cloud computing. *Artificial Intelligence Review,* 54, 3599–3637.

Abdulredha, M. N., Bara'a, A. A. & Jabir, A. J. 2021. An evolutionary algorithm for task scheduling problem in the cloud-fog environment. *Journal of Physics: Conference Series*, IOP Publishing, 012044.

Abdulkhaleq, M. T., Rashid, T. A., Alsadoon, A., Hassan, B. A., Mohammadi, M., Abdullah, J. M., Chhabra, A., Ali, S. L., Othman, R. N., Hasan, H. A. & Azad, S. 2022. Harmony search: current studies and uses on healthcare systems. *Artificial Intelligence in Medicine*, 131, 102348.

Alkayal, E. S., Jennings, N. R. & Abulkhair, M. F. 2016. Efficient task scheduling multi-objective particle swarm optimization in cloud computing. *2016 IEEE 41st Conference on Local Computer Networks Workshops (LCN Workshops)*, IEEE, 17–24.

Atlam, H. F., Walters, R. J. & Wills, G. B. 2018. Fog computing and the internet of things: a review. *Big Data and Cognitive Computing,* 2, 10.

Attiya, I. & Zhang, X. 2017. D-choices scheduling: a randomized load balancing algorithm for scheduling in the cloud. *Journal of Computational and Theoretical Nanoscience,* 14, 4183–4190.

Bacanin, N., Bezdan, T., Tuba, E., Strumberger, I., Tuba, M. & Zivkovic, M. 2019. Task scheduling in cloud computing environment by grey wolf optimizer. *2019 27th telecommunications forum (TELFOR),*. IEEE, 1–4.

Ben Alla, H., Ben Alla, S., Touhafi, A. & Ezzati, A. 2018. A novel task scheduling approach based on dynamic queues and hybrid meta-heuristic algorithms for cloud computing environment. *Cluster Computing,* 21, 1797–1820.

Bhalaji, N. 2019. Delay diminished efficient task scheduling and allocation for heterogeneous cloud environment. *Journal of trends in Computer Science and Smart technology (TCSST),* 1, 51–62.

Cao, Z. & Deng, X. 2023. Dependent task offloading in edge computing using GNN and deep reinforcement learning. *arXiv preprint arXiv:2303.17100.*

Chamkoori, A. & Katebi, S. 2021. Robustness of the storage in cloud data centers based on simple swarm optimization algorithm. *Security and Communication Networks,* 2021, 1–17.

Chen, X., Cheng, L., Liu, C., Liu, Q., Liu, J., Mao, Y. & Murphy, J. 2020. A WOA-based optimization approach for task scheduling in cloud computing systems. *IEEE Systems Journal,* 14, 3117–3128.

Chhabra, A., Sahana, S. K., Sani, N. S., Mohammadzadeh, A. & Omar, H. A. 2022. Energy-aware bag-of-tasks scheduling in the cloud computing system using hybrid oppositional differential evolution-enabled whale optimization algorithm. *Energies,* 15, 4571.

Chhabra, A., Singh, G. & Kahlon, K. S. 2021. Multi-criteria HPC task scheduling on IaaS cloud infrastructures using meta-heuristics. *Cluster Computing,* 24, 885–918.

Dubey, K. & Sharma, S. C. 2021. A hybrid multi-faceted task scheduling algorithm for cloud computing environment. *International Journal of System Assurance Engineering and Management,* 1–15.

Fu, B., Huo, Y. & Zhao, H. 2023. Sublinear approximation schemes for scheduling precedence graphs of bounded depth. *arXiv preprint arXiv:2302.00133.*

Hassan, B.A., 2021. CSCF: a chaotic sine cosine firefly algorithm for practical application problems. *Neural Computing and Applications*, 33(12), 7011–7030.

Houssein, E. H., Gad, A. G., Wazery, Y. M. & Suganthan, P. N. 2021. Task scheduling in cloud computing based on meta-heuristics: review, taxonomy, open challenges, and future trends. *Swarm and Evolutionary Computation,* 62, 100841.

Huang, X., Li, C., Chen, H. & An, D. 2020. Task scheduling in cloud computing using particle swarm optimization with time varying inertia weight strategies. *Cluster Computing,* 23, 1137–1147.

Ibrahim, M., Nabi, S., Hussain, R., Raza, M. S., Imran, M., Kazmi, S. A., Oracevic, A. & Hussain, F. 2020. A comparative analysis of task scheduling approaches in cloud computing. *2020 20th IEEE/ACM International Symposium on Cluster, Cloud and Internet Computing (CCGRID)*, IEEE, 681–684.

Jaber, S., Ali, Y. & Ibrahim, N. 2022. An automated task scheduling model using a multi-objective improved Cuckoo optimization algorithm. *International Journal of Intelligent Engineering and Systems,* 15, 295–304.

Karaboga, D. 2005. An idea based on honey bee swarm for numerical optimization. Technical report-tr06, Erciyes university, engineering faculty, computer.

Karaja, M., Ennigrou, M. & Said, L. B. 2020. Budget-constrained dynamic bag-of-tasks scheduling algorithm for heterogeneous multi-cloud environment. *2020 International Multi-Conference on: "Organization of Knowledge and Advanced Technologies" (OCTA)*, IEEE, 1–6.

Li, J., Ma, T., Tang, M., Shen, W. & Jin, Y. 2017. Improved FIFO scheduling algorithm based on fuzzy clustering in cloud computing. *Information,* 8, 25.

Li, X. & Engelbrecht, A. P. 2007. Particle swarm optimization: an introduction and its recent developments. *Proceedings of the 9th Annual Conference Companion on Genetic and Evolutionary Computation*, 3391–3414.

Manikandan, N. & Pravin, A. 2019. LGSA: hybrid task scheduling in multi objective functionality in cloud computing environment. *3D Research,* 10, 1–16.

Moori, A., Barekatain, B. & Akbari, M. 2022. LATOC: an enhanced load balancing algorithm based on hybrid AHP-TOPSIS and OPSO algorithms in cloud computing. *The Journal of Supercomputing,* 78, 4882–4910.

Nguyen, B. M., Thi Thanh Binh, H., The Anh, T. & Bao Son, D. 2019. Evolutionary algorithms to optimize task scheduling problem for the IoT based bag-of-tasks application in cloud–fog computing environment. *Applied Sciences,* 9, 1730.

Pradeep, K. & Jacob, T. P. 2018. CGSA scheduler: a multi-objective-based hybrid approach for task scheduling in cloud environment. *Information Security Journal: A Global Perspective,* 27, 77–91.

Prasanna Kumar, K. & Kousalya, K. 2020. Amelioration of task scheduling in cloud computing using crow search algorithm. *Neural Computing and Applications,* 32, 5901–5907.

Rashedi, E., Nezamabadi-Pour, H. & Saryazdi, S. 2009. GSA: a gravitational search algorithm. *Information Sciences,* 179, 2232–2248.

Sujan, S. & Kanniga Devi, R. 2016. An efficient task scheduling scheme in cloud computing using graph theory. *Proceedings of the International Conference on Soft Computing Systems: ICSCS 2015*, Volume 2. Springer, 655–662.

Tekawade, A. & Banerjee, S. 2023. A cost effective reliability aware scheduler for task graphs in multi-cloud system. *2023 15th International Conference on COMmunication Systems & NETworkS (COMSNETS)*, IEEE, 295–303.

Yang, X.-S. & Deb, S. 2009. Cuckoo search via Lévy flights. *2009 World Congress on Nature & Biologically Inspired Computing (NaBIC)*, IEEE, 210–214.

Zhang, P. & Zhou, M. 2017. Dynamic cloud task scheduling based on a two-stage strategy. *IEEE Transactions on Automation Science and Engineering,* 15, 772–783.

Zhao, W., Zhang, Z. & Wang, L. 2020. Manta ray foraging optimization: an effective bio-inspired optimizer for engineering applications. *Engineering Applications of Artificial Intelligence,* 87, 103300.

6 Exploring the Potential of Modified Metaheuristic Optimized Long Short-term Memory Neural Networks for Earthquake Magnitude Forecasting

Nebojsa Bacanin, Luka Jovanovic, Milos Dobrojevic, Miodrag Zivkovic, Milos Antonijevic and Mohamed Salb

6.1 INTRODUCTION

The intricate interplay of stress and strain occurring between tectonic plates, coupled with various geological phenomena, culminates in the occurrence of earthquakes, a natural disaster of profound impact. These seismic events are particularly pronounced in regions where tectonic plates converge or interact, resulting in a higher frequency of such occurrences. The specialized field of seismology is dedicated to comprehensively studying earthquakes, focusing on both the seismic waves they generate and the underlying geological processes. Seismologists play a crucial role in monitoring and analyzing seismic events globally, meticulously recording the intensity and geographical locations of these occurrences. The amassed wealth of data provides a fertile ground for the application of data-driven methodologies, offering the promise of developing robust earthquake forecasting techniques (N. Yang, Li, Zhao, Zhang, & Zhao, 2023).

However, despite the abundance of seismic data, challenges persist in achieving uniformity in data formats, collection methods, and presentation standards. This lack of consistency poses a hurdle in accurately forecasting earthquake events, adding complexity to an already intricate task. The urgency for an effective forecasting system becomes evident considering the devastating impact of earthquakes (Karataş,

DOI: 10.1201/9781003601555-6

Ateş, Alptekin, Dal, & Yakar, 2023). Nevertheless, the inherently nonlinear nature of this challenge further complicates efforts to devise a comprehensive and reliable forecasting methodology. Addressing these complexities is essential for advancing our ability to predict and mitigate the destructive consequences of earthquakes.

Advancements in artificial intelligence (AI) in recent times have created new opportunities to tackle the problems related to earthquake forecasting. AI in seismology has the potential to significantly increase forecast efficiency and accuracy, which will improve our capacity to lessen the effects of seismic occurrences. Since earthquake precursors are frequently subtle and intricate (Pulinets, Ouzounov, Karelin, & Boyarchuk, 2023), advanced pattern detection skills are needed. Machine Learning (ML) algorithms may be trained to recognize minute signals or abnormalities in seismic data that can be signs of an impending earthquake. Incorporating AI and a time series forecasting models for seismic activity can help scientists and emergency responses workers make more informed decisions. Additionally, the detection of aftershocks (Im & Avouac, 2023) can prove vital to preserving lives in disaster scenarios.

The challenge of proper hyperparameter selection further complicates the application of ML algorithms in earthquake forecasting. Many algorithms are initially designed with a focus on achieving good general performance across various tasks. However, adapting these algorithms to specific applications with desired outcomes requires careful adjustment of hyperparameters (Bischl et al., 2023). The task of hyperparameter tuning becomes particularly daunting due to the nature of emerging algorithms, which introduce an increasing number of parameters. This process can be considered NP-hard, and traditional approaches often involve a trial-and-error method. In recent years, researchers have increasingly turned to metaheuristic algorithms to address the complexity of hyperparameter selection. These innovative approaches aim to optimize the algorithm's performance by exploring the hyperparameter space more efficiently, providing a promising avenue for overcoming challenges in the customization of ML models for earthquake prediction.

This work examines the prospect of Long Short-term Memory (LSTM) (Cahuantzi, Chen, & Güttel, 2023) neural networks for earthquake magnitude forecasting. Recognizing that the performance of LSTM networks heavily relies on appropriate hyperparameter choice, an altered variant of the Sine Cosine Algorithm (SCA) (Mirjalili, 2016) is proposed specifically tailored to meet the requirements of this study. The proposed approach is evaluated using real-world, publicly available seismological data, and a comparative analysis is executed with several state-of-the-art optimization methods. The contributions of the conducted research can be summarized as follows:

- the proposal of a robust timeserver-based approach for earthquake prediction based on LSTM networks;
- the incorporation of metaheuristic optimizer algorithms for tuning LSTM parameters to improve performance; and
- the introduction of an altered version of SCA that builds on the excellent results of the original.

The remainder of this chapter follows the structure outlined below: Section 6.2 reviews prior work that provided support and inspiration for this research. Section 6.3 describes the methodology, encompassing both the original and modified algorithms introduced in this work. The experimental setup and obtained results are detailed in Sections 6.4 and 6.5, respectively. Lastly, Section 6.6 summarizes the research and puts forth suggestions for future research endeavors.

6.2 RELATED WORK

This section commences by giving the background on the topic of the earthquake prediction problem. Afterwards, a brief overview of Recurrent Neural Networks (RNNs) for time series prediction is provided, followed by metaheuristic optimization. Finally, this section introduces LSTM networks utilized in the simulations.

6.2.1 Seismology and Earthquake Prediction

Seismology has witnessed numerous recent contributions, with notable examples, including the development and integration of fiber optic technology for sensing, offering high-quality data and wide-bandwidth monitoring. Further contributions have been achieved with advancements in satellite-based seismic monitoring, enabling researchers to track ground movements and enhance global seismic studies. The applications of ML techniques to earthquake signal detection have demonstrated remarkable success, extending their application from traditional seismic data and events to innovative methods, such as earthquake identification from digitized images of analog seismograms (Jia & Ye, 2023; Mousavi & Beroza, 2023). Additionally, researchers have employed ML in earthquake detection using smartphone data (Kong, Inbal, Allen, Lv, & Puder, 2019) and for discerning low-frequency earthquakes and tremors (Thomas, Inbal, Searcy, Shelly, & Bürgmann, 2021).

RNNs, and LSTM networks in particular, are very appropriate for detecting earthquakes, as they can process sequential data and identify temporal dependencies. These are vital to the analysis of seismic signals (Berhich, Belouadha, & Kabbaj, 2023; Bilal, Wang, Ji, Akhter, & Liu, 2023). Moreover, LSTM models are capable of detecting patterns in seismic waves, thus allowing the very accurate identification and classification of earthquakes and other tectonic events like active volcanoes (Kavianpour, Kavianpour, Jahani, & Ramezani, 2023; Titos et al., 2023). Such methods may be implemented as part of real-time earthquake tracking and early warning systems that may provide important early alerts to minimize the impacts (A. Wang et al., 2023). This diversification underscores the versatility of ML approaches in advancing earthquake detection across various data sources and phenomena.

6.2.2 Recurrent Neural Networks

Emerging recurrent models have demonstrated remarkable proficiency when applied to time series forecasting in various complex nonlinear applications (Masini, Medeiros, & Mendes, 2023). Notable examples include forecasting in finance and stock markets, as well as cloud system forecasting (Predić et al., 2023). While the potential of recurrent models is widely recognized, it is crucial to acknowledge that simple RNNs may encounter limitations (Fang, Ma, Pan, Yang, & Arce, 2023). The LSTM was developed as a promising solution to address some of these challenges.

Finding the optimal hyperparameters poses a significant challenge in neural networks in general. The quest for identifying the most suitable hyperparameters for each dataset or problem presents a computationally complex NP-hard challenge (Hassan & Rashid, 2021; Jovanovic, Jovanovic, et al., 2023). Traditional or manual methods are prohibitively time-demanding in this context, prompting the exploration of alternative methods. On the other hand, metaheuristic algorithms continuously demonstrate proficiency in solving complex problems, especially those classified as NP-hard. Furthermore, the No Free Lunch (NFL) theorem (Wolpert & Macready, 1997) underscores that there is no universal solution applicable to every single optimization task, highlighting the necessity for experimentation and evaluation of different algorithms tailored to each specific task.

6.2.3 Metaheuristic Optimization

With the diverse array of metaheuristic methods, each characterized by distinct behaviors or patterns found in nature, it is possible to classify them into four main families: swarm intelligence, evolutionary, physics-based, and human-inspired methods. Some methods have gained prominence in the research community thanks to their consistently robust performance. Methods like particle swarm optimization (D. Wang, Tan, & Liu, 2018) and the Genetic Algorithm (GA) (Mirjalili & Mirjalili, 2019), are perhaps the two most well-known algorithms. Others are inspired by collaborative behavioral patterns expressed by groups of individuals, such as the Artificial Bee Colony (ABC) (Karaboga & Basturk, 2008), Firefly Algorithm (FA) (X.-S. Yang, 2010), and the more recent Reptile Search Algorithm (Abualigah, Abd Elaziz, Sumari, Geem, & Gandomi, 2022). Moreover, abstract concepts also served as inspiration for algorithms, like the Self-Adaptive Step-Size Search (Azad & Hasançebi, 2014) and COLSHADE (Gurrola-Ramos, Hernàndez-Aguirre, & Dalmau-Cedeño, 2020) algorithms. All these algorithms comprise of two distinctive phases, namely exploration and exploitation, and to achieve good performance (Hassan et al., 2024), it is vital to determine the appropriate balance between these phases (Ahmed et al., 2024).

Utilizing optimization metaheuristics yielded considerable benefits in addressing a variety of complex challenges. Often employed for tuning hyperparameters of other methods, metaheuristic techniques have demonstrated remarkable success across a broad spectrum of application domains, such as environmental sciences (Jovanovic, Bacanin, Simic, et al., 2023; Jovanovic et al., 2022), green energy production and distribution (Bacanin, Jovanovic, et al., 2023; Damaševičius et al., 2024; Stoean et al., 2023), financial aspects (Bacanin, Zivkovic, Jovanovic, Ivanovic, & Rashid, 2022; Todorovic et al., 2023), sentiment analysis (Abdalla et al., 2023), and the medical domain (Abdulkhaleq et al., 2022, 2023; Jovanovic, Bacanin, Zivkovic, et al., 2023; Majidpour et al., 2024; Zivkovic, Bacanin, et al., 2022). Hybridization is distinguished as a prevalent strategy in the realm of optimization, seeking to overcome the constraints associated with existing optimizers. Extensive literature emphasizes that the integration of multiple methods often yields outcomes that surpass the performance achieved by simple baseline models (Hassan, 2021). This approach, where ML models have been paired with metaheuristic algorithms, has been widely applied over different domains, including networks and computer security (Dobrojevic et al., 2023; Salb et al., 2023), intrusion detection (AlHosni et al., 2022; Savanović et al., 2023; Zivkovic, Jovanovic, et al., 2022), cloud and edge systems (Bacanin, Zivkovic, Bezdan, Venkatachalam, & Abouhawwash, 2022; Predić et al., 2023), and spam detection (Bacanin, Zivkovic, Stoean, et al., 2022; Salb et al., 2022). Medicine in particular is observed as a promising area (Bacanin, Zivkovic, Al-Turjman, et al., 2022; Jovanovic, Kljajic, et al., 2023; Minic et al., 2023; Pilcevic et al., 2023; Zivkovic, Bacanin, et al., 2022, 2021; Zivkovic, K, et al., 2021), where significant enhancements have been attained by applying hybrid methods.

6.2.4 Long Short-term Memory

The LSTM architecture was devised as an improved variant of RNNs (Salehinejad, Sankar, Barfett, Colak, & Valaee, 2017), aiming to efficiently handle the input in the form of sequential time series by recognising and preserving long-term connections (Hochreiter, 1991). Opposite to the baseline RNN, the LSTM model is efficient in handling the issue of a vanishing or exploding gradient (Hochreiter & Schmidhuber, 1997). Moreover, it is considerably adaptable to time series input data. Finally, LSTM overcomes the drawback of the baseline RNN, where it was observed that the input signals fade away while moving along the propagation paths, decreasing the network's ability to efficiently manage past events.

This approach employs an advanced four-layer gating process that facilitates the dynamic and precise management, adjustment, and removal of cell inputs to allow the maintenance of long-term memory. Inside the LSTM model, the cell state functions as a storage unit designated to keep essential information. The forget gate, represented by the sigmoid function, has a pivotal role in deciding which data should be disregarded, efficiently eliminating unnecessary data from the cell state:

$$f_t = \sigma(W_f x_t + U_f h_{t-1} + b_f) \tag{1}$$

where f_t denotes the forget gate, x_t denotes its inputs in time step t, h_{t-1} denotes the previous hidden state, W_f and U_f are the weight coefficients associated with the inputs, and b_f denotes the vector of biases.

The decision about which incoming information is to be kept in the cell state is enabled by a specific sigmoid function known as the input gate (i_t):

$$i_t = \sigma(W_i x_t + U_i h_{t-1} + b_i) \tag{2}$$

where W_i and U_i denote the appropriate weight factors, and b_i represents the bias.

An additional collection of candidate solutions is produced through the application of a hyperbolic tangent (*tanh*) layer:

$$\tilde{C}_t = tanh(W_c x_t + U_c h_{t-1} + b_c) \tag{3}$$

In this process, it is vital to take into account the weight coefficients, namely W_c, U_c, and the bias term denoted by b_c.

Using an element-wise product represented by ⊙ alongside the forget gate f_t, the preceding cell state C_{t-1} will be dropped to make room for the new cell state C_t, filled with the current data. The newly obtained candidate values $\tilde{C}_t$ are multiplied by the input gate i_t and then combined with this result:

$$Ct = ft \odot Ct - 1 + it \odot \tilde{C}t \tag{4}$$

The initial sigmoid output o_t undergoes further manipulation, relying on the cell state:

$$ot = \sigma(Wo\ xt + Uo\ ht{-}1 + bo) \tag{5}$$

and the resultant calculation is then sent through the *tanh* layer, finally resulting in producing the fresh hidden state:

$$h_t = o_t \odot tanh(C_t) \tag{6}$$

where W_o and U_o represent the weight coefficients, and b_o denotes the bias.

LSTM networks have found numerous applications for different time series prediction tasks, including cryptocurrencies (Mizdrakovic et al., 2024), pollution monitoring (Bacanin, Sarac, et al., 2022), oil prices (Jovanovic et al., 2022), gold prices (Stankovic et al., 2023), and cloud load prediction (Bacanin, Simic, Zivkovic, Alrasheedi, & Petrovic, 2023). Medicine is also a significant application domain, where LSTM has been used to predict arrhythmia (Alamatsaz et al., 2024; Tiwari et al., 2023), epilepsy (X. Wang, Wang, Liu, Wang, and Wang, 2023), and autism (Lakhan, Mohammed, Abdulkareem, Hamouda, & Alyahya, 2023), among other successful use cases. LSTMs are also widely implemented in applications like language modeling and speech recognition as they are capable of capturing complex temporal patterns.

6.3 METHODS

This section sheds light on the elementary version of the SCA, and discusses the drawbacks of this basic variant. Afterwards, a modified variant of the algorithm is suggested to address these constraints.

6.3.1 Basic SCA

The original variant of SCA metaheuristics was devised and presented by Mirjalili et al. (Mirjalili, 2016), belonging to the group of population-based methods that draw inspiration from the mathematical behavior of elementary trigonometry functions. The algorithm commences by synthesizing a random set of solutions. Later, the position of each solution in the population is updated by employing sine and cosine functions:

$$Y_i^{t+1} = Y_i^t + r_1 \times sin(r_2) \cdots \left| r_3 \cdots P_i^t - Y_i^t \right| \tag{7}$$

$$Y_i^{t+1} = Y_i^t + r_1 \times cos(r_2) \cdots \left| r_3 \cdots P_i^t - Y_i^t \right| \tag{8}$$

where the i-th component belonging to the observed solution is represented by Y_i^t during round t and Y_i^{t+1} denotes the freshly updated solution in the following round. Control variables of the algorithm that guide the motion of the solutions are labeled r_1, r_2, and r_3. The location of the destination is denoted by P_i^t.

An arbitrary value labeled r_4 is utilized to determine if the sine or cosine equation shall be employed with respect to the following condition:

$$Y_i^{i+1} = \begin{cases} Y_i^{t+1} = Y_i^t + r_1 \cdot sin\left(r_2\right) \cdot \left| r_3 \cdot P_i^{*t} - Y_i^t \right|, & r_4 < 0.5 \\ Y_i^{t+1} = Y_i^t + r_1 \cdot cos\left(r_2\right) \cdot \left| r_3 \cdot P_i^{*t} - Y_i^t \right|, & r_4 \geq 0.5 \end{cases} \tag{9}$$

where $r_4 \in [0,1]$.

These two trigonometric functions permit exploitation in the proximity of the current solutions. The balance between the exploration and exploitation phases is allowed through the control variable r_1, which is updated as defined by

$$r1 = a - t\frac{a}{T}, \tag{10}$$

where the ongoing iteration is represented by t, T is the maximum count of iterations allowed in a single execution of the algorithm, and a represents a constant value.

The value of r_2 is inside the limits $[0,2\pi]$ and r_3 represents an arbitrary value in the range $[-2,2]$.

6.3.2 Hybrid Adaptive Sine Cosine Algorithm

SCA performs better than many other algorithms that are currently in use. Even though it's a relatively new addition to optimization techniques, trials are essential to find possible enhancements. Its efficiency has been increased by a number of changes. To improve variety, the main change consists of integrating Quasi-Reflexive Learning (QRL) (Basha et al., 2021) initialization. QLR initialization helps to improve the convergence rate and solution quality by allowing the algorithm to commence close to the promising areas of the search space, and consequently enhances the variety of solutions and overall algorithmic efficacy and performance. To clarify, the first half of the population is normal initialized, whereas the second half is created according to the implementation

$$A_Z^{qr} = rand\left(\frac{lb_z + ub_z}{2}, a_z\right) \tag{11}$$

where *rad* represents a random value inside the specified interval, and *lb* and *ub* designate the search space's lower and upper bounds, respectively. Previous research has demonstrated that the QRL mechanism enhances exploration.

Hybridization in an adaptable way using the Firefly Algorithm (FA) (X.-S. Yang & He, 2013b) handles additional boosts to convergence, as FA is renowned for its powerful search. Hybridization allows taking the best of both SCA and FA, while canceling out their individual drawbacks. The mechanism of firefly search is explained by

$$X_i(t+1) = X_i(t) + \beta e^{-\gamma r_{ij}^2}\left(X_j(t) - X_i(t)\right) + \alpha \epsilon_i(t), \tag{12}$$

where β_0 represents the attractiveness at $r = 0$. However, Eq. (12) is commonly swapped for the following equation to improve computational performance:

$$\beta(r) = \frac{\beta_0}{\left(1 + \gamma \times r^2\right)} \tag{13}$$

where r_{ij} indicates the current location of agent j during the corresponding iteration t and $X_i(t)$ indicates the current position of agent i at a particular iteration t. The distance between agents indexed as i and j, which serves as a measure of their mutual attraction, is indicated by the parameter β. The agent attraction coefficient is β; the light absorption coefficient is γ; the degree of randomness is α; and a stochastic vector is $\epsilon_i(t)$. In this hybridization, the FA's function is to increase exploitation in the last few optimization rounds.

An empirically determined value is chosen for an introduced parameter $\theta = 0.8$, and a random value between 0 and 1 is picked for value ψ in each optimization iteration. FA search is performed if $\psi > \theta$; if not, SCA search is performed. After every iteration, θ is reduced by 0.4, which raises the likelihood that FA will be utilized. With this strategy, the updated version of the algorithm may profit from the early exploration boost from QRL while concentrating on convergence in subsequent iterations.

The updated method is formulated using a hybridization and adaptive approach, resulting in an optimizer known as the hybrid adaptive SCA (HASCA). Algorithm 1 presents the pseudocode for this optimizer.

Algorithm 1 Pseudocode for the proposed HASCA

Initialize $\frac{1}{2}$ of the population P
Apply QRL to generate latter half of agents.
Set $\theta = 0.8$ **while** $T > t$ **do**
  Evaluate population fitness
  Generate a random value for ψ
  if $\psi > \theta$ **then**
    Apply FA search **else**
    Apply SCA search
  end if $\theta = \theta - 0.4$
end while

6.4 EXPERIMENTAL SETUP

To validate the suggested methodology, a publicly available seismology dataset is utilized that was last accessed on the Earthquake Hazards Program website[1] on February 7, 2024. The datasets cover all earthquakes occurring over a 30-day basis from January 5 to February 4. The intensity of earthquakes is logged in Richter local magnitude scale. The data are converted into a time series format with 6 lags (samples provided as inputs) in the hope of predicting the magnitude one, two, and three earthquakes ahead. Such an approach provides a good basis for detecting aftershocks, and allowing communities and individuals to plan disaster relief and recovery. A visualization of the dataset is provided in Figure 6.1.

Models are constructed and trained with the initial 70% of the available data, with an additional 10% serving for validation and the last 20% withheld for testing. Several metaheuristic optimizers are included in a comparative analysis with the introduced HASCA. The algorithms included in the evaluations are the original SCA (Mirjalili, 2016), Genetic Algorithm (GA) (Mirjalili & Mirjalili, 2019), Bat Algorithm (BA) (X.-S. Yang & He, 2013a), ABC (Karaboga & Basturk, 2008), and Fitness Dependent Optimizer (FDO) (Abdullah & Ahmed, 2019). Optimization algorithms are independently implemented in Python, using the default hyperparameters suggested in the works that originally introduced each respective algorithm. Each algorithm is allocated a population size of five agents and allowed a total of five iterations to enhance the results. Additionally, simulations are carried out through 30 independent runs to ensure a fair comparison is presented despite the randomness associated with the heuristic approach. Optimizers are tasked with selecting optimal network architectures and training parameter form ranges shown in Table 6.1

To evaluate the performance of the proposed method, several metrics are monitored during the experiments. These include the mean absolute error (MAE), mean squared error (MSE), root MSE (RMSE), coefficient of determination (R^2) and the index of alignment (IA):

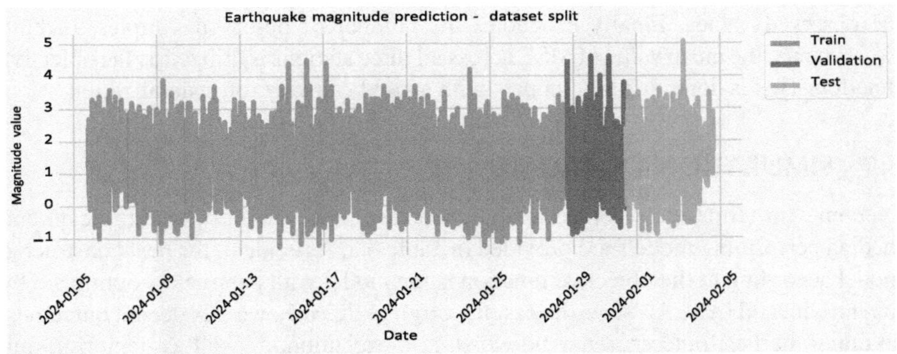

FIGURE 6.1 Dataset visualization.

TABLE 6.1
LSTM optimizer parameters and their respective ranges

Parameter	Lower bound	Upper bound
Learning rate	0.0001	0.01
Dropout	0.05	0.2
Epochs	300	600
Layers	1	3
Layer neurons	100	200

$$MAE = \frac{1}{n}\sum_{i=1}^{n}\left|b_i - \hat{b}_i\right| \tag{14}$$

$$MSE = \frac{1}{n}\sum_{i=1}^{n}\left(b_i - \hat{b}_i\right)^2 \tag{15}$$

$$RMSE = \sqrt{\frac{1}{n}\sum_{i=1}^{n}\left(b_i - \hat{b}_i\right)^2} \tag{16}$$

$$R^2 = 1 - \frac{\sum_{i=1}^{n}\left(b_i - \hat{b}_i\right)^2}{\sum_{i=1}^{n}\left(b_i - \bar{b}\right)^2} \tag{17}$$

$$IA = 1 - \frac{\sum_{i=1}^{n}\left(b_i - \hat{b}_i\right)^2}{\sum_{i=1}^{n}\left(\left|b_i - \bar{b}\right| + \left|b_i - \hat{b}_i\right|\right)^2} \tag{18}$$

where b_i denotes the observed values, $\hat{b}_i$ the forecast value, and $\bar{b}$ the mean of the observed values. Finally n denotes the number of observed samples. In both experiments, the mean value of MSE across all three stations is utilized as the objective function. This is done to attain models with a good capacity for generalization.

6.5 SIMULATION OUTCOMES

Outcomes in terms of the objective function results for the best, worst, mean, and median performing models are provided in Table 6.2. As evident, the best constructed model according to the objective function is the model with parameters optimized by the introduced HASCA. Nevertheless, the original SCA showcases decent outcomes, attaining the best outcomes for the worst case execution, as well as demonstrating admirable stability. The BA algorithm showcases good outcomes in the mean and median executions, despite not attaining the best outcomes.

TABLE 6.2
Objective function comparisons in terms of the best worst mean and median constructed models

Method	Best	Worst	Mean	Median	Std	Var
LSTM-HASCA	**0.018787**	0.020068	0.019428	0.019428	0.000640	4.10E-07
LSTM-SCA	0.019071	**0.019094**	0.019083	0.019094	**0.000011**	**1.26E-10**
LSTM-GA	0.018864	0.020623	0.019626	0.018864	0.000872	7.60E-07
LSTM-BA	0.018828	0.019271	**0.019020**	**0.018828**	0.000220	4.84E-08
LSTM-ABC	0.019300	0.019735	0.019518	0.019518	0.000217	4.72E-08
LSTM-FDO	0.019110	0.019546	0.019328	0.019328	0.000218	4.74E-08

Stability comparisons can further be compared in Figure 6.2, where the admirable consistency of the SCA is evident, despite the algorithm not attaining the best outcomes. Additionally, swarm plots for the best agents in each iteration are shown in Figure 6.3.

Detailed metric comparisons for the best performing models in terms of overall performance for all three prediction steps are provided in Table 6.3. As evident, the best outcomes in terms of R^2, MSE, and MAE are showcased by the models optimized by the introduced HASCA. The best outcomes in terms of MSE are showcased by the BA and the best IA is attained by the ABC algorithm.

Detailed metrics for each individual prediction step are shown in Table 6.4. A clear dominance for all three prediction steps is evident in the results showcased by the introduced algorithm attaining the best outcomes in terms of MSE, RMSE, MAE, and IA for one step ahead, and R^2, MSE, RMSE, and MAE for two steps ahead; however, only the best outcomes in terms of R^2 and MAE are attained for three steps ahead. The GA demonstrates good performance for longer-term predictions, with the best MSE and RMSE outcomes for predictions three steps ahead.

An additional consideration when assessing optimizers is the convergence rates demonstrated during optimization. Comparisons in terms of convergence are demonstrated in Figure 6.4. A clear improvement over the original SCA is evident, with the original algorithm showcasing a slower convergence rate, resulting in a local minima being located. The introduced algorithm showed rapid convergence toward the best attained outcome, suggesting that the introduced modifications demonstrate a sufficient basis for improvement. Finally, to support the easier reproducibility of the experiments, the parameter selections obtained by all metaheuristics for the respective highest-achieving models are showcased in Table 6.5.

The simulation outcomes show the considerable potential of LSTM models tuned by the metaheuristic algorithms, which may be integrated into earthquake detection and early warning systems in the future. The described LSTM networks may enhance the accuracy and responsiveness of these systems by providing the capability to analyze complex seismic data in real time. Seismic data typically include a variety of time series signals originating from different sensors, which can be fed to the pretrained models deployed for real-time earthquake tracking. These models can then process incoming data, identify potential earthquakes, and trigger early warning systems.

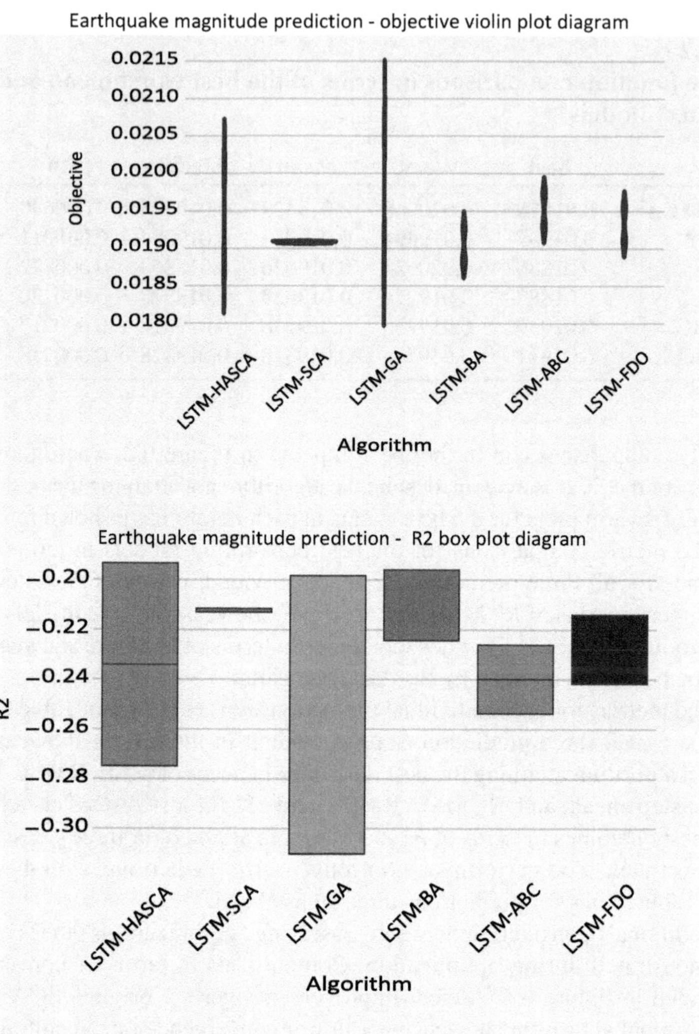

FIGURE 6.2 Objective and indicator distribution plots.

Challenges that need to be addressed before employing these models in real-time systems include quality of data (which can be noisy and affect the performance of the model), additional computational resources required for real-time data processing, and integration into existing infrastructure, which may necessitate considerable modifications.

6.6 CONCLUSION

This study delved into the crucial realm of earthquake prediction, recognizing the significant impact of natural disasters on communities and countries. The unpredictability

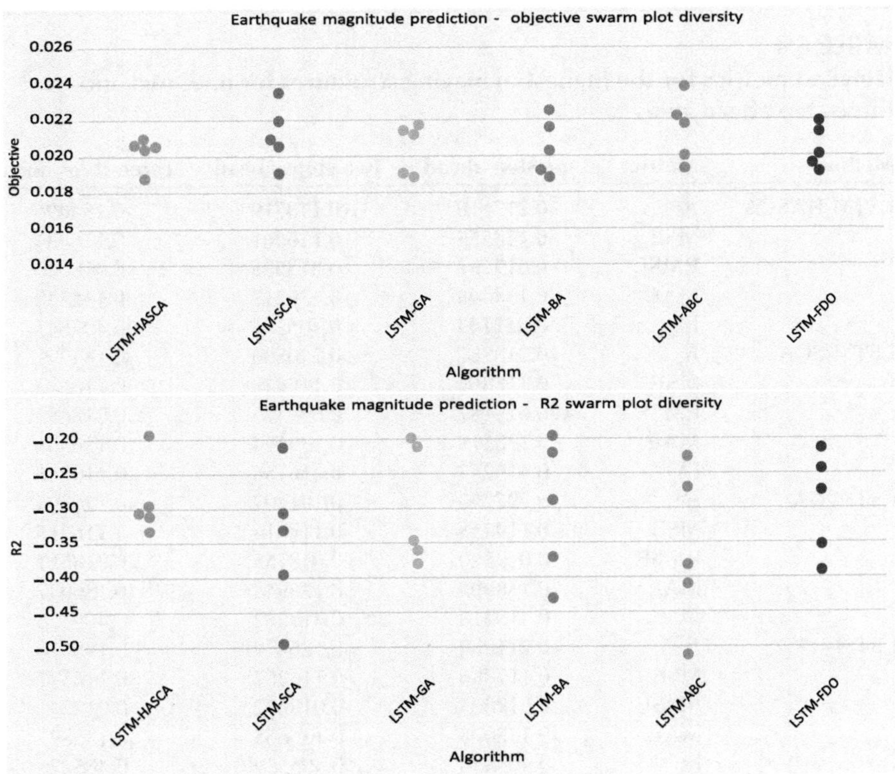

FIGURE 6.3 Objective and indicator function swarm plots.

TABLE 6.3
Overall outcomes for three-step ahead forecasts

Method	R^2	MSE	RMSE	MAE	IA
LSTM-HASCA	**-0.193475**	0.117033	**0.018787**	**0.137067**	0.408592
LSTM-SCA	-0.211495	0.118319	0.019071	0.138098	0.414423
LSTM-GA	-0.198330	0.117429	0.018864	0.137346	0.411955
LSTM-BA	-0.196023	**0.116942**	0.018828	0.137213	0.405988
LSTM-ABC	-0.226061	0.118951	0.019300	0.138926	**0.415379**
LSTM-FDO	-0.213977	0.118428	0.019110	0.138240	0.413007

of these events poses a formidable challenge, leading to immense property damage, loss of livelihoods, and tragically, human lives. The need for a robust prediction system has become evident, aiming to mitigate the devastating effects of earthquakes by providing timely warnings for evacuations and preparations, as well as predicting aftershock occurrences to aid rescue and recovery efforts. The study focused on harnessing the potential of LSTM models for earthquake magnitude prediction.

TABLE 6.4
Detailed metrics for the highest-achieving structures for one, two and three-step ahead steps

Method	Metric	One step ahead	Two steps ahead	Three steps ahead
LSTM-HASCA	R^2	-0.217610	**-0.178719**	**-0.184096**
	MSE	**0.118555**	**0.116061**	0.116484
	RMSE	**0.019167**	**0.018555**	0.018640
	MAE	**0.138446**	**0.136217**	**0.136528**
	IA	**0.412749**	0.405384	0.407643
LSTM-SCA	R^2	-0.218881	-0.230806	-0.184799
	MSE	0.118864	0.119448	0.116643
	RMSE	0.019187	0.019375	0.018651
	MAE	0.138518	0.139194	0.136568
	IA	0.416257	**0.416156**	0.410856
LSTM-GA	R^2	-0.227290	-0.191407	-0.176294
	MSE	0.119258	0.117014	**0.116015**
	RMSE	0.019320	0.018755	**0.018517**
	MAE	0.138995	0.136948	0.136077
	IA	0.415218	0.411282	0.409366
LSTM-BA	R^2	**-0.211080**	-0.186479	-0.190510
	MSE	0.117886	0.116367	0.116572
	RMSE	0.019065	0.018677	0.018741
	MAE	0.138074	0.136665	0.136897
	IA	0.406530	0.405559	0.405876
LSTM-ABC	R^2	-0.254984	-0.219057	-0.204143
	MSE	0.120689	0.118527	0.117636
	RMSE	0.019756	0.019190	0.018955
	MAE	0.140555	0.138528	0.137678
	IA	0.419106	0.414577	0.412453
LSTM-FDO	R^2	-0.222958	-0.208962	-0.210012
	MSE	0.119000	0.118058	0.118226
	RMSE	0.019252	0.019031	0.019048
	MAE	0.138750	0.137954	0.138014
	IA	0.414286	0.411495	**0.413241**

TABLE 6.5
Parameter selection for the highest-achieving networks made by each optimizer

Approach	Learning rate	Dropout	Epochs	Layers	Neurons L1	Neurons L2
LSTM-HASCA	0.003497	0.166512	600	1	145	N/a
LSTM-SCA	0.001278	0.193595	304	1	139	N/a
LSTM-GA	0.002494	0.119767	493	2	111	134
LSTM-BA	0.003723	0.129873	438	2	195	144
LSTM-ABC	0.004034	0.139697	372	2	165	137
LSTM-FDO	0.001730	0.123893	320	1	100	N/a

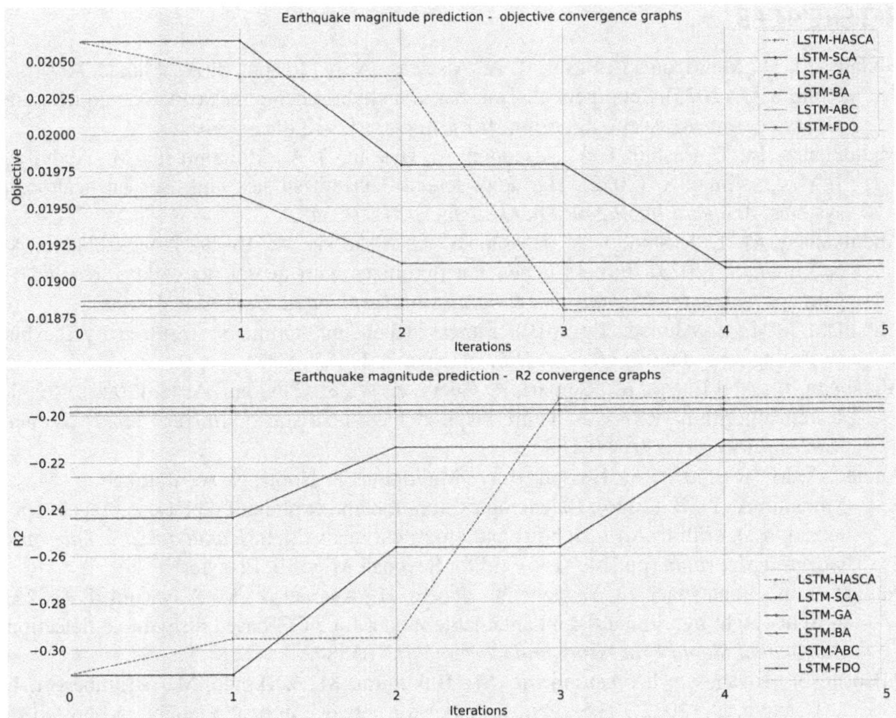

FIGURE 6.4 Objective and indicator function convergence.

Leveraging the wealth of data available from seismology institutes worldwide, the research employed metaheuristic optimizers to calibrate the performance of LSTM models. A modified variant of sine cosine metaheuristics was suggested to address the specific demands of the presented methodology, demonstrating its superiority in comparison to several contemporary optimizers.

Modest yet promising outcomes were achieved by the best models constructed through the introduced algorithm. These results underscore the potential effectiveness of LSTM models in earthquake prediction, especially when optimized with the proposed metaheuristic approach. Nevertheless, it is crucial to state the constraints of this study. One notable limitation lies in the complexity of the seismic data and the dynamic nature of earthquake events, which may impact the generalizability of the models. Additionally, the study's reliance on historical data may not fully capture the evolving nature of seismic patterns. Future work in this field should aim to address these limitations by incorporating real-time data and refining the model to adapt to changing seismic conditions. Furthermore, collaborative efforts with seismologists and domain experts could enhance the accuracy and applicability of the predictive models.

NOTE

1 https://earthquake.usgs.gov/earthquakes/feed/v1.0/csv.php

REFERENCES

Abdalla, M. H., Majidpour, J., Rasul, R. A., Alsewari, A. A., Rashid, T. A., Ahmed, A. M., ... Salih, S. Q. (2023). Sentiment analysis based on hybrid neural network techniques using binary coordinate ascent algorithm. *IEEE Access*, *11*, 134087–134099.

Abdulkhaleq, M. T., Rashid, T. A., Alsadoon, A., Hassan, B. A., Mohammadi, M., Abdullah, J. M., ... Vimal, S. (2022). Harmony search: Current studies and uses on healthcare systems. *Artificial Intelligence in Medicine*, *131*, 102348.

Abdulkhaleq, M. T., Rashid, T. A., Hassan, B. A., Alsadoon, A., Bacanin, N., Chhabra, A., & Vimal, S. (2023). Fitness dependent optimizer with neural networks for covid-19 patients. *Computer Methods and Programs in Biomedicine Update*, *3*, 100090.

Abdullah, J. M., & Ahmed, T. (2019). Fitness dependent optimizer: Inspired by the bee swarming reproductive process. *IEEE Access*, *7*, 43473–43486.

Abualigah, L., Abd Elaziz, M., Sumari, P., Geem, Z. W., & Gandomi, A. H. (2022). Reptile search algorithm (RSA): A nature-inspired meta-heuristic optimizer. *Expert Systems with Applications*, *191*, 116158.

Ahmed, A. M., Rashid, T. A., Hassan, B. A., Majidpour, J., Noori, K. A., Rahman, C. M., ... Mohammed, N. B. (2024). Balancing exploration and exploitation phases in whale optimization algorithm: An insightful and empirical analysis. In *Handbook of whale optimization algorithm* (pp. 149–156). Editor Seyedali Mirjalili. Elsevier.

Alamatsaz, N., Tabatabaei, L., Yazdchi, M., Payan, H., Alamatsaz, N., & Nasimi, F. (2024). A lightweight hybrid cnn-lstm explainable model for ECG-based arrhythmia detection. *Biomedical Signal Processing and Control*, *90*, 105884.

AlHosni, N., Jovanovic, L., Antonijevic, M., Bukumira, M., Zivkovic, M., Strumberger, I., ... Bacanin, N. (2022). The xgboost model for network intrusion detection boosted by enhanced sine cosine algorithm. In International Conference on Image Processing and Capsule Networks, Springer (pp. 213–228).

Azad, S. K., & Hasançebi, O. (2014). An elitist self-adaptive step-size search for structural design optimization. *Applied Soft Computing*, *19*, 226–235.

Bacanin, N., Jovanovic, L., Zivkovic, M., Kandasamy, V., Antonijevic, M., Deveci, M., & Strumberger, I. (2023). Multivariate energy forecasting via metaheuristic tuned long-short term memory and gated recurrent unit neural networks. *Information Sciences*, *642*, 119122.

Bacanin, N., Sarac, M., Budimirovic, N., Zivkovic, M., AlZubi, A. A., & Bashir, A. K. (2022). Smart wireless health care system using graph lstm pollution prediction and dragonfly node localization. *Sustainable Computing: Informatics and Systems*, *35*, 100711.

Bacanin, N., Simic, V., Zivkovic, M., Alrasheedi, M., & Petrovic, A. (2023). Cloud computing load prediction by decomposition reinforced attention long short-term memory network optimized by modified particle swarm optimization algorithm. *Annals of Operations Research*, 1–34.

Bacanin, N., Zivkovic, M., Al-Turjman, F., Venkatachalam, K., Trojovský, P., Strumberger, I., & Bezdan, T. (2022). Hybridized sine cosine algorithm with convolutional neural networks dropout regularization application. *Scientific Reports*, *12*(1), 6302.

Bacanin, N., Zivkovic, M., Bezdan, T., Venkatachalam, K., & Abouhawwash, M. (2022). Modified firefly algorithm for workflow scheduling in cloud-edge environment. *Neural Computing and Applications*, *34*(11), 9043–9068.

Bacanin, N., Zivkovic, M., Jovanovic, L., Ivanovic, M., & Rashid, T. A. (2022). Training a multilayer perception for modeling stock price index predictions using modified whale optimization algorithm. In Computational Vision and Bio-inspired Computing: Proceedings of *ICCVBIC* 2021 (pp. 415–430). Springer.

Bacanin, N., Zivkovic, M., Stoean, C., Antonijevic, M., Janicijevic, S., Sarac, M., & Strumberger, I. (2022). Application of natural language processing and machine learning boosted with swarm intelligence for spam email filtering. *Mathematics*, *10*(22), 4173.

Basha, J., Bacanin, N., Vukobrat, N., Zivkovic, M., Venkatachalam, K., Hubálovský, S., & Trojovský, P. (2021). Chaotic Harris Hawks optimization with quasi-reflection-based learning: An application to enhance cnn design. *Sensors*, *21*(19), 6654.

Berhich, A., Belouadha, F.-Z., & Kabbaj, M. I. (2023). An attention-based lstm network for large earthquake prediction. *Soil Dynamics and Earthquake Engineering*, *165*, 107663.

Bilal, M. A., Wang, Y., Ji, Y., Akhter, M. P., & Liu, H. (2023). Earthquake detection using stacked normalized recurrent neural network (SNRNN). *Applied Sciences*, *13*(14), 8121.

Bischl, B., Binder, M., Lang, M., Pielok, T., Richter, J., Coors, S., ... Lindauer, M. (2023). Hyperparameter optimization: Foundations, algorithms, best practices, and open challenges. *Wiley Interdisciplinary Reviews: Data Mining and Knowledge Discovery*, *13*(2), e1484.

Cahuantzi, R., Chen, X., & Güttel, S. (2023). A comparison of LSTM and GRU networks for learning symbolic sequences. Springer. In *Science and Information Conference* (pp. 771–785).

Damaševičius, R., Jovanovic, L., Petrovic, A., Zivkovic, M., Bacanin, N., Jovanovic, D., & Antonijevic, M. (2024). Decomposition aided attention-based recurrent neural networks for multistep ahead time-series forecasting of renewable power generation. *PeerJ Computer Science*, *10*.

Dobrojevic, M., Zivkovic, M., Chhabra, A., Sani, N. S., Bacanin, N., & Amin, M. M. (2023). Addressing internet of things security by enhanced sine cosine metaheuristics tuned hybrid machine learning model and results interpretation based on shap approach. *PeerJ Computer Science*, *9*, e1405.

Fang, Z., Ma, X., Pan, H., Yang, G., & Arce, G. R. (2023). Movement forecasting of financial time series based on adaptive lstm-bn network. *Expert Systems with Applications*, *213*, 119207.

Gurrola-Ramos, J., Hernàndez-Aguirre, A., & Dalmau-Cedeño, O. (2020). Colshade for realworld single-objective constrained optimization problems. In *2020 IEEE Congress on Evolutionary Computation (CEC), IEEE.* (pp. 1–8).

Hassan, B. A. (2021). CSCF: A chaotic sine cosine firefly algorithm for practical application problems. *Neural Computing and Applications*, *33*(12), 7011–7030.

Hassan, B. A., & Rashid, T. A. (2021). A multidisciplinary ensemble algorithm for clustering heterogeneous datasets. *Neural Computing and Applications*, *33*(17), 10987–11010.

Hassan, B. A., Rashid, T. A., Ahmed, A. M., Qader, S. M., Majidpour, J., Abdalla, M. H., ... Ramadhan, A. A. (2024). Equitable and fair performance evaluation of whale optimization algorithm. Editor Seyedali Mirjalili In *Handbook of whale optimization algorithm* (pp. 157–168). Elsevier.

Hochreiter, S. (1991). Studies on dynamic neural networks. *Master's thesis, Institute for Computer Science, Technical University, Munich*, *1*, 1–150.

Hochreiter, S., & Schmidhuber, J. (1997, November). Long short-term memory. *Neural Computing*, *9*(8), 1735–1780.

Im, K., & Avouac, J.-P. (2023). Cascading foreshocks, aftershocks and earthquake swarms in a discrete fault network. *Geophysical Journal International*, *235*(1), 831–852.

Jia, J., & Ye, W. (2023). Deep learning for earthquake disaster assessment: Objects, data, models, stages, challenges, and opportunities. *Remote Sensing*, *15*(16), 4098.

Jovanovic, L., Bacanin, N., Simic, V., Mani, J., Zivkovic, M., & Sarac, M. (2023). Optimizing machine learning for space weather forecasting and event classification using modified metaheuristics. *Soft Computing*, 1–20.

Jovanovic, L., Bacanin, N., Zivkovic, M., Antonijevic, M., Jovanovic, B., Sretenovic, M. B., & Strumberger, I. (2023). Machine learning tuning by diversity oriented firefly metaheuristics for industry 4.0. *Expert Systems*, e13293.

Jovanovic, L., Jovanovic, D., Bacanin, N., Jovancai Stakic, A., Antonijevic, M., Magd, H., ... Zivkovic, M. (2022). Multi-step crude oil price prediction based on lstm approach tuned by salp swarm algorithm with disputation operator. *Sustainability*, *14*(21), 14616.

Jovanovic, L., Jovanovic, G., Perisic, M., Alimpic, F., Stanisic, S., Bacanin, N., ... Stojic, A. (2023). The explainable potential of coupling metaheuristics-optimized-xgboost and shap in revealing vocs' environmental fate. *Atmosphere*, *14*(1). Retrieved from www.mdpi.com/2073-4433/14/1/109

Jovanovic, L., Kljajic, M., Mizdrakovic, V., Marevic, V., Zivkovic, M., & Bacanin, N. (2023). Predicting credit card churn: Application of xgboost tuned by modified sine cosine algorithm. In *2023 3rd International Conference on Smart Data Intelligence (ICSMDI). Springer.* (pp. 55–62).

Karaboga, D., & Basturk, B. (2008). On the performance of artificial bee colony (abc) algorithm. *Applied Soft Computing*, *8*(1), 687–697.

Karataş, L., Ateş, T., Alptekin, A., Dal, M., & Yakar, M. (2023). A systematic method for post-earthquake damage assessment: Case study of the antep castle, tu¨rkiye. *Advanced Engineering Science*, *3*, 62–71.

Kavianpour, P., Kavianpour, M., Jahani, E., & Ramezani, A. (2023). A CNN-BiLSTM model with attention mechanism for earthquake prediction. *The Journal of Supercomputing*, *79*(17), 19194–19226.

Kong, Q., Inbal, A., Allen, R. M., Lv, Q., & Puder, A. (2019). Machine learning aspects of the myshake global smartphone seismic network. *Seismological Research Letters*, *90*(2A), 546–552.

Lakhan, A., Mohammed, M. A., Abdulkareem, K. H., Hamouda, H., & Alyahya, S. (2023). Autism spectrum disorder detection framework for children based on federated learning integrated cnn-lstm. *Computers in Biology and Medicine*, *166*, 107539.

Majidpour, J., Rashid, T. A., Thinakaran, R., Batumalay, M., Dewi, D. A., Hassan, B. A., ... Arabi, H. (2024). NSGA-II-DL: Metaheuristic optimal feature selection with deep learning framework for her2 classification in breast cancer. *IEEE Access* (2), pp. 38885-38898

Masini, R. P., Medeiros, M. C., & Mendes, E. F. (2023). Machine learning advances for time series forecasting. *Journal of Economic Surveys*, *37*(1), 76–111.

Minic, A., Jovanovic, L., Bacanin, N., Stoean, C., Zivkovic, M., Spalevic, P., ... Stoean, R. (2023). Applying recurrent neural networks for anomaly detection in electrocardiogram sensor data. *Sensors*, *23*(24), 9878.

Mirjalili, S. (2016). SCA: A sine cosine algorithm for solving optimization problems. *Knowledgebased Systems*, *96*, 120–133.

Mirjalili, S., & Mirjalili, S. (2019). Genetic algorithm. In *Evolutionary algorithms and neural networks: Theory and applications* Editor Seyedali Mirjalili. Springer. (pp. 43–55).

Mizdrakovic, V., Kljajic, M., Zivkovic, M., Bacanin, N., Jovanovic, L., Deveci, M., & Pedrycz, W. (2024). Forecasting bitcoin: Decomposition aided long short-term memory based time series modelling and its explanation with shapley values. *Knowledge-Based Systems*, 112026.

Mousavi, S. M., & Beroza, G. C. (2023). Machine learning in earthquake seismology. *Annual Review of Earth and Planetary Sciences*, *51*, 105–129.

Pilcevic, D., Jovicic, M. D., Antonijevic, M., Bacanin, N., Jovanovic, L., Zivkovic, M., ... Bisevac, P. (2023). Performance evaluation of metaheuristics-tuned recurrent neural networks for electroencephalography anomaly detection. *Frontiers in Physiology*, *14*.

Predić, B., Jovanovic, L., Simic, V., Bacanin, N., Zivkovic, M., Spalevic, P., ... Dobrojevic, M. (2023). Cloud-load forecasting via decomposition-aided attention recurrent neural network tuned by modified particle swarm optimization. *Complex & Intelligent Systems*, 1–21.

Pulinets, S., Ouzounov, D., Karelin, A., & Boyarchuk, K. (2023). *Earthquake precursors in the atmosphere and ionosphere: New concepts*. Springer Nature.

Salb, M., Jovanovic, L., Bacanin, N., Antonijevic, M., Zivkovic, M., Budimirovic, N., & Abualigah, L. (2023). Enhancing internet of things network security using hybrid CNN and xgboost model tuned via modified reptile search algorithm. *Applied Sciences*, *13*(23), 12687.

Salb, M., Jovanovic, L., Zivkovic, M., Tuba, E., Elsadai, A., & Bacanin, N. (2022). Training logistic regression model by enhanced moth flame optimizer for spam email classification. In *Computer Networks and Inventive Communication Technologies: Proceedings of Fifth ICCNCT 2022* (pp. 753–768). Springer.

Salehinejad, H., Sankar, S., Barfett, J., Colak, E., & Valaee, S. (2017). Recent advances in recurrent neural networks. *arXiv preprint arXiv:1801.01078.*

Savanović, N., Toskovic, A., Petrovic, A., Zivkovic, M., Damaševičius, R., Jovanovic, L., ... Nikolic, B. (2023). Intrusion detection in healthcare 4.0 internet of things systems via metaheuristics optimized machine learning. *Sustainability*, *15*(16), 12563.

Stankovic, M., Bacanin, N., Budimirovic, N., Zivkovic, M., Sarac, M., & Strumberger, I. (2023). Bi-directional long short-term memory optimization by improved teaching-learning based algorithm for univariate gold price forecasting. In *2023 International Conference on Inventive Computation Technologies (ICICT)Springer* (pp. 1650–1657).

Stoean, C., Zivkovic, M., Bozovic, A., Bacanin, N., Strulak-Wójcikiewicz, R., Antonijevic, M., & Stoean, R. (2023). Metaheuristic-based hyperparameter tuning for recurrent deep learning: Application to the prediction of solar energy generation. *Axioms*, *12*(3), 266.

Thomas, A. M., Inbal, A., Searcy, J., Shelly, D. R., & Bürgmann, R. (2021). Identification of low-frequency earthquakes on the san andreas fault with deep learning. *Geophysical Research Letters*, *48*(13), e2021GL093157.

Titos, M., Gutiérrez, L., Benítez, C., Rey Devesa, P., Koulakov, I., & Ibáñez, J. M. (2023). Multi-station volcano tectonic earthquake monitoring based on transfer learning. *Frontiers in Earth Science*, *11*, 1204832.

Tiwari, S., Jain, A., Sapra, V., Koundal, D., Alenezi, F., Polat, K., ... Nour, M. (2023). A smart decision support system to diagnose arrhythymia using ensembled convnet and convnet-lstm model. *Expert Systems with Applications*, *213*, 118933.

Todorovic, M., Stanisic, N., Zivkovic, M., Bacanin, N., Simic, V., & Tirkolaee, E. B. (2023). Improving audit opinion prediction accuracy using metaheuristics-tuned xgboost algorithm with interpretable results through shap value analysis. *Applied Soft Computing*, *149*, 110955.

Wang, A., Li, S., Lu, J., Zhang, H., Wang, B., & Xie, Z. (2023). Prediction of pga in earthquake early warning using a long short-term memory neural network. *Geophysical Journal International*, *234*(1), 12–24.

Wang, D., Tan, D., & Liu, L. (2018). Particle swarm optimization algorithm: An overview. *Soft Computing*, *22*, 387–408.

Wang, X., Wang, Y., Liu, D., Wang, Y., & Wang, Z. (2023). Automated recognition of epilepsy from eeg signals using a combining space–time algorithm of CNN-LSTM. *Scientific Reports*, *13*(1), 14876.

Wolpert, D. H., & Macready, W. G. (1997). No free lunch theorems for optimization. *IEEE Transactions on Evolutionary Computation*, *1*(1), 67–82.

Yang, N., Li, G., Zhao, P., Zhang, J., & Zhao, D. (2023). Porosity prediction from pre-stack seismic data via a data-driven approach. *Journal of Applied Geophysics*, *211*, 104947.

Yang, X.-S. (2010). Firefly algorithm, stochastic test functions and design optimisation. *International Journal of Bio-inspired Computation*, *2*(2), 78–84.

Yang, X.-S., & He, X. (2013a). Bat algorithm: Literature review and applications. *International Journal of Bio-inspired Computation*, *5*(3), 141–149.

Yang, X.-S., & He, X. (2013b). Firefly algorithm: Recent advances and applications. *International Journal of Swarm Intelligence*, *1*(1), 36–50.

Zivkovic, M., Bacanin, N., Antonijevic, M., Nikolic, B., Kvascev, G., Marjanovic, M., & Savanovic, N. (2022). Hybrid CNN and XGBoost model tuned by modified arithmetic optimization algorithm for covid-19 early diagnostics from x-ray images. *Electronics*, *11*(22), 3798.

Zivkovic, M., Bacanin, N., Venkatachalam, K., Nayyar, A., Djordjevic, A., Strumberger, I., & Al-Turjman, F. (2021). Covid-19 cases prediction by using hybrid machine learning and beetle antennae search approach. *Sustainable Cities and Society*, *66*, 102669.

Zivkovic, M., Jovanovic, L., Ivanovic, M., Bacanin, N., Strumberger, I., & Joseph, P. M. (2022). XGBoost hyperparameters tuning by fitness-dependent optimizer for network intrusion detection. In *Communication and Intelligent Systems: Proceedings of ICCIS 2021* (pp. 947–962). Springer.

Zivkovic, M., K, V., Bacanin, N., Djordjevic, A., Antonijevic, M., Strumberger, I., & Rashid, T. A. (2021). Hybrid genetic algorithm and machine learning method for covid-19 cases prediction. In *Proceedings of International Conference on Sustainable Expert Systems: ICSES 2020. Springer* (pp. 169–184).

7 Backtracking Search Optimization Algorithm

A Practical Tutorial

Bryar A. Hassan, Alla A. Hassan, Aram M. Ahmed, Tarik A. Rashid and Abdulhady A. Abdulla

7.1 OVERVIEW

Over the last few decades, there has been a growing interest in biologically inspired algorithms that rely on evolutionary computing theory. This category of algorithms is developed based on the natural theory of survival of fittest, and it is generally referred to as an Evolutionary Algorithm (EA) [1]. Referencing [2], the premise of an EA is very plain. Initialization, selection, genetic operators, and termination are the four phases that make up an EA. These phases offer obvious approaches to modularize this algorithm implementation and roughly correlate to a particular component of natural selection. In an EA, more physically fit individuals endure and spread. As under natural selection, however, defective individuals disappear and do not add to the gene pool of subsequent generations. The main components of a simple EA are depicted in Figure 7.1.

1. **Initialization:** To start our algorithm, we should construct a population of solutions. A random number of potential answers to the issue, often known as representatives, make up the population. If any prior information about the job is known, it is also created generally based on what is regarded ideal or at random (within the bounds of the problem). Since it represents a gene pool, the population must have a wide range of solutions. We should attempt to have many distinct genes present if we wish to investigate a variety of choices throughout the process.
2. **Selection:** Once a population has been generated, it is critical to evaluate its members in light of a fitness trait. A fitness function is a function that incorporates a member's characteristics and gives a numerical assessment of how practical a solution is. Finding a decent fitness function that truly represents the data can be challenging at times and depends much on the particular challenge at hand. Now, we evaluate everyone's level of fitness and choose some of the top-scoring competitors.

DOI: 10.1201/9781003601555-7

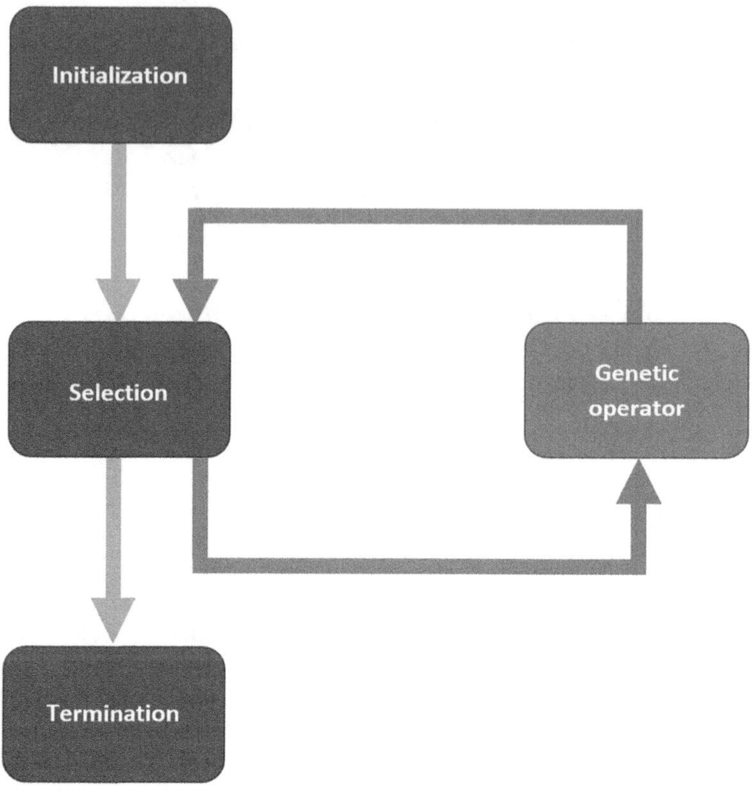

FIGURE 7.1 Simple EA.

3. **Genetic operator:** This stage often has two sub-steps: crossover and mutation. After choosing the best members, the algorithm builds the following generation using these members (typically the top 2, but this number can vary). New offspring are created by combining the traits of the parents, using the traits of the chosen parents. This can frequently be challenging depending on the type of data, but in combinatorial problems, it is typically easy to combine combinations and obtain valid combinations from these inputs. Now the generation must assimilate the new genetic material. This crucial step must be taken since failing to do so would cause us to fall victim to the local extreme and provide subpar outcomes. In this mutation phase, we simply alter a small portion of the offspring such that they no longer exactly reflect particular gene subsets from the parents. The possibility of a mutation occurring in a kid and the severity of the mutation are often determined by a probability distribution.
4. **Termination:** There should be an end to the algorithm. There are two ways in which this often happens: either the method has achieved some maximum runtime or the algorithm has hit some output threshold. It is now time to choose and deliver the ideal answer.

7.2 BACKTRACKING SEARCH OPTIMIZATION ALGORITHM AT A GLANCE

The Backtracking Search Optimization Algorithm (BSA) is a contemporary population-based evolutionary method that was proposed by P. Civicioglu to address issues with numerical optimization. BSA utilizes a single control parameter to construct an experimental population, unlike other EAs. The method may successfully and efficiently tackle issues involving numerical optimization thanks to this control parameter. Two more crossover and mutation operators are included in BSA's method for building a trial population. Strong capabilities for exploration and exploitation are provided by BSA's methods for creating trial populations and monitoring the amplitude of the search-direction matrix and search-space bounds.

BSA is a population-based EA that relies on recombination and Differential Evolution (DE) mutation [3]. The six phases of BSA include initialization, selection-I, mutation, crossover, selection-II, and fitness evaluation, according to [4, 5]. Figure 7.2 depicts the BSA flowchart. The following subsections provide descriptions of the stages in the flowchart.

7.2.1 Steps of BSA

BSA consists of five steps. These steps are described below.

1. **Initialization:** The initial population (P) of BSA is created at random and consists of D variables and N people:

$$P_{ij} \sim U\left(low_j, up_j\right) \tag{1}$$

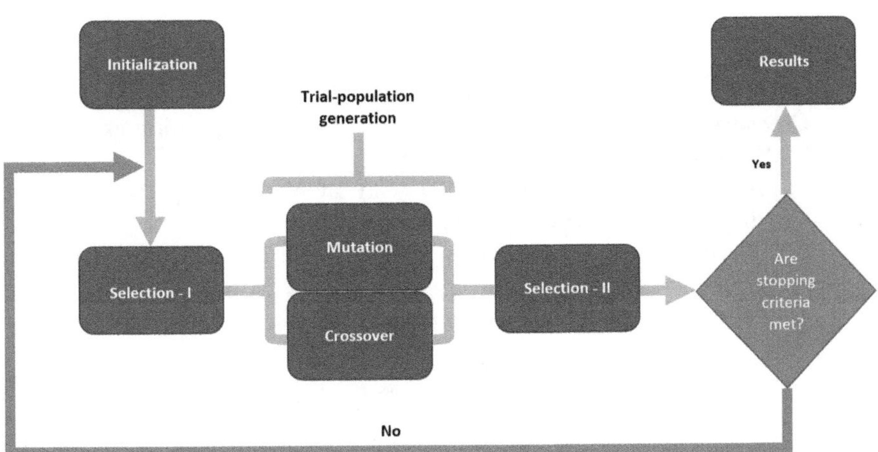

FIGURE 7.2 Basic steps of BSA (reprinted from [4]).

for $i = 1, 2,..., N$ and $n = 1, 2,..., D$, where

- N is the population size,
- D is the problem dimension,
- U is the uniform distribution.

2. **Selection-I.** The historical population (P^{old}) is obtained using the pre-selection operator, which is then utilized to determine the search's direction. The value of $P^{old}$is calculated using the following three steps:
    1. Produce the starting historical population at random:

$$P_{ij} \sim U\left(low_j, up_j\right) \tag{2}$$

    2. Redefine P^{old} at the start of each repetition:

$$P_{ij}^{old} \sim U\left(low_j, up_j\right) \tag{3}$$

if $a < b$ then $P^{old} := P \mid a, b \sim U(0,1)$,

where:

- = is the update operation,
- a and b are random numbers.

    3. Reorder its members after obtaining the historical population:

$$P^{old} := Permuting\left(P^{old}\right) \tag{4}$$

A function that randomly shuffles data is the permutation function.

3. **Mutation.** The following equation demonstrates how BSA uses the mutation operator to create the initial trial population (mutant):

$$Mutant = P + F.\left(P^{old} - P\right) \tag{5}$$

where the search direction matrix's amplitude $\left(P^{old} - P\right)$ is controlled by F.

4. **Crossover.** The trial population (T) takes on its ultimate form as a result of BSA's crossover. There are two steps to it.
    1. A binary-valued matrix (map) with an $N \times D$ map size is created, displaying the members of the trail population T.
    2. The binary integer matrix is initially set to 1 as its value, where $n \in \{1,2,..., N\}$ and $m \in \{1,2,..., D\}$.

The following equation is used to update T's value:

$$T_{n,m} := P_{n,m} \tag{6}$$

5. **Selection-II.** This is referred to as BSA greedy selection. When the trial population *T*'s fitness values surpass those of the population *P*, the trial population *T*'s members are swapped out for the population *P*'s members. The person with the greatest fitness value overall is chosen as the best overall solution.
6. **Fitness evaluation.** A group of individuals is evaluated during the fitness assessment. Fitness is the result, using the list of populations as the input.

7.2.2 Key Success Strategy of BSA

Based on the discussion in the literature [6], the following are the main success strategy and a few elements that could be responsible for BSA's greater success than other EAs.

- In each generation, the trial populations are generated quite efficiently using the mutation and crossover strategies of BSA.
- Due to the control parameter, *F*, BSA adequately balances the correlation between discovery (global search) and extraction (local search). This parameter accurately and fairly directs the direction of both local and international searches. This equilibrium may generate a wide range of numerical numbers needed for both local and international searches. For global searches, a wide range of numerical values is created, but for local searches, a smaller range is provided. BSA can therefore solve a number of difficulties;
- Using the historical population, the quest path of BSA is determined (oldP). In other words, compared to older generations, the historical populations are equal. This operation allows BSA to deliver the populations in the courtroom effectively.
- Compared with DE's crossover, BSA's crossover is a dynamic operation. BSA's crossover operator generates trial individuals in each generation to boost the capacity of BSA to cope with problems.
- In BSA, demographic diversity can be accomplished by a boundary management system to ensure successful searches.
- BSA utilizes historical and present populations, unlike other metaheuristic algorithms, since it is considered a dual-population algorithm.
- BSA's flexibility can make it simpler for researchers and professionals.

7.2.3 Notational Pseudocode of BSA

BSA is an iterative, population-based EA designed to be a global minimizer. Breaking down BSA's activities into the following five processes—initialization, selection-I, mutation, crossover, and selection-II—helps to clarify how it works similarly to other EAs. Algorithm 1 demonstrates that BSA may be recognized by its general design.

Algorithm 1. BSA algorithm

Set the initial values of population size N, *mat_itr*, *mix_rate*
For $i = 1; i \leq N; i$ ++
  For $j = 1; j \leq D; j$ ++
  Set $P_{ij} = rnd.(up_j - low_j) + low_j$
  Set $P_{ij}^{old} = rnd.(up_j - low_j) + low_j$
  End
Evaluate each solution in the population P by calculating its fitness function $f(P)$
End
Set $t = 0$
Repeat
  Generate random numbers a, b, where $a, b \in [0, 1]$
  If $(a < b\ P|a, b \sim U(0, 1))$ **Then**
    Set $P^{old} := P$
  End
  $P^{old} := permuting(P^{old})$ant
  Set$P^{mutant} = P + 3 \bullet rndn \bullet (P^{old} - P)$
  Apply the crossover operator on the solutions inP^{mutant}
  If individuals are beyond the search space **Then**
    Regenerate the individuals
  End
  Evaluate each solution in the trail population T, $f(T)$
  For $i = 1; i \leq N; i$ ++
    If $f(T_i) < f(P_i)$ **Then**
      Set $f(P_i) = f(T_i)$
      Set $P_i = T_i$
    End
  End
  $P^{best} = \min f(P)$
  If $P^{best} < global\_min$ **Then**
    $global\_min = P^{best}$
  End
  Set $t = t + 1$
  Until $t < max\_itr$
Produce the best obtained solution

Table 7.1 also provides a notational explanation of BSA.

7.3 BSA SOURCE CODE

All the studies and examples presented in the literature have been utilized via several functions of MATLAB, Java, and C++ programs.

7.3.1 MATLAB Program

The MATLAB source code and examples of BSA are freely accessible in [7]. The source code and files for this project are shown in the project files section to ensure

TABLE 7.1
Notational description of BSA's algorithm

Line #	Notation (description)
1	Parameter initialisation
2 to 8	Population initialisation and evaluation
11 to 15	Selection I
16 to 21	Mutation
22 to 27	Crossover
28 to 32	Selection II
34	Producing the best-obtained solution

TABLE 7.2
MATLAB functions for implementing BSA

Functions	Parameters	Description
bsa	fnc, mydata, popsize, dim, DIM_RATE, low, up, epoch	Running BSA from the first step until it terminates
GeneratePopulation	popsize, dim, low, up	Generating a random population of BSA
BoundaryControl	pop, low, up	Controlling the boundary of the population
get_scale_factor		Changing the generation strategy of the scale factor (F)
fitcircle	population, data	Fitting the population
Imageclustering	population, data	Generating clusters of a population from image for circle fitting
Plotcircle	a, b, r, color	Plotting a specific set of the population
run_circle_fitting_demo	-	Generating a population for circle fitting
run_image_clustering_demo	-	Creating population groups for circle fitting
FncNumber	-	Initializing the function number
settingOfBenchmarkFnc	fname, fnc, low, up, dim	Running the algorithm based on the five parameters

that the source code mentioned meets the requirements of scholars. Table 7.2 explains the functions and parameters used for BSA.

The same project uses five benchmark issues. These issues are displayed in Table 7.3.

7.3.2 Java Program

The Java source code for BSA was developed in [8]. This Java code is a pure and simple version of BSA without any modification, extension, or hybridization. The

TABLE 7.3
Sample of benchmarking problems

Function name	MATLAB function	Parameters
griewank	settingOfBenchmarkFnc(FunctionNo)	FunctionNo=1
rastrigin		FunctionNo=2
rosenbrock		FunctionNo=3
FoxHoles		FunctionNo=4
ackley		FunctionNo=5

TABLE 7.4
Java classes for implementing BSA

Java classes	Methods	Description
f_xj	func	Specifying the problem name
backtracking_search	backtracking_search, getminval_index, randperm, shuffle, rndintD, initialize, solution and toStringnew	A class for assessing BSA based on multidimensional optimization benchmark functions.

code developer mentioned that this program is not intended for marketing purposes, only to direct users who study metaheuristic algorithms. It is worth mentioning that the code contains multidimensional benchmarking optimization functions to evaluate the efficiency of the proposed system. These Java methods are listed in Table 7.4.

The above Java classes have several methods. These methods are described in Table 7.5.

7.3.3 C++ Program

A C++ program is available in [9] for developing a BSA trial population strategy to minimize six boundary-constrained benchmark optimization problems. In this coding program, a user is required to enter the demanded values and make certain choices of the functions. In addition, the code automatically generates the desired result within a few seconds or minutes, depending on the number of cycles. The population size, dimension, and other parameters are entered by the user. The results solely correspond to the implemented BSA algorithm and not any brute force or other mechanisms. Table 7.6 presents a summary of the contents of each file that make up the BSA application.

The main requirements to run the BSA application using C++ are as follows:

1. Use the latest updated version of Visual Studio 2013 with all the Service Packs installed in it.

TABLE 7.5
Java methods for implementing BSA

Java methods	Parameters	Description
func	x[]	Specifying the problem name
backtracking_search	iff, iN, iLower, iUpper, imaxiter	Constructor for initializing the class
getminval_index	a	Returning the minimum value of the population
randperm	a [][]	Population's random variation
shuffle	a []	Shuffling the population
rndintD	-	Generating and returning a random number
initialize	-	Initializing the population
solution	-	Problem solver using the steps of BSA
toStringnew	-	Casting from double to string

TABLE 7.6
C++ files for implementing the BSA application

Files	Description
BSA.cpp	This is the application's primary source file.
BSA.vcxproj	This is the primary project file for VC++ projects created with an Application Wizard. It includes details on the project platforms, configurations, and features the user chooses in the Application Wizard, as well as the version of Visual C++ that was used to create the file.
BSA.vcxproj.filters	This file serves as a filter for VC++ projects created by application wizards. It includes details on how the files in the user's project are related to the filters. The IDE uses this relationship to show a collection of files with related extensions under a certain node (e.g., ".cpp" files are associated with the "Source Files" filter).
StdAfx.h and StdAfx.cpp	These files produce the precompiled type file StdAfx.obj and the precompiled header file BSA.pch.
Other standard files	AppWizard uses "TODO:" comments to highlight sections of the source code that require code additions or modifications.

2. The code works on any machine with VS2013 installed on it.
3. The code debugs in both Win32 and x64 configurations without any discrepancies.

7.4 NUMERICAL EXAMPLE

To illustrate and trace the functions of BSA in detail, we use the Griewank problem in a simple numerical example. The Griewank function is

TABLE 7.7
Initialization parameters of BSA

Parameters	Value
Search dimension	2 × 2
Number of iterations	7

$$f(x)=\sum_{i=1}^{d}\frac{x_i^2}{4000}-\prod_{i=1}^{d}cos\left(\frac{x_i}{\sqrt{i}}\right)+1 \quad (7)$$

Table 7.7 lists the initial parameters of BSA for solving Eq. (7).

Moreover, Table 7.8 presents the values acquired by BSA for different variables during its initial eight iterations while solving Eq. (7).

7.5 GENERAL FRAMEWORK OF BSA VARIANTS

Using the standard BSA, several optimization issues in the real world and on benchmarks have been resolved effectively. Additionally, in order to get around the existing BSA's purported speed and efficiency restrictions, academics have been asked to improve its performance. While some scholars have made modifications to the original BSA, others have been inspired to apply it to diverse challenges in the real world. [6] divides the three categories of BSA variations into BSA alterations, BSA hybridization, and BSA extensions. Figure 7.3 provides a conceptual structure for the various BSA expansions and their expansion procedures.

These three BSA variations all operate differently from one another. The BSA modification variation may be based on the control factor (F), a routing issue, historical data, or a combination of BSA and another metaheuristic algorithm. It may also take the form of binary BSA, embedded BSA, or BSA that is used in a microprocessor. The hybridization of the BSA variation, in contrast, often relies on guided and memetic BSA, control factor (F), local and global searches, neural networks, fuzzy logic, and the hybridization of BSA with other algorithmic and mathematical functions. Combining BSA with Burger's chaotic map with wavelet-based mutation and support vector machine (SVM) are a few examples of this hybridization. BSA has now been expanded to include restricted optimization, multi-objective optimization, and single-objective optimization (including binary and numerical optimizations). The relationships between BSA and three constraint handlers, a novel mutation technique, network structure, and crossover operation are the main topics of these extensions.

TABLE 7.8
Numerical example illustrating the operation of BSA via the use of Eq. (7)

Iteration	P	fitnessP	Eq. (3)	oldP	F ~ N(0,3)	Mutant	Map	T	fintnessT	Global min.
Initial	-485.235 396.817	99.306	0	-	-	-	-	-	-	Infinity
step	-331.388 356.853	60.317	0	-	-	-	-	-	-	
1	-485.235 396.817	99.306	0	288.586 -520.583	3.432	2.17 -2.751	1 1	-485.235 396.817	99.306	60.317
	-331.388 356.853	60.317	0	550.803 206.469	3.432	2.696 -159.304	1 1	-331.388 356.853	60.317	
2	-485.235 396.817	99.306	0	288.586 -520.583	-2.67	-2.551 396.817	1 0	-485.235 396.817	99.306	60.317
	-331.388 356.853	60.317	0	550.803 206.469	-2.67	-331.388 758.412	0 1	-331.388 356.853	60.317	
3	-485.235 396.817	99.306	1	-331.388 356.853	77.85	-473.258 396.817	1 0	-485.235 396.817	96.121	60.317
	-331.388 356.853	60.317	1	-485.235 396.817	77.85	-343.365 356.853	1 0	-331.388 356.853	60.317	
4	-473.258 396.817	96.121	1	-473.258 396.817	2.603	-473.258 396.817	0 1	-473.258 396.817	96.121	60.317
	-331.388 356.853	60.317	1	-331.388 356.853	2.603	-331.388 356.853	1 0	-331.388 356.853	60.317	
5	-473.258 396.817	96.121	1	-331.388 356.853	-160.592	-473.258 403.235	0 1	-473.258 396.817	96.121	59.11
	-331.388 356.853	60.317	1	-473.258 396.817	-160.592	-331.388 350.435	0 1	-331.388 356.853	59.11	
6	-473.258 396.817	96.121	0	-331.388 356.853	3.437	-473.258 259.42	0 1	-473.258 396.817	73.965	59.11
	-331.388 350.435	59.11	0	-473.258 396.817	3.437	-819.136 350.435	1 0	-331.388 350.435	59.11	
7	-473.258 259.42	73.965	1	-473.258 259.42	-2.009	-473.258 259.42	1 1	-473.258	73.965	59.11
	-331.388 350.435	59.11	1	-331.388 350.435	-2.009	-331.388 350.435	0 1	259.4205	59.11	
								-331.388 350.435		
8	-473.258 259.42	73.965	0	-473.258 259.42	-6.575	-473.258 259.42	0 1	-473.258 259.42	73.965	59.11
	-331.388 350.435	59.11	0	-331.388 350.435	-6.575	-331.388 350.435	0 1	-331.388 350.435	59.11	

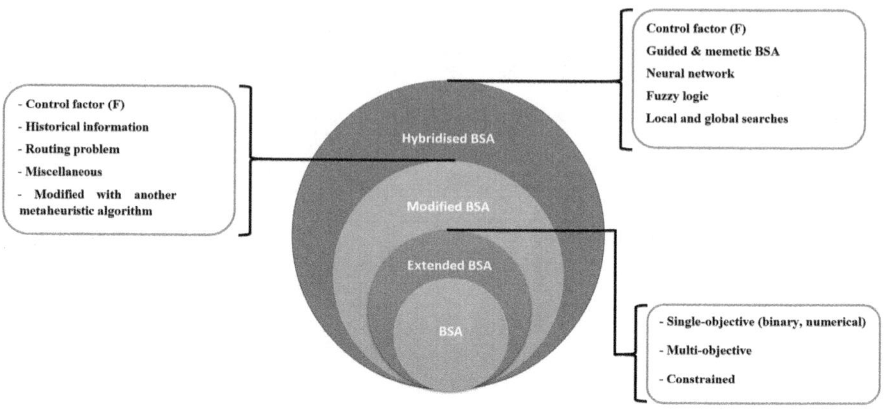

FIGURE 7.3 General framework of BSA variants.

7.6 CONCLUSION

This chapter provided a tutorial introduction to BSA to pique readers' interest in the subject and promote the wider application of the approach. EAs were introduced first, along with their artificial implementation. The fundamental steps of BSA were then discussed. Next, the applications' current source code needed to implement BSA was shown. Then, an illustrative example was provided to demonstrate how BSA operates concerning a numerical optimization issue. Finally, using a generic framework of BSA variations, a broad overview (extended explanation) of the method was demonstrated based on the potential of BSA for minimizing various optimization problems. Compared to other algorithms, BSA appears to be better suited to handle real-valued optimization because it does not require any recoding (discretization) of the variables or redefinition of the operators. However, BSA is also very adaptable and can be used to solve binary- or discrete-coded problems. Hopefully, this chapter will provide researchers working on BSA with useful resources and suggestions for improving it. In future work, we will prepare a practical tutorial about the chaotic sine cosine firefly algorithm [10], non-dominated sorting genetic algorithm optimizer [11], modified-improved fitness dependent optimizer [12], and modified bat algorithm [13]. For future reading, the authors advise the reader to optionally read the following research works: [10–24].

ACKNOWLEDGMENTS

The University of Kurdistan Hewler provided the facilities and ongoing assistance necessary to carry out this work, for which the authors sincerely thank them.

REFERENCES

[1] A. Konar, Ed., *Evolutionary Computing Algorithms BT – Computational Intelligence: Principles, Techniques and Applications*, Springer Berlin Heidelberg, 2005, pp. 323–351. doi: 10.1007/3-540-27335-2_12.

[2] A. E. Eiben and J. E. Smith, "What is an evolutionary algorithm?" *Introduction to Evolutionary Computing*, Springer, 2015, pp. 25–48.

[3] P. Civicioglu and E. Besdok, "A+ Evolutionary search algorithm and QR decomposition based rotation invariant crossover operator," *Expert Syst Appl*, vol. 103, pp. 49–62, 2018.

[4] P. Civicioglu, "Backtracking search optimization algorithm for numerical optimization problems," *Appl Math Comput*, vol. 219, no. 15, pp. 8121–8144, 2013.

[5] Y. Sheoran, V. Kumar, K. P. S. Rana, P. Mishra, J. Kumar, and S. S. Nair, "Development of backtracking search optimization algorithm toolkit in LabView™," *Procedia Comput Sci*, vol. 57, pp. 241–248, 2015.

[6] B. A. Hassan and T. A. Rashid, "Operational framework for recent advances in backtracking search optimisation algorithm: a systematic review and performance evaluation," *Appl Math Comput*, p. 124919, 2019.

[7] P. Civicioglu, "Backtracking Search Optimization Algorithm," MATLAB Central File Exchange. Accessed: Mar. 24, 2021. [Online]. Available: www.mathworks.com/matlabcentral/fileexchange/44842-backtracking-search-optimization-algorithm

[8] O. Turgut, "Backtracking Search Algorithm java code," Sep. 20, 2015. doi: 10.13140/RG.2.1.3212.3367.

[9] J. Sharma, "Backtracking Search Optimization Algorithm C++ Code." Accessed: Mar. 23, 2021. [Online]. Available: https://github.com/imjayanti/Backtracking-Search-Optimization-Algorithm

[10] B. A. Hassan, "CSCF: a chaotic sine cosine firefly algorithm for practical application problems," *Neural Comput Appl*, vol. 33, no. 12, 2021, doi: 10.1007/s00521-020-05474-6.

[11] T. A. Rashid *et al.*, "NSGA-II-DL: metaheuristic optimal feature selection with deep learning framework for HER2 classification in breast cancer," *IEEE Access*, vol. 12, pp. 38885–38898, 2024, doi: 10.1109/ACCESS.2024.3374890.

[12] H. K. Hamarashid, B. A. Hassan, and T. A. Rashid, "Modified-improved fitness dependent optimizer for complex and engineering problems," *Knowl Based Syst*, vol. 300, p. 112098, Sep. 2024, doi: 10.1016/J.KNOSYS.2024.112098.

[13] S. U. Umar, T. A. Rashid, A. M. Ahmed, B. A. Hassan, and M. R. Baker, "Modified Bat Algorithm: a newly proposed approach for solving complex and real-world problems," *Soft comput*, pp. 1–16, Jul. 2024, doi: 10.1007/S00500-024-09761-5/METRICS.

[14] B. A. Hassan, N. B. Tayfor, A. A. Hassan, A. M. Ahmed, T. A. Rashid, and N. N. Abdalla, "From A-to-Z review of clustering validation indices," *Neurocomputing*, p. 128198, Jul. 2024, doi: 10.1016/J.NEUCOM.2024.128198.

[15] R. K. Muhammed *et al.*, "Comparative analysis of AES, Blowfish, Twofish, Salsa20, and ChaCha20 for Image Encryption," *Kurdistan J Appl Res*, vol. 9, no. 1, pp. 52–65, May 2024, doi: 10.24017/SCIENCE.2024.1.5.

[16] A. M. Ahmed *et al.*, "Balancing exploration and exploitation phases in whale optimization algorithm: an insightful and empirical analysis," *Handbook of Whale Optimization Algorithm: Variants, Hybrids, Improvements, and Applications*, pp. 149–156, Jan. 2024, doi: 10.1016/B978-0-32-395365-8.00017-8.

[17] B. A. Hassan *et al.*, "Equitable and fair performance evaluation of whale optimization algorithm," *Handbook of Whale Optimization Algorithm: Variants, Hybrids, Improvements, and Applications*, pp. 157–168, Jan. 2024, doi: 10.1016/B978-0-32-395365-8.00018-X.

[18] M. H. Abdalla *et al.*, "Sentiment analysis based on hybrid neural network techniques using binary coordinate ascent algorithm," *IEEE Access*, vol. 11, pp. 134087–134099, 2023, doi: 10.1109/ACCESS.2023.3334980.

[19] B. A. Hassan, "Ontology learning using formal concept analysis and WordNet," Nov. 2023, Accessed: Jul. 24, 2024. [Online]. Available: https://arxiv.org/abs/2311.14699v1
[20] T. A. Rashid *et al.*, "Awareness requirement and performance management for adaptive systems: a survey," *Journal of Supercomputing*, vol. 79, no. 9, pp. 9692–9714, Jun. 2023, doi: 10.1007/S11227-022-05021-1/METRICS.
[21] M. T. Abdulkhaleq *et al.*, "Fitness dependent optimizer with neural networks for COVID-19 patients," *Computer Methods and Programs in Biomedicine Update*, vol. 3, p. 100090, Jan. 2023, doi: 10.1016/J.CMPBUP.2022.100090.
[22] B. A. Hassan and T. A. Rashid, "Artificial Intelligence algorithms for natural language processing and the semantic web ontology learning," 2021.
[23] T. A. R. Bryar and A. Hassan, "Formal context reduction in deriving concept hierarchies from corpora using adaptive evolutionary clustering algorithm," *Complex Intell Syst*, 2021.
[24] B. A. Hassan and T. A. Rashid, "A multidisciplinary ensemble algorithm for clustering heterogeneous datasets," *Neural Comput Appl*, 2021, doi: 10.1007/s00521-020-05649-1.

8 Optimizing LPB Algorithms using Simulated Annealing

Dana Rasul Hamad and Tarik A. Rashid

8.1 INTRODUCTION

In the realm of optimization, many algorithms display significant promise; however, they require greater efficiency and robustness to effectively solve complex problems and obtain better results. The Learner Performance-based Behavior (LPB) algorithm shows potential, while, with its exploitation and exploration, there is room to improve or enhance its overall performance. Metaheuristics are sophisticated approaches for addressing problems that have been created to address a variety of optimization issues. These algorithms have demonstrated efficacy in managing intricate and insolvable scenarios, providing an approximate resolution within a brief timeframe. In many disciplines, including engineering, medicine, corporate planning, and energy management, optimization is essential. In addition, metaheuristic algorithms are capable of making intelligent choices among a wide range of potential solutions [1]. Natural systems are the source of inspiration for some popular metaheuristic algorithms, such as the Genetic Algorithm (GA) and Particle Swarm Optimization (PSO).

The LPB algorithm was recently introduced by Rahman and Rashid (2021) [2] and is intended to handle both single and multi-objective situations. Rahman and Rashid's study illustrates how to apply LPB in a methodical manner to maximize and acquire optimal results. Researchers interested in creating, improving, or hybridizing the method can use the publication as a guide as well.

The primary solution focuses on the improvement of the LPB algorithm to address its limitations. To do this, we introduce LPB using Simulated Annealing (LPBSA), which integrates the strengths of LPB with the powerful technique of Simulated Annealing (SA) to enhance its performance. By striking a balance between exploring new possibilities and exploiting known ones, LPBSA provides the best approach as the algorithm converges more effectively to the optimal solution. The proposed algorithm is an improvement on LPB that expands on its capabilities. Utilizing SA as a potent method, LPBSA outperforms well-known algorithms like GAs and PSO in terms of LPB optimization. By applying crossover and mutation operations, splitting a main population into Good and Bad populations, and using SA with a Metropolis Acceptance Criterion (MAC) formula, the LPBSA technique selects only the best

DOI: 10.1201/9781003601555-8

candidates for the subsequent iteration. By improving the population, this technique raises the algorithm's efficiency and improves its performance.

Optimization algorithms are crucial in terms of solving complex or high-dimensional problems effectively. Among those algorithms, LPB has obtained impressive results by drawing inspiration from the academic progress of students, especially students that are currently moving to university from high school. They are new students from college and have chosen a new field in which they may find it difficult to improve their performance, so they need to share information with other students in order to improve their behavior. To this end, the LPB algorithm tries to use its methods and techniques to improve the performance of students. LPB still has room to improve its performance, which can be achieved by integrating an additional metaheuristic technique to address its limitation. One of the main objectives of developing the LPB algorithm is to enhance students' behavior with the goal of improving their overall experience across academic disciplines. Consequetly, progression in the LPB algorithm directly contributes to the enhancement of students' knowledge in various fields. Therefore, algorithms can be developed in order to iteratively refine their search processes, leading to more efficient solutions.

In order to strengthen the LPB, simulating annealing was chosen suitably because it is a popular metaheuristic algorithm that was inspired by the metallurgical annealing process. In this process, materials are gradually cooled to obtain an optimal crystalline structure. Similarly, SA helps the algorithm to escape local optima, making it extremely effective in improving the exploration phase of the optimization process. Additionally, SA increases the possibility of discovering the goal optimum because it helps in broadening the search space and probabilistically accepting worse solutions.

8.2 LITERATURE REVIEW

Optimization algorithms play an important role for complex problem solutions in a wide range of fields. For example, their application in education has garnered pivotal attention, as these algorithms provide potential solutions for improving learning outcomes, supporting decision-making, and personalizing the learning experience. Among these optimization approaches, GA stands out due to its capability to effectively search large solution spaces, inspired by the principles of natural selection. GA has been established as an essential approach in enhancing several tasks, from resource allocation to curriculum design, highlighting its versatility and importance in educational contexts. This algorithm has also been applied to resource allocation and optimization scheduling in educational settings. For instance, GAs have been used to mimic adaptive systems for learning, which tailors educational content to the need of a specific learner [3]. PSO is another optimization algorithm that was inspired by both "fish schooling" and the "social behavior of flocking birds", and it has been utilized for optimization tasks in numerus fields such as education. It is also absolutely useful in e-learning systems, especially for some significant facilities related to these systems, including improving the collaborative learning area and developing intelligent tutorials. One of PSO's aspects that makes it suitable for "real-time educational applications" is that this algorithm has a high ability to converge fast to optimal solutions [4]. Moreover, [5] presents differential evolution as another simple and effective optimization algorithm that has been utilized in solving many problems in the education field, particularly for some relevant educational facilities like managing resources, scheduling exams, and designing curriculums. Simplicity and robustness are the two essential aspects that make this algorithm a popular choice for educational tasks.

Some approaches like "data-driven analytics", "machine learning models", and "education data mining" are aiming to identify trends in student data for developing their involvement targets. [6] states that one of the key links that will improve a student's achievement and engagement is personalized learning by addressing student learning styles, preferences, and wants. In addition, by using real-time data, "adaptive learning systems" can be significantly utilized to alter the contents and learning environment to meet the learner student's needs. Moreover, optimization algorithms are employed by these systems to modify feedback, assessment, and instructional resources dynamically. Furthermore, the systems created based on these algorithms can lead to greater student fulfillment and outcomes for improved learning [7]. Next, more of a focus will be on existing gaps not addressed properly in previous studies.

In this section, we highlight some significant gaps that we focus on in this study, especially regarding personalized learning and optimization algorithms. One of the crucial gaps that has not been addressed properly is the challenge of admitting students into university programs for which they may lack foundational knowledge. This problem highlights the need for more effective algorithms that help the students to improve their skills (performance) by selecting their specific learning needs. By refining these algorithms, better support will be available for the students in overcoming knowledge gaps and enhancing their academic progress. One of the most important areas for tracking and exposing previous algorithms is "scalability". Many optimization algorithms struggle during applications to large educational datasets. With existing algorithms, there is a need to provide actionable and interpretable insights for educators. This is another gap that should be under focus. Another gap is that, so far, there is a lack of focus on integrating optimization algorithms for real-world educational systems and practices. LPBSA, as a novel contribution, provides a scalable, interpretable, and integrative approach to address these gaps to optimize learner performance. LPBSA is inspired by the process of admitting students to colleges or universities where they are encouraged to adapt and enhance their study skills or behaviors. This process aligns with the essential principles of personalized and adaptive learning, which focus on tailoring education to meet the student's needs. By the incorporation of SA, which strengthens the LPB results, LPBSA makes the algorithm more reliable and, more importantly, it is also designed to be adaptable and responsive for further use in the future, making it essential for optimization processes and developing outcomes in various fields like healthcare and education.

8.3 LPB

The phrases required in learner acceptance are mirrored in the LPB algorithm, which draws inspiration from the university admissions process for recent high school graduates. These procedures include grouping students according to their cumulative rates and subsequently enhancing their behavior and performance after admission.

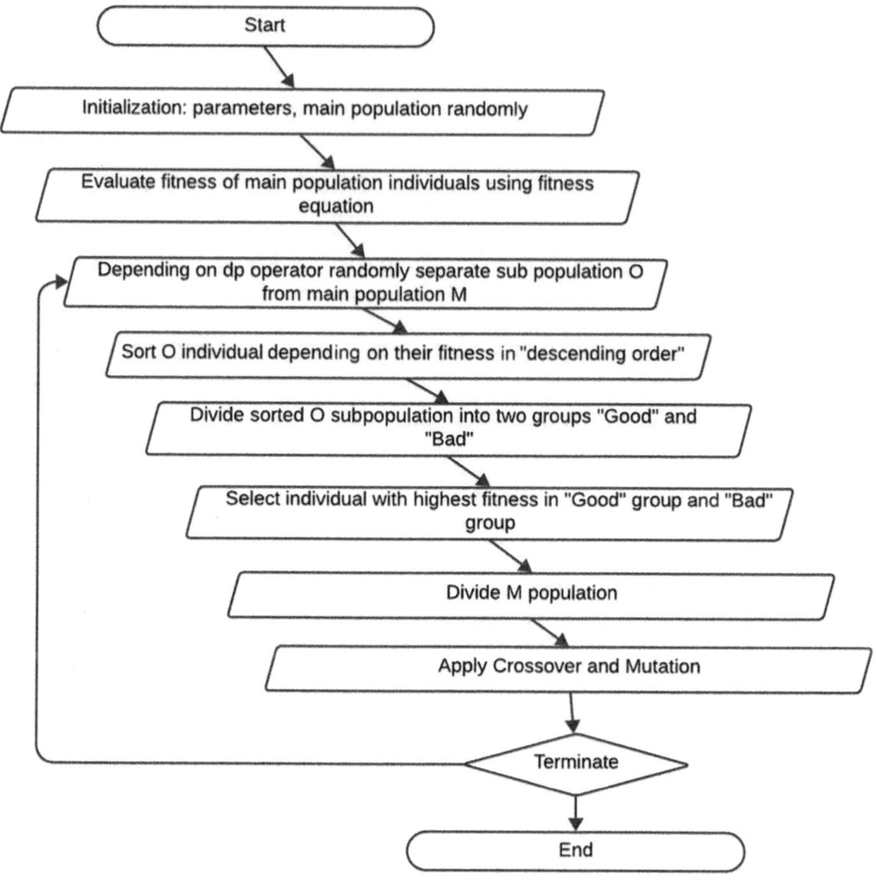

FIGURE 8.1 LPB flowchart.

These techniques also help to improve individuals' conduct and performance level when they are admitted to a particular department. According to [8–10], given the necessity for learners to adopt new study habits as they transition from junior high to college, in order to solve this, the algorithm first chooses a subset of the population. These individuals are then separated into smaller groups, from which the most qualified applications are selected in accordance with their qualifications. Then, through cooperative efforts that resemble cooperation and facilitate information exchange during study sessions, their conduct and performance are further improved (referred to as crossover). Furthermore, this approach introduces mutation, or unpredictable impacts, on their behavior. LPB incorporates mutation and crossover strategies similar to those used in GA. Researchers used the diagram shown in Figure 8.1 as a flowchart to develop this algorithm.

It is important to represent how the LPB algorithm works in the next paragraph.

Initialization is the first step in the process, in which a group of people (learners) is separated into two sub-groups based on their fitness. Subsequently, work is targeted at enhancing behavior through promoting working as a team and information transfer within the sub-groups. This approach, along with the development of effective study habits, is aimed at improving individual performance and learning outcomes. Later, we have to select the best individuals from each sub-group based on their improved behaviors.

Then we start to exchange information between selected individuals, called "crossover", to create new solutions. This is similar to exchanging information between parents to produce offspring. When crossover is complete, the "mutation" operation starts, which introduces random changes to some individuals to explore new potential solutions, and assess the fitness of the new individuals to determine their performance. Then, the process is repeated to select the next iteration. The process stops when a certain condition is met.

8.4 SA

The SA algorithm was introduced in 1983 by Kirkpatrick et al. to solve the "Traveling Salesman Problem (TSP)" [11]. SA is a probabilistic way to estimate a function's global optimum. In addition, it is a metaheuristic specifically designed to approximate global optimization for an optimization problem across a broad search space, and it is an optimization technique that maintains constant control limits while systematically modifying the variables, and upper and lower truncation limits in a stochastic way [12].

SA algorithms draw inspiration from metallurgy's annealing process, which involves rapidly heating a metal followed by gradual cooling [13]. At high temperatures, atoms within the metal exhibit fast movement, but as the temperature decreases, their kinetic energy diminishes. Ultimately, this process leads to a more ordered atomic arrangement, resulting in a material that is more ductile and easier to manipulate [13]. Numerous optimization issues, including TSP, protein folding, graph partitioning, and job-shop scheduling, have been effectively solved with SA [14–16]. The capacity of SA to break free from local minima and converge to a global minimum is its primary benefit [12]. Furthermore, SA does not require prior knowledge of the search space and is quite simple to implement. Starting with an initial solution, the SA method iteratively improves the existing solution by randomly perturbing it and accepting the perturbation with a predetermined probability [17]. With more iterations, there is a progressive drop in the probability of accepting a worse answer from its original high level [17]. Figure 8.2 offers a through flowchart that illustrates each phase of SA.

As mentioned, SA is an optimization method inspired by the annealing process in metallurgy. It involves heating a material and then cooling it slowly to remove defects, analogous to exploring and refining solutions in optimization.

8.4.1 Key Concepts Behind SA

Some key concepts have a significant role to play in developing the techniques of the SA algorithm. "Energy states and cost functions" is one of those concepts. In optimization, the energy concept in a system is analogous to the cost function, which

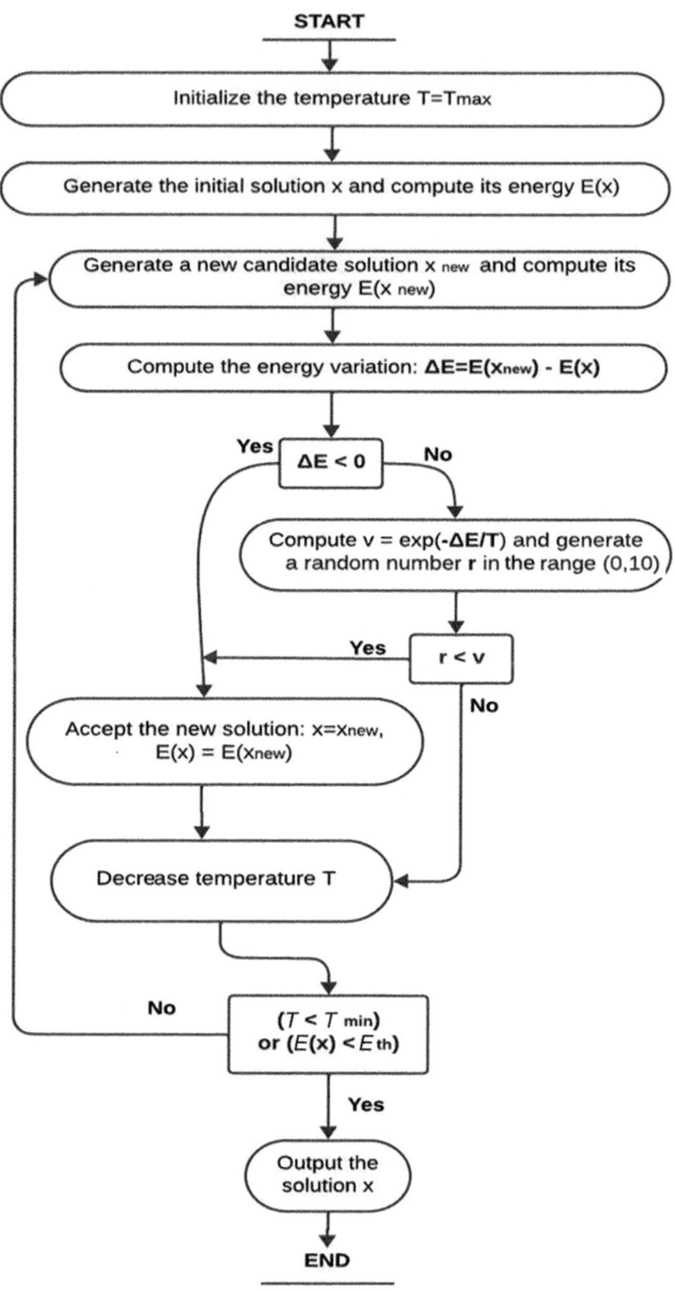

FIGURE 8.2 SA algorithm flowchart.

shows the value required to be minimized or maximized. Each acceptable solution to the optimization problem corresponds to a state of the system. Additionally, based on the cost function, the quality of that solution is evaluated. The temperature and cooling schedule are two other key concepts, where temperature acts as a controlling parameter to determine the likelihood of accepting a less optimal solution across the algorithm's progress, allowing the exploration of a broader solution space. Means with a higher probability of a higher temperature allow the algorithm to accept worse solutions, promoting exploration of the solution space. Based on a predefined cooling schedule, the temperature is regularly reduced. As the temperature reduces, the possibility of accepting a worse solution is reduced, guiding the algorithm toward convergence to the optimal solution. The MAC is another concept and an important key to SA's effectiveness that allows the acceptance of worse solutions developed on the basis of a probability that decreases with temperature assisting the algorithm in escaping local minima. LPBSA combines the robust exploration capabilities of SA with the structured approach of LPB. So, by incorporating the MAC, LPBSA enhances its capacity to avoid local minima and explore the solution space more effectively.

8.4.2 How the Algorithm Works

In the context of the algorithm's work process, here is a step-by-step presentation of how the algorithm works.

Initialization

Start the algorithm with a high temperature or cooling rate. This is known as the initial solution. Then the cost of the initial solution will be evaluated.

Iterative Process

The following steps are repeated until the specified number of iterations is reached or the stopping criteria are met:

- Generate a new solution: The existing solution is disrupted to produce a new candidate solution and then the cost of the new solution is evaluated.
- Evaluate the new solution: The difference between the new solution and the current solution in cost (ΔE) is calculated.
- Apply the MAC: The MAC determines the likelihood of accepting a solution with a higher cost, enabling the algorithm to explore a broader solution space.
- Update the temperature: According to the cooling schedule, reduce the temperature.

Convergence

The proposed algorithm concentrates on refining the solution discovered during the temperature decrease and then the algorithm becomes less inclined to accept worse solutions. When the temperature meets the stopping condition or gets close to zero, the algorithm converges to an optimum solution.

8.5 LPBSA

LPBSA is a hybrid optimization technique that blends SA and LPB. However, it is known as an improvement approach that focuses more on the LPB algorithm's optimization. By increasing convergence, resilience, and adaptability in the solution of challenging optimization problems, LPBSA seeks to improve on established LPB approaches. By comparing LPBSA with LPB and other well-known algorithms, such as PSO and GA, LPBSA performs better through assessments utilizing benchmark test functions. LPBSA plays a crucial role in improving the efficiency of intelligent systems by fine-tuning parameters, lowering the complexity of computation, and improving performance overall. This algorithm is especially helpful in dynamic settings where the optimal solution could alter over time. Because of its adaptive qualities, LPBSA can modify its search strategies to change conditions or circumstances.

8.5.1 Practical Implementation using LPBSA to Optimize Learner Performance

In this section, the capabilities of LPBSA are integrated into real-world environments. Essentially, given the purpose of personalized learning pathways, this algorithm provides promising chances in real-world educational settings to improve learner performance over personalized learning pathways by analyzing individual performance data, wherever students are struggling, and tailoring learning activities to their precise needs. The algorithm can identify such areas, which will have a great effect in terms of improving learning experiences. There is significant potential for the performance of optimizing learners available in various education settings for LPBSA. It also has the ability to be effectively applied in real-world environments to provide personalized learning pathways. LPBSA identifies areas where students are facing challenges and tailors learning activities to address their specific requirements based on analyzing individual performance data. Additionally, this targeted approach improves learning skills and helps to enhance overall academic outcomes. One of the most prominent implementation considerations is "data collection", which means it is very significant to collect comprehensive data on student performance, such as grades, attendance, or any other relevant metrics. Optimized resource sharing and adaptive learning systems, and identifying students that are currently at-risk, can be beneficial for LPBSA, which is crucial because it will adjust the difficulty of lessons based on real-time feedback from the algorithm, dynamically ensuring that each student is challenged appropriately. In terms of implementation considerations, LPBSA should be integrated with learning management systems (LMS) in order to adjust learning materials automatically.

The primary objectives of this altered algorithm are to improve convergence, adaptability, and robustness. By integrating LPB with SA, the approach improves global exploration, which serves to avoid local optima. Additionally, introducing temperature-controlled evolution allows the acceptance of less optimal solutions when beneficial and combines SA's exploration abilities with adaptive learning to create an effective optimization strategy, which ensures robustness throughout several problem domains. The following section explains how LPBSA works.

LPBSA starts by initializing a population of solutions and calculating the fitness of each individual. It then selects a subset of individuals, sorts them into Good and Bad groups based on fitness, and assigns each individual to a group. Next, it selects individuals for crossover, ensuring the ideal group is not empty, and performs crossover to create new individuals. Mutation is applied to introduce random changes. The algorithm uses the MAC to accept or reject new individuals based on a random number and temperature rate. The population is updated with the accepted individuals and the process is repeated for a specified number of iterations or until a termination condition is met. The optimal solution is returned at the end. Figure 8.3 shows the flowchart of LPBSA.

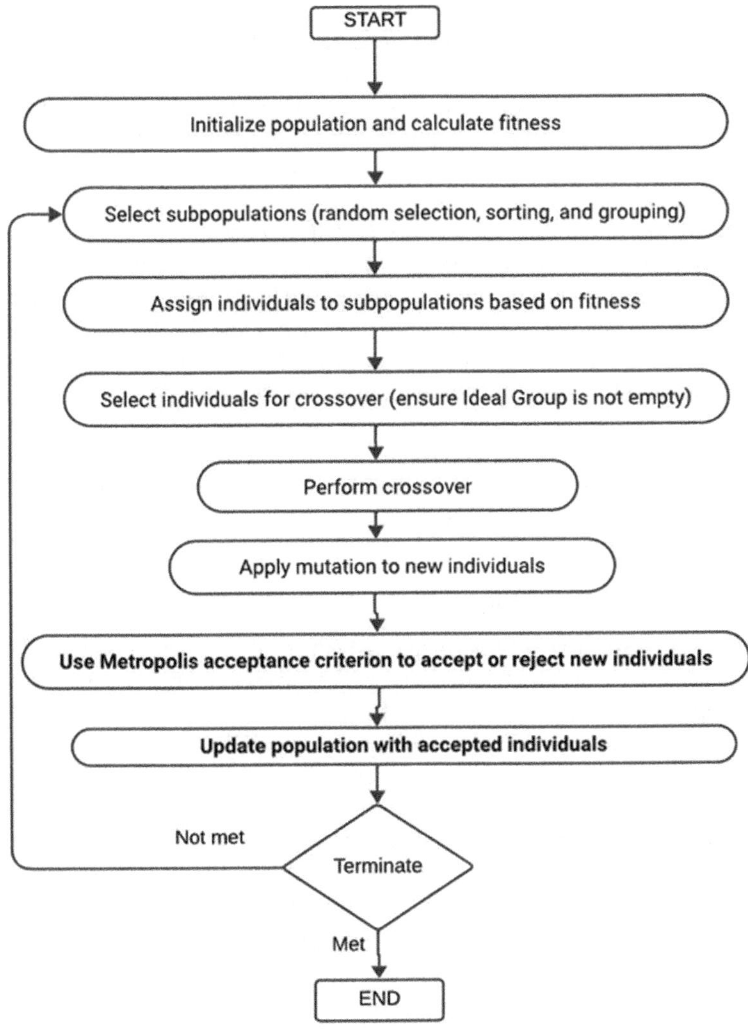

FIGURE 8.3 LPBSA flowchart.

The MAC is applied before proceeding to the next iteration in LPBSA during the mutation process. The criterion is defined as

$$P(accept\ worse)\exp\left(-\frac{Cost(new)-Cost(current)}{temperature}\right) \quad (1)$$

By calculating the probability of accepting the worse solution, this equation helps to prevent premature convergence and maintain diversity within the population by calculating the probability of accepting less optimal solutions.

8.6 KEY STEPS OF CROSSOVER AND MUTATION IN LPBSA

There are some reasons behind crossover and mutation operators being used with LPBSA, such as "exploration and exploitation". Crossover mainly enables exploration by sharing information from diverse regions in the search space. On the other hand, mutation facilitates exploitation by implementing random and small modifications, which fine-tune solutions. Hence, these two operators together balance exploitation and exploration, which is very significant in order to provide effective optimization. [18] states that, properly balancing two modes like exploitation and exploration increases the chances of achieving robust results. Additionally, diversity maintenance is important for preserving genetic diversity within the population by leveraging both crossover and mutation, which allows the algorithm to reduce the risk of getting trapped in local optima. So, this diversity is crucial for achieving global optima, which is the ultimate goal of the optimization process.

Regarding "adaptability", LPBSA can adapt to various optimization landscapes by combining crossover and mutation. The algorithm's capability to fine-tune solutions within the fitness environment is ensured through the mutation process.

8.6.1 Crossover Operator

In LPBSA the crossover operator is inspired by the process of integrating the behaviors of individuals to deliver a new individual with possibly superior characteristics. The crossover operator is utilized in LPBSA to combine information from two solutions to create a new candidate solution, helping to explore the search space more effectively. Two parents are selected for crossover:

i. Parents are introduced as a decimal number that is converted to binary.
ii. The left half of the binary number for the first parent is merged with the right half of the binary number for the second parent, and the right half of the binary number for the first parent is integrated with the left half of the binary number for the second parent.
iii. Two new offspring are produced.

8.6.2 Mutation Operator

In LPBSA, the mutation operator provides a random change to the individual solution, which means more improvement in selected individuals. This promotes variety within the population and aids the avoidance of premature convergence. The mutation process is performed simply as presented below.

i. New offspring from crossover are selected for mutation.
ii. A binary number is utilized for modification.
iii. One point of the binary number is randomly changed, like one 0 point changed to 1.
iv. New and improved offspring are delivered.

8.7 CASE STUDY

Consider the following function: $f(x) = x_1^2 + x_2^2$, for integer x, $0 \le x_1 \le 8000$ and $0 \le x_2 \le 8000$.

Step 1: Create a random population (*Bi*) of between 0 and 8000 individuals and then provide an average for the **fitness of $X_1^2$and X_2^2**, as shown in Table 8.1.

TABLE 8.1
Main population (*Bi*)

Bi	X_1	X_2	X_1^2	X_2^2	Fitness of $X_1^2$and X_2^2
B1	4320	3120	18662400	9734400	28396800
B2	1233	4523	1520289	20457529	21977818
B3	5100	3209	26010000	10297681	36307681
B4	4355	5210	18966025	27144100	46110125
B5	2331	4266	5433561	18198756	23632317
B6	2040	2755	4161600	7590025	11751625
B7	5043	1977	25431849	3908529	29340378
B8	3460	4781	11971600	22857961	34829561
B9	1920	5510	3686400	30360100	34046500
B10	4222	3741	17825284	13995081	31820365
B11	5401	1740	29170801	3027600	32198401
B12	3351	2850	11229201	8122500	19351701
B13	5201	4989	27050401	24890121	51940522
B14	2188	3477	4787344	12089529	16876873
B15	3409	1877	11621281	3523129	15144410
B16	4560	2776	20793600	7706176	28499776

Average: 28889053

TABLE 8.2
***K* population**

Bi	*Ki*	Fitness	Groups
B3	*K1*	**36307681**	Good
B8	*K2*	34829561	
B11	*K3*	32198401	
B1	*K4*	28396800	
B5	*K5*	**23632317**	Bad
B12	*K6*	19351701	
B15	*K7*	15144410	
B6	*K8*	11751625	

Highest fitness for the Good group = **36307681**

Highest fitness for the Bad group = **23632317**

TABLE 8.3
Comparison of *Bi* individuals with Ki highest Good and Bad fitness

Bi	Fitness	Fit (*Bi*)<= Fit (*Ki5*)	Fit (*Bi*)<= Fit(*Ki1*)	Ideal group
B1	28396800		Good	
B2	21977818	Bad		
B3	36307681		Good	
B4	46110125			Ideal
B5	23632317	Bad		
B6	11751625	Bad		
B7	29340378		Good	
B8	34829561		Good	
B9	34046500		Good	
B10	31820365		Good	
B11	32198401		Good	
B12	19351701	Bad		
B13	51940522			Ideal
B14	16876873	Bad		
B15	15144410	Bad		
B16	28499776		Good	

Step 2: Create subpopulation (s) randomly from the main population by selecting eight individuals and put them in descending order. Divide those eight individuals into two groups (Good) and (Bad). Select individuals from highest to lowest from those groups, which is shown in Table 8.2.

Step 3: Compare the main population (*Bi*) individuals to the highest fitness of the Good and Bad group in Ki, which is presented in Table 8. 3.

TABLE 8.4
Selected individuals

Proper individuals are selected to crossover from Table 8.3		
B13	51,940,522	Ideal
B4	46,110,125	Ideal
B3	36,307,681	Good
B8	34,829,561	Good

- If the fitness values of *Bi* <= highest fitness of the Bad group, move *Bi* to the Bad group;
- else if the fitness value of *Bi* <= highest fitness of the Good group, move *Bi* to the Good group;
- else if the fitness value of *Bi* > highest fitness of the Good group, move *Bi* to the Ideal group.

Step 4: In this step, we have to select four individuals from the main population and ensure that the Ideal group is not empty. If it is, choose individuals from the Good group. If the Good group is also empty, select individuals from the Bad group. displays the selected individuals from, these four individuals are used for the next phase (crossover stage). Ensure that the number of selected individuals matches the specified requirement $N = 4$, as stated in the initial step.

Step 5: Table 8.5 presents the applied crossover between the individuals selected from Table 8.4 with *Ki* individuals from Table 8.2.

Step 6: For the new individuals (children), apply mutation and convert one 0 to 1 for each individual randomly in order to maximize the function. This process is shown in Table 8.6.

New individuals are shown in Table 8.7.

Step 7: Before passing to the next iteration, apply the MAC formula shown in Eq. (1). In this chapter, a random number = 0.6 and temperature rate = 100 are provided.

For each cost, compare the acceptance probability with the random number to determine acceptance or rejection:

For C1: Acceptance probability = exp(-(7276 - 5228)/100) ≈ 0.6703. Since 0.6 < 0.6703, C1 is **accepted**.
For C2: Acceptance probability = exp(-(7121 - 5073)/100) ≈ 0.6775. Since 0.6 < 0.6775, C2 is **accepted**.
For C3: Acceptance probability = exp(-(3204 - 2180)/100) ≈ 0.7408. Since 0.6 < 0.7408, C3 is **accepted**.
For C4: Acceptance probability = exp(-(7939 - 6915)/100) ≈ 0.6525. Since 0.6 < 0.6525, C4 is **accepted**.

TABLE 8.5
Crossover operation

IdCh	NInd	X_1	X_2	Binary (X_1)	Binary(X_2)
E1	B13	5201	4989	**101000**1010001	**100110**1111101
E2	K1	5100	3209	100111**1101100**	110010**001001**
C1	New	5228	2441	1010001101100	100110001001
C2	New	5073	6525	1001111010001	1100101111101
E3	B4	4355	5210	**100010**0000011	**101000**1011010
E4	K2	3460	4781	110110**000100**	100101**0101101**
C3	New	2180	5165	100010000100	1010000101101
C4	New	6915	4826	1101100000011	1001011011010
E5	B3	5100	3209	**100111**1101100	**110010**001001
E6	K3	3351	2850	110100**010111**	101100**100010**
C5	New	2519	3234	100111010111	110010100010
C6	New	6764	2825	1101001101100	101100001001
E7	B8	3460	4781	**110110**000100	**100101**0101101
E8	K4	4320	3120	100001**1100000**	110000**110000**
C7	New	7008	2416	1101101100000	100101110000
C8	New	1028	6189	10000000100	1100000101101

TABLE 8.6
Apply mutation

IdCh	X_1	X_2	Binary (X_1)	Binary(X_2)
C1	5228	2441	1010001101100	100110001001
C2	5073	6525	1001111010001	1100101111101
C3	2180	5165	100010000100	1010000101101
C4	6915	4826	1101100000011	1001011011010
C5	2519	3234	100111010111	110010100010
C6	6764	2825	1101001101100	101100001001
C7	7008	2416	1101101100000	100101110000
C8	1028	6189	10000000100	1100000101101

For C5: Acceptance probability = exp(-(3543 - 2519)/100) ≈ 0.7288. Since 0.6 < 0.7288, C5 is **accepted**.

For C6: Acceptance probability = exp(-(7788 - 6764)/100) ≈ 0.6742. Since 0.6 < 0.6742, C6 is **accepted**.

For C7: Acceptance probability = exp(-(7024 - 7008)/100) ≈ 0.5391. Since 0.6 > 0.5391, C7 is **rejected**.

For C8: Acceptance probability = exp(-(1540 - 1028)/100) ≈ 0.6428. Since 0.6 < 0.6428, C8 is **accepted**.

TABLE 8.7
New individuals after mutation

IdCh	X_1	X_2	Binary (X_1)	Binary(X_2)
C1	7276	3465	1110001101100	110110001001
C2	7121	7549	1101111010001	1110101111101
C3	3204	7213	110010000100	1110000101101
C4	7939	6874	1111100000011	1101011011010
C5	3543	3746	110111010111	111010100010
C6	7788	3849	1111001101100	111100001001
C7	7024	3440	1101101110000	110101110000
C8	1540	7213	11000000100	1110000101101

TABLE 8.8
Accepted individuals

IdCh	X_1	X_2	Binary (X_1)	Binary(X_2)
C1	7276	3465	1110001101100	110110001001
C2	7121	7549	1101111010001	1110101111101
C3	3204	7213	110010000100	1110000101101
C4	7939	6874	1111100000011	1101011011010
C5	3543	3746	110111010111	111010100010
C6	7788	3849	1111001101100	111100001001
C8	1540	7213	11000000100	1110000101101

TABLE 8.9
Calculate accepted individuals

IdCh	X_1	X_2	X_1^2	X_2^2	Fitness of X_1^2 and X_2^2
C1	7276	3465	52940176	12006225	64946401
C2	7121	7549	50708641	56987401	107696042
C3	3204	7213	10265616	52027369	62292985
C4	7939	6874	63027721	47251876	110279597
C5	3543	3746	12552849	14032516	26585365
C6	7788	3849	60652944	14814801	75467745
C8	1540	7213	2371600	52027369	54398969

Therefore, based on the comparison with the random number (0.6), C7 is rejected, whereas all other costs are accepted. Table 8.8 shows the individuals that are accepted and used for the next iteration.

Step 8: Calculate the new accepted individuals, which are displayed in Table 8.9.

TABLE 8.10
Provide an average

IdCh	X_1	X_2	X_1^2	X_2^2	Fitness of X_1^2 and X_2^2
B13	5201	4989	27050401	24890121	51940522
K1	5100	3209	26010000	10297681	36307681
B4	4355	5210	18966025	27144100	46110125
K2	3460	4781	11971600	22857961	34829561
B3	5100	3209	26010000	10297681	36307681
K3	3351	2850	11229201	8122500	19351701
B8	3460	4781	11971600	22857961	34829561
K4	4320	3120	18662400	9734400	28396800
C1	7276	3465	52940176	12006225	64946401
C2	7121	7549	50708641	56987401	107696042
C3	3204	7213	10265616	52027369	62292985
C4	7939	6874	63027721	47251876	110279597
C5	3543	3746	12552849	14032516	26585365
C6	7788	3849	60652944	14814801	75467745
C8	1540	7213	2371600	52027369	54398969

Average: 52649382

TABLE 8.11
Second K subpopulation

Bi	*Ki*	Fitness	Groups
B4	*K1*	**46110125**	Good
B9	*K2*	34046500	
B10	*K3*	31820365	
B1	*K4*	28396800	
B2	*K5*	**21977818**	Bad
B12	*K6*	19351701	
B14	*K7*	16876873	
B6	*K8*	11751625	

Highest fitness for the Good group = **46110125**

Highest fitness for the Bad group = **21977818**

Step 9: Calculate the average for accepted individuals in Table 8.9, the individuals in Table 8.4, and (Good group) individuals in Table 8.2. The results are shown in Table 8.10.

Steps 2–9 are repeated until the desired number of iterations or the termination condition is satisfied, at which point the optimal solution is provided. Let's follow Steps 2–9 for the next iteration.

TABLE 8.12
Compare *Bi* individuals with Ki highest Good and Bad fitness

Bi	Fitness	Fit (*Bi*)<= Fit (*Ki5*)	Fit (*Bi*)<= Fit(*Ki1*)	Ideal group
B1	28396800		Good	
B2	21977818		Good	
B3	36307681		Good	
B4	46110125		Good	
B5	23632317		Good	
B6	11751625	Bad		
B7	29340378		Good	
B8	34829561		Good	
B9	34046500		Good	
B10	31820365		Good	
B11	32198401		Good	
B12	19351701	Bad		
B13	51940522			Ideal
B14	16876873	Bad		
B15	15144410	Bad		
B16	28499776		Good	

Step 2: From the main population *B*, create a new subpopulation *K*. The results are illustrated in Table 8.11.

Step 3: Compare the main population (*Bi*) individuals to the highest fitness of the Good group in *Ki* and highest fitness of the Bad group in *Ki*, presented in Table 8.12.

- If the fitness values of *Bi* <= highest fitness of the Bad group, move *Bi* to the Bad group;
- else if the fitness value of *Bi* <= highest fitness of the Good group, move *Bi* to the Good group;
- else if the fitness value of *Bi* > highest fitness of the Good group, move *Bi* to the Ideal group.

Step 4: In this step, we have to select four individuals from the main population and ensure that the Ideal group is not empty. If it is, choose individuals from the Good group. If the Good group is also empty, select individuals from the Bad group. The selected individuals are used in the next phase (Crossover Operator). The results are shown in Table 8.13.

Ensure that the number of the selected individuals matches the specified requirement $N = 4$, as stated in the initial step.

Step 5: Apply crossover between the individuals selected from Table 8.13 and Good group individuals from Table 8.11. The results of this operation are shown in Table 8.14.

TABLE 8.13
Selected individuals

Proper individuals are selected to crossover from Table 8.13		
B13	51940522	Ideal
B4	46110125	Good
B3	36307681	Good
B8	34829561	Good

TABLE 8.14
Crossover operation

IdCh	NInd	X_1	X_2	Binary (X_1)	Binary(X_2)
E1	B13	5201	4989	**101000**1010001	**100110**1111101
E2	K1	4355	5210	100010**0000011**	101000**1011010**
C1	New	5123	4954	1010000000011	1001101011010
C2	New	4433	5245	1000101010001	1010001111101
E3	B4	4355	5210	**100010**0000011	**101000**1011010
E4	K2	1920	5510	111100**00000**	101011**0000110**
C3	New	1088	5126	10001000000	1010000000110
C4	New	7683	5594	1111000000011	1010111011010
E5	B3	5100	3209	**101011**0000110	**110010**001001
E6	K3	4222	3741	100000**1111110**	111010**011101**
C5	New	5630	3229	1010111111110	110010011101
C6	New	4102	3721	1000000000110	111010001001
E7	B8	3460	4781	**110110**000100	**100101**0101101
E8	K4	4320	3120	1000011**100000**	110000**110000**
C7	New	3488	2416	110110100000	100101110000
C8	New	4292	6189	1000011000100	1100000101101

TABLE 8.15
Apply mutation

IdCh	X_1	X_2	Binary (X_1)	Binary(X_2)
C1	5123	4954	1010000000011	1001101011010
C2	4433	5245	1000101010001	1010001111101
C3	1088	5126	10001000000	1010000000110
C4	7683	5594	1111000000011	1010111011010
C5	5630	3229	1010111111110	110010011101
C6	4102	3721	1000000000110	111010001001
C7	3488	2416	110110100000	100101110000
C8	4292	6189	1000011000100	1100000101101

TABLE 8.16
New individuals after mutation

IdCh	X_1	X_2	Binary (X_1)	Binary(X_2)
C1	7171	7002	1110000000011	1101101011010
C2	6481	7293	1100101010001	1110001111101
C3	1600	7174	11001000000	1110000000110
C4	7939	7642	1111100000011	1110111011010
C5	7678	3741	1110111111110	111010011101
C6	6150	3977	1100000000110	111110001001
C7	4000	3440	111110100000	110101110000
C8	6340	7213	1100011000100	1110000101101

Step 6: For the new individuals (children), apply mutation and convert one 0 to 1 for each individual randomly in order to maximize the function. This process is shown in Table 8.15.

New individuals are shown in Table 8.16.

Step 7: Before passing to the next iteration, apply the MAC formula shown in Eq. (1). In this chapter, we provide random number = 0.6 and temperature rate = 100. The results are the same as below:

For C1: Acceptance probability = exp(-(7171 - 5123)/100) ≈ 0.6703. Since 0.6 < 0.6703, C1 is **accepted**.
For C2: Acceptance probability = exp(-(6481 - 4433)/100) ≈ 0.6703. Since 0.6 < 0.6703, C2 is **accepted**.
For C3: Acceptance probability = exp(-(1600 - 1088)/100) ≈ 0.6065. Since 0.6 > 0.6065, C3 is **rejected**.
For C4: Acceptance probability = exp(-(7939 - 7683)/100) ≈ 0.6311. Since 0.6 > 0.6311, C4 is **rejected**.
For C5: Acceptance probability = exp(-(7678 - 5630)/100) ≈ 0.7061. Since 0.6 < 0.7061, C5 is **accepted**.
For C6: Acceptance probability = exp(-(6150 - 4102)/100) ≈ 0.6703. Since 0.6 < 0.6703, C6 is **accepted**.
For C7: Acceptance probability = exp(-(4000 - 3488)/100) ≈ 0.6055. Since 0.6 > 0.6055, C7 is **rejected**.
For C8: Acceptance probability = exp(-(6340 - 4292)/100) ≈ 0.6703. Since 0.6 < 0.6703, C8 is **accepted**.

Based on the comparison with the random number (0.6), C3, C4, and C7 are rejected, whereas all other costs are accepted. Table 8.17 shows the accepted individuals.

Step 8: Calculate the new accepted individuals after applying simulated annealing the results are shown in Table 8.18.

TABLE 8.17
Accepted individuals

IdCh	X_1	X_2	Binary (X_1)	Binary(X_2)
C1	7171	7002	1110000000011	1101101011010
C2	6481	7293	1100101010001	1110001111101
C5	7678	3741	1110111111110	111010011101
C6	6150	3977	1100000000110	111110001001
C8	6340	7213	1100011000100	1110000101101

TABLE 8.18
Calculate accepted individuals

IdCh	X_1	X_2	X_1^2	X_2^2	Fitness of X_1^2 and X_2^2
C1	7171	7002	51423241	49028004	100451245
C2	6481	7293	42003361	53187849	95191210
C5	7678	3741	58951684	13995081	72946765
C6	6150	3977	37822500	15816529	53639029
C8	6340	7213	40195600	52027369	92222969

TABLE 8.19
Provide sum, average, and max

IdCh	X_1	X_2	X_1^2	X_2^2	Fitness of X_1^2 and X_2^2
E1	5201	4989	27050401	24890121	51940522
E2	4355	5210	18966025	27144100	46110125
E3	4355	5210	18966025	27144100	46110125
E4	1920	5510	3686400	30360100	34046500
E5	5100	3209	26010000	10297681	36307681
E6	4222	3741	17825284	13995081	31820365
E7	3460	4781	11971600	22857961	34829561
E8	4320	3120	18662400	9734400	28396800
C1	7171	7002	51423241	49028004	100451245
C2	6481	7293	42003361	53187849	95191210
C5	7678	3741	58951684	13995081	72946765
C6	6150	3977	37822500	15816529	53639029
C8	6340	7213	40195600	52027369	92222969

Average: 55693299

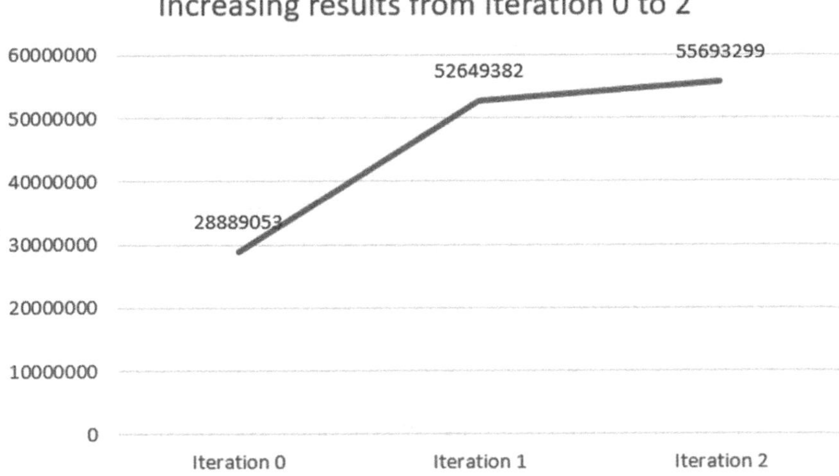

FIGURE 8.4 Average results of three iterations.

TABLE 8.20
Comparison of the average

Iteration *N*	Average
Iteration 0	28889053
Iteration 1	52649382
Iteration 2	55693299

Step 9: Provide the sum, average, and max for the accepted individuals in Table 8.18, as well as the selected individuals from Table 8. 13. These results are displayed in Table 8.19.

Therefore, it is evident from the outcomes that the population has been enhanced, leading to increased efficiency among individuals. Table 8.20 presents a comparison of the average for each iteration.

Figure 8.4 shows the average results for each iteration and how they increased from iteration 0 to 2.

8.8 COMPARATIVE ANALYSIS OF LPBSA WITH OTHER ALGORITHMS

As mentioned, LPBSA claims to outperform other algorithms. Therefore, it is crucial to present a detailed comparison analysis utilizing particular benchmarks. In this section, more focus is on the comparative analysis, the rationale behind the selected benchmarks, and the detailed outcomes that were obtained.

8.8.1 Metrics for Comparison

The metrics discussed in this section have played a major role in determining the performance of LPBSA. The average value is one of those metrics that can be utilized to represent the cost function average value that was obtained over multiple runs, signifying the algorithm's capability to deliver promised solutions consistently. To reflect the algorithm's reliability and stability, as well as measure the cost function value variability, standard deviation is utilized as another metric. Additionally, the next metric is the best value that represents the lowest cost function value to assess the algorithm's capability to find the optimal solution.The worst value is used as another metric to obtain the highest cost function value to display the worst case of the algorithm's performance.

8.8.2 Benchmark Test Functions

To evaluate the performance of LPBSA, 19 benchmark test functions were utilized. These functions are commonly used to measure the efficiency of optimization algorithms due to their potential characteristics, such as dimensionality, separability, and multimodality. For example, test function one, which is a sphere function, is a simple unimodal function with a single global minimum, and test function two is a Rastrigin function used as a multimodal function with numerous local minima, inspiring the algorithm's exploration capability.

These test functions provide a comprehensive evaluation of the algorithm's performance. Different aspects of those algorithms are tested via those functions, such as their ability to converge to the global minima, avoid the local minima, and maintain stability. Table 8.21 shows a summary of evaluating LPBSA with other algorithms. The table entries represent the average and standard deviation of the cost function value over 30 runs for each entry.

8.9 RESULTS AND DISCUSSION

The results in Table 8.21 show that LPBSA consistently achieved better performance than the other algorithms, like PSO, across the majority of benchmark functions, GA, and the original LPB. In addition, LPBSA achieved a lower standard deviation and superior average solution, and it also showed its ability to deliver better solutions more consistently. Table 8.21 presents a whole map for all the tests. The best tests are highlighted and the table shows the best outcomes of LPBSA. Some test function results are illustrated in the following sentences. For example, in TF1, LPBSA recorded an average of 3.87E-04 and standard deviation of 7.25E-4, outperforming LPB, which achieved an average of 1.88E-03 and standard deviation of 2.09E-03. PSO and GA provided lower results. PSO achieved 4.20E-18, but with a high standard deviation, and GA recorded an average of 748.60.

In TF5, LPBSA obtained an average of 4.7676233 and standard deviation of 2.7775532. By recording these results, LPBSA outperformed all the other proposed algorithms in this test function. Moreover, LPBSA recorded the highest average and standard deviation compared to other algorithms in TF9,TF14,TF15,TF16,TF18, and

TABLE 8.21
Result of 19 benchmark test functions to compare LPBSA with other algorithms

	LPBSA		LPB		DA		PSO		GA	
TF	AVA	STD	AVA	STD	AVA	STD	AVA	STD	AVA	STD
TF1	3.86896E-04	7.25127E-04	0.001877	0.0020936	**2.85E-18**	7.16E-18	4.20E-18	4.31E-18	748.5972	324.926
TF2	3.9134E-03	2.67553E-03	0.005238	0.0036525	1.49E-05	3.76E-05	0.003154	0.009811	5.971358	1.53310
TF3	15.5732633	9.35452E+00	36.47488	29.224155	1.29E-06	2.10E-06	0.001891	0.003311	1949.003	994.273
TF4	0.15603627	3.49740E-02	0.393866	0.135818	0.000988	0.002776	0.001748	0.002515	21.16304	2.60540
TF5	**4.76762333**	**2.77755315**	16.76919	22.19251	7.600558	6.786473	63.45331	80.12726	133307.1	85007.6
TF6	0.001353802	1.83590E-03	0.002031	0.0027832	4.17E-16	1.32E-15	**4.36E-17**	**1.38E-16**	563.8889	229.699
TF7	0.002900520	**0.001495889**	0.004975	0.002965	0.010293	0.010293	0.005973	0.003583	0.166872	0.07257
TF8	-3723.968593	191.566968	-3747.65	**189.0206**	-2857.58	383.6466	-7.10E+11	1.2E+12	**-3407.25**	164.478
TF9	**0.00067658**	**0.0007894**	0.001567	0.001842	16.01883	9.479113	10.44724	7.879807	25.51886	6.66936
TF10	0.01168585	6.82155E-03	0.017933	0.013532	0.23103	0.487053	0.280137	0.601817	9.498785	1.27139
TF11	0.062534300	2.58492E-02	0.066355	0.030973	0.193354	0.073495	0.083463	0.035067	7.719959	3.62607
TF12	3.06720E-05	5.69906E-05	2.79E-05	3.84E-05	0.031101	0.098349	**8.57E-11**	**2.71E-10**	1858.502	5820.21
TF13	2.63645E-04	7.24801E-04	0.000309	0.000512	0.002197	0.004633	0.002197	0.004633	68047.23	87736.7
TF14	**0.998000000**	**4.51681E-16**	0.998004	1.26E-11	103.742	91.24364	150	135.4006	130.0991	21.3203
TF15	**0.001032395**	**6.76267E-04**	0.002358	0.003757	193.0171	80.6332	188.1951	157.2834	116.0554	19.1935
TF16	**-1.031600000**	**6.77522E-16**	-1.03163	2.46E-06	458.2962	165.3724	263.0948	187.1352	383.9184	36.6053
TF17	2.705400000	**1.35504E-15**	**0.397888**	3.16E-06	596.6629	171.0631	466.5429	180.9493	503.0485	35.7940
TF18	**3.000000000**	**0.00000E+00**	3.000142	0.000283	229.9515	184.6095	136.1759	160.0187	118.438	51.0018
TF19	**-3.862800**	**3.16177E-15**	-3.86278	9.61E-07	679.588	199.4014	741.6341	206.7296	544.1018	13.3016

TF19. Even in TF17, LPBSA provided the highest standard deviation of 1.35504E-15, and achieved the highest average by recording 2.705400000 compared to other algorithms, except LPB, which recorded an average of 1.35504E-15. These results show the precision, consistency, robustness, and superior performance of LPBSA, showcasing the algorithm's enhanced capability to find the optimal solution with great stability.

8.10 CONCLUSION

LPBSA offers a promising approach to optimizing solutions in various domains. By integrating SA and LPB, LPBSA effectively balances exploration and exploitation, allowing for the discovery of high-quality solutions. LPBSA's ability to adapt to changing environments and its efficient search capabilities make it a valuable tool for solving complex optimization problems. In addition, LPBSA outperformed the LPB algorithm in terms of LPB optimization, indicating a notable improvement over the LPB algorithm. A total of 19 benchmark test functions were utilized and tested to compare the proposed algorithms. LPBSA beat well-known algorithms like PSO and GAs by utilizing SA as a potent optimization strategy. This study's systematic methodology, which includes mutation, crossover, population division, and SA techniques, successfully increased the population and boosted overall efficiency. Thus, these outcomes highlight LPBSA's potential as a useful tool for solving challenging issues in a variety of fields.

REFERENCES

[1] K. Rajwar, K. Deep, and S. Das, "An Exhaustive Review of the Metaheuristic Algorithms for Search and Optimization: Taxonomy, Applications, and Open Challenges," *Artificial Intelligent Review*, vol. 56, no. 11, pp. 13187–13257, Nov. 2023, doi: 10.1007/s10462-023-10470-y.

[2] S. Ahmed and T. A. Rashid, "Learner Performance-based Behavior Optimization Algorithm: A Functional Case Study," 2022, doi: 10.21203/rs.3.rs-1688246/v2.

[3] J. H. Holland, *Adaptation in Natural and Artificial Systems*. Mit Press, 1992.

[4] J. Kennedy and R. Eberhart, "Particle Swarm Optimization." In *Proceedings of ICNN'95-International Conference on Neural Networks*, IEEE, 1995.

[5] R. Storn and K. Price, "Differential Evolution-A Simple and Efficient Heuristic for Global Optimization over Continuous Spaces," J*ournal of* Glob*al* Optimi*zation*, vol. 11, pp. 341–359. Kluwer Academic Publishers, 1997. doi: 10.1023/A:1008202821328.

[6] J. F. Pane, E. D. Steiner, M. D. Baird, and L. S. Hamilton, "Continued Progress: Promising Evidence on Personalized Learning," 2015. [Online]. Available: www.rand.org/t/RR1365z2.

[7] J. M. Keller, "Development and Use of the ARCS Model of Instructional Design", *Journal of Instructional Development*, vol. 10, no. 3, pp. 2–10, 1987. doi: 10.1007/BF02905780.

[8] A. Qayyum, "Student Help-Seeking Attitudes and Behaviors in a Digital Era," *International Journal of Educational Technology in Higher Education*, vol. 15, no. 1, Dec. 2018, doi: 10.1186/s41239-018-0100-7.

[9] S. L. Chew, "Improving Classroom Performance by Challenging Student Misconceptions About Learning," *APS Observer*, vol. 23, 2010.
[10] L. Cao, "College Students' Metacognitive Awareness of Difficulties in Learning the Class Content Does Not Automatically Lead to Adjustment of Study Strategies 1," *Australian Journal of Educational & Developmental Psychology*, vol. 7, pp. 31–46, 2007.
[11] S. Kirkpatrick, C. D. Gelatt, and M. P. Vecchi, "Optimization by Simulated Annealing," 1983. [Online]. Available: www.sciencemag.org, doi: 10.1126/science.220.4598.671
[12] D. Delahaye, S. Chaimatanan, and M. Mongeau, "Simulated Annealing: From Basics to Applications," vol. 272, p. 978, 2019, doi: 10.1007/978-3-319-91086-4_1ï.
[13] C. Venkateswaran, M. Ramachandran, R. Kurinjimalar, P. Vidhya, and G. Mathivanan, "Application of Simulated Annealing in Various Field," *Materials and its Characterization*, vol. 1, no. 1, pp. 01–08, Feb. 2022, doi: 10.46632/mc/1/1/1.
[14] J. Stastny, V. Skorpil, Z. Balogh, and R. Klein, "Job Shop Scheduling Problem Optimization by means of Graph-Based Algorithm," *Applied Sciences (Switzerland)*, vol. 11, no. 4, pp. 1–16, Feb. 2021, doi: 10.3390/app11041921.
[15] L. Hernández-Ramírez, J. Frausto-Solis, G. Castilla-Valdez, J. Javier González-Barbosa, D. Terán-Villanueva, and M. L. Morales-Rodríguez, "A Hybrid Simulated Annealing for Job Shop Scheduling Problem," *International Journal of Combinatorial Optimization Problems and Informatics*, vol. 10, no. 1, pp. 6–15, doi: 10.1007/s10878-017-0148-0
[16] F. P. Agostini, D. D. O. Soares-Pinto, M. A. Moret, C. Osthoff, and P. G. Pascutti, "Generalized Simulated Annealing Applied to Protein Folding Studies," *Journal of Computational Chemistry*, vol. 27, no. 11, pp. 1142–1155, Aug. 2006, doi: 10.1002/jcc.20428.
[17] G. Rajeswarappa and S. Vasundra, "Red Deer and Simulation Annealing Optimization Algorithm-Based Energy Efficient Clustering Protocol for Improved Lifetime Expectancy in Wireless Sensor Networks," *Wireless Personal Communication*, vol. 121, no. 3, pp. 2029–2056, Dec. 2021, doi: 10.1007/s11277-021-08808-2.
[18] A. M. Ahmed *et al.*, "Balancing Exploration and Exploitation Phases in Whale Optimization Algorithm: An Insightful and Empirical Analysis," In *Handbook of Whale Optimization Algorithm*. Academic Press, pp. 149–156, 2024.

9 A Systematic Study of GOOSE Algorithms

Rebwar Kh. Hamad and Tarik A. Rashid

9.1 INTRODUCTION

The goal of an optimization problem is to maximize or minimize a function with respect to a set, which often represents the range of options accessible under a given circumstance. This feature makes it possible to compare the options and decide which could be the "best". Metaheuristics play a key role in determining solutions to real-world problems. Recently, metaheuristic optimization approaches have gained popularity as a means of solving complex optimization problems. In order to address issues in important fields including data science, engineering, economics, and health, optimization algorithms are required. Depending on how they conduct searches, optimization algorithms may be categorized as deterministic (precise) or nondeterministic, or stochastic (heuristic). The primary characteristic of the deterministic or predictable approach is that the entire search space is examined rather than just an algorithm's approximation of the result.

Natural systems, biological systems, chemical systems, and physical systems also serve as the basis for the development of optimization algorithms, such as the Krill Herd (Gandomi and Alavi, 2012), FOX Algorithm (Mohammed, 2022), Fitness-Dependent Optimizer Algorithm (Abdullah and Ahmed, 2019), Dragonfly Algorithm (Mirjalili, 2016), Whale Optimization Algorithm (Mirjalili and Lewis, 2016), GOOSE Algorithm (Hamad and Rashid, 2024), Chaotic Sine Cosine Firefly (Hassan, 2021), Sparrow Search Algorithm (SSA) (Xue and Shen, 2020), Particle Swarm Optimization (Kennedy and Eberhart, 1995), Evolutionary Clustering Algorithm (Hassan and Rashid, 2021), Gravitational Search Algorithm (Rashedi, Nezamabadi-pour and Saryazdi, 2009), Artificial Bee Colony (Karaboga and Basturk, 2007), and Shuffled Frog Leaping Algorithm (Maaroof *et al.*, 2022).

Consequently, a GOOSE Algorithm is given for the purpose of resolving a wide variety of real-world issues that arise in many aspects of our life. The GOOSE Algorithm is based on geese's behavior during rest and foraging (Hamad and Rashid, 2024). The main contribution of this study is the manual optimization and acquisition of optimum solutions through the use of GOOSE Algorithm case studies. Thus, straightforward, step-by-step directions are given. The chapter can also be used by researchers to improve, expand, or combine these techniques with others (Hamad and Rashid, 2023).

 DOI: 10.1201/9781003601555-9

This chapter is organized into several sections. Section 9.2 provides an introduction to the pseudocode, flowchart, and the GOOSE Algorithm. Section 9.3 illustrates the mathematical implementation of the GOOSE Algorithm. In Section 9.4, the GOOSE Algorithm is shown. Section 9.5 provides an explanation of the case study of the GOOSE Algorithm and the conclusion.

9.2 GOOSE

A unique metaheuristic algorithm, the GOOSE Algorithm is based on how geese behave when they are at rest and while they are foraging. The goose balances and stands on one leg to observe and defend the other birds in the flock. This method begins by creating a search space and randomly selecting a sample of individuals from the entire population. The fitness function is examined to ascertain the optimal score and location. A flowchart and pseudocode of the GOOSE Algorithm is shown in Figure 9.1 and Algorithm 1.

Algorithm 1 GOOSE Algorithm

```
1:  Initialize the goose population X_i (i =1, 2,..., n)
2:  While Loop < Max_IT
3:  Generate D_S_T, T_o_A_S, BestX, D_G, M_T, alpha, BestFitness
4:  Calculate the object function of each search agent
5:  Select BestX and BestFitness among the goose population (X) in each iteration
6:  If_1 for checking each agent
7:  Update the current position of the search agent
8:  Endif_1
9:  For
10: Calculate rnd, pro, and coe
11: Find S_W using Eq. (1)
12: Calculate time randomly using Eqs. (2) and (3)
13: Calculate T_T and T_A using Eqs. (4) and (5)
14:   If2 rnd >= 0.5
15:     If3 pro > 0.2 and W_S >= 12
16:               Find F_F_S using Eq. (6)
17:               Calculate D_S_T using Eq. (7)
18:               Calculate D_G using Eq. (8)
19:               Find X_(it+1) using Eq. (9)
20:       Else
21:               Find F_F_S using Eq. (10)
22:               Calculate D_S_T using Eq. (7)
23:               Calculate D_G using Eq. (8)
24:               Find X_(it+1) using Eq. (11)
25:     EndIf3
26:   Else
27:     Find alpha using Eq. (12)
28:   Explore X_(it+1) using Eq. (13)
29:     EndIf2
30:     If they go over the restrictions, they are adapted X.
```

31: Evaluate search agents by their fitness
32: Update BestX
33: **EndFor**
34: $Loop = Loop + 1$
35: **End while**
36: Return BestX and BestFitness
37: Initialize the goose population X_i $(i = 1, 2, \ldots, n)$
38: While $Loop < Max\_IT$

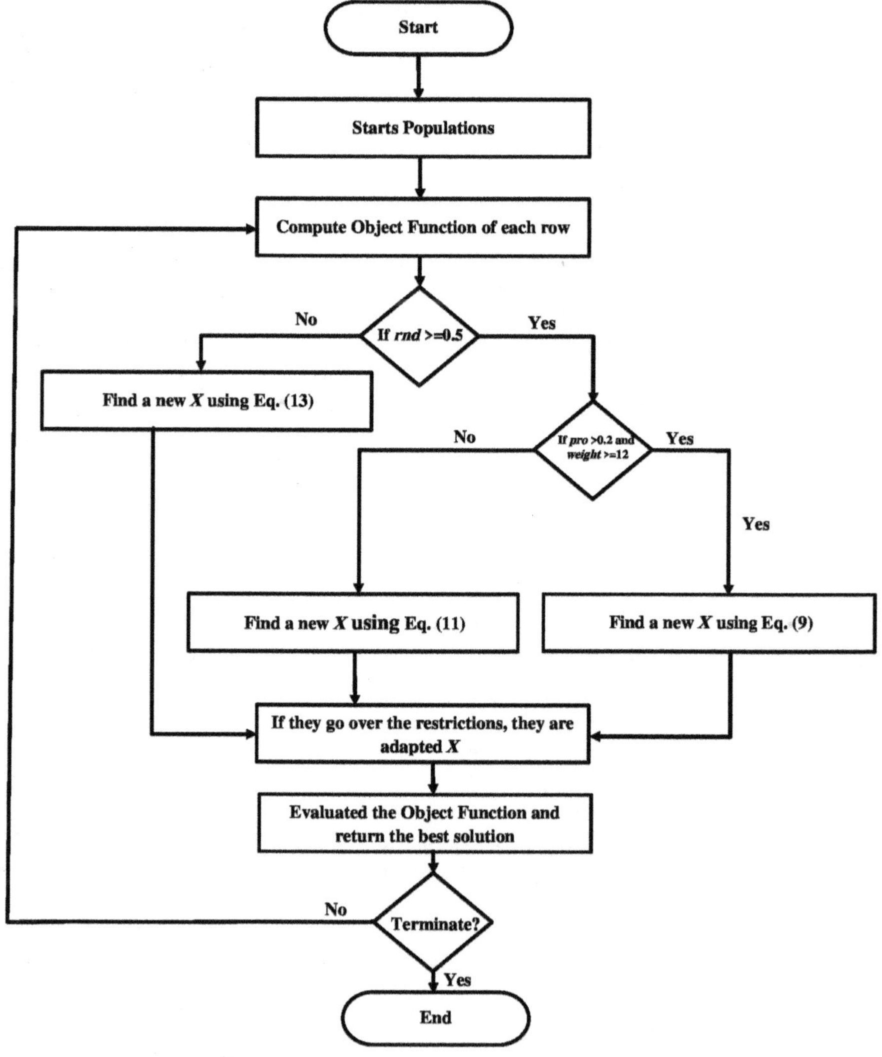

FIGURE 9.1 Flowchart of the GOOSE Algorithm.

9.3 A CASE STUDY OF THE GOOSE ALGORITHM

Examine the following minimization function, $f(x)$, where $f(x) = X_1^2 + X_2^2$. For integers X_1 and X_2, $0 \leq X_1 \leq 10$ and $0 \leq X_2 \leq 10$.

9.3.1 Calculating the First Iteration

Step 1: Assume for the moment that five agents and a population were randomly generated.

To obtain the optimal score and location in the first iteration, the fitness function was computed, as shown in Table 9.1.

After creating the population size, the algorithm reviews the agents that have exceeded the boundary and adjusts them to the limited boundaries of the population. The outcome of the fitness function is sorted from the lowest to the highest value. The Best score = 13, Worse score =181, and Best position = 3 2. Some of the variables that are randomly assigned between 0 and 1 are *pro*, *rnd*, and *coe*. However, the weights of the stones were randomly assigned between 5 and 25. As a result, we assign some variables at this stage. In the initial iteration, for each of the five agents, for instance, *pro* = 0.3, *rnd* = 0.5, *coe* = 0.16, dim = 2, S_S = 343.2, gravity (g) = 9.81, and *M_T* = inf.

9.3.1.1 These are the Positions for the First Agent: $x_1 = 3$ and $x_2 = 2$.

Step 2: Using Eq. (1), we estimate the weight of the stone that the goose keeps in its foot to be between 5 and 25 g. For each iteration, we randomly determine the weight of the stone. The number of iterations is indicated by the *it* variable.

$$S\_W_{it} = rand(5,25) \tag{1}$$

Suppose S_W_{it}=12.

Step 3: Next, we calculate the time ($T\_o\_A\_O_{it}$) required for the stone to descend to Earth using Eq. (2). Every loop iteration is produced at random between one and the entire number of dimensions.

TABLE 9.1
Create an agent population at random and examine the fitness of each individual

Agent	X_1	X_2	X_1^2	X_2^2	$F(x) = X_1^2 + X_2^2$
1	3	2	9	4	**13**
2	2	4	4	16	**20**
3	6	5	36	25	**61**
4	4	6	16	36	**52**
5	9	10	81	100	**181**

$$T\_o\_A\_O_{it} = rand(1, dim) \quad (2)$$

$$T\_o\_A\_O_{it} =$$

0.87 0.21

Step 4: When an object strikes the ground, sound is produced and sent to each individual goose in the flock. We calculate the duration of the sound travel $T\_o\_A\_S_{it}$ using Eq. (3) from the guardian geese to the other geese.

$$T\_o\_A\_S_{it} = rand(1, dim) \quad (3)$$

$$T\_o\_A\_S_{it} =$$

0.16 0.35
0.13 0.08
0.42 0.78
0.13 0.11
0.73 0.82

Step 5: The following formula determines the total amount of time needed for each iteration of the sound to travel through the air to each individual goose in the flock. We split the entire time by the dimensions, as shown in Eq. (4).

$$T\_T = sum(T\_o\_A\_S_{it}, T\_o\_A\_O_{it})/dim \qquad t = 10, \quad (4)$$

Step 6: We compute the average time of arrival of the sound to groups. We split the total time into two to obtain the average amount of time required. These stages are explained in Eq. (5).

$$T - A = \frac{T - T}{2} \quad (5)$$

Suppose $T\_A = 0.8$.

Step 7: As we discussed in the previous sections, there is a random variable (*rnd*), which is responsible for the distribution of the exploitation and exploration phases. The value of a variable (*pro*) is randomly selected from the range [0, 1]. Considering that the value of the variable *pro* is greater than 0.2 and S_W_{it} is greater than or equal to 12, Eq. (6) is applied, where $T\_o\_A\_O_{it}$ is multiplied by the square root of S_W_{it} and then multiplied by the object's acceleration (g) at 9.81 meters per square second, m/s². These equations should be used to protect and awaken the individual in the group.

We use Eq. (6) to calculate the values of the free fall speed F_F_S:

$$\mathrm{F\_F\_S} = T\_o\_A\_O_{it} * \left(sqrt\left(S\_W_{it}\right)\right) * g \tag{6}$$

$$\mathrm{F\_F\_S} =$$

29.57 7.14

Step 8: In Eq.(7), the distance of sound travel, $D\_S\_T_{it}$, must be the speed of sound S_S in the air multiplied by the time of sound travel $T\_o\_A\_S_{it}$; the speed of sound is 343.2 meters per second in the air. :

$$D\_S\_T_{it} = S\_S * T\_o\_A\_S_{it} \tag{7}$$

$$D\_S\_T_{it} =$$

54.91 120.12

Step 9: In this step, we find D_G_{it}, the distance between the guard goose and another goose that is resting or feeding in the herd. Each individual in the flock was at a different distance from the guardian's goose. This equation can be used to determine the distance between the individuals in a flock. In Eq. (8), we use the distance of sound travel $D\_S\_T_{it}$ multiplied by 1/2 or 0.5 because we only need the time for the sound to travel, not the time for the sound to return:

$$D\_G_{it} = 0.5 * D\_S\_T_{it} \tag{8}$$

27.46 60.06

Step 10: In other words, to wake up the individual in the flock, we must find a $BestX_{it}$, as demonstrated in Eq. (9). This equation is composed of the $Current - position\, or\,(Xt)$, with the speed of the falling object $\_F\_S$ subtracted, plus the distance of the goose D_G_{it}, multiplied by the average of the time squaredT_A, to determine a new X in the population:

$$X_{it} = Xt\_F\_F\_S + D\_G_{it} * T\_A^2 \tag{9}$$

$$X_{it} =$$

-9.00 33.30

9.3.1.2 These are the Positions for the Second Agent: $x_1 = 2$ and $x_2 = 4$.

Step 1: By contrast, if both variables are the weight of the stone S_W_{it} and *pro*, together less than 12 and less than or equal to 0.2, find the new X, as shown in Eq. (11). To obtain the speed of a falling object F_F_S, multiply the time $T\_o\_A\_O_{it}$ taken to arrive at the object by the weight of the stone S_W_{it} multiplied by gravity. In addition, to determine the distance of sound travel $D\_S\_T_{it}$ and the distance of the goose D_G_{it}, we reuse the previous Eqs. (7) and (8).

We use Eq. (10) to calculate the values of the free fall speed F_F_S. Suppose $S\_W = 11$:

$$F\_F\_S = T\_o\_A\_O_{it} * S\_W_{it} * g \tag{10}$$

$$F\_F\_S =$$

93.88 22.66

Step 2: In this step, we compute the distance of sound travel in the air by multiplying the speed of sound by the time of arrival of the sound:

$$D\_S\_T_{it} = S\_S * T\_o\_A\_S_{it} \tag{7}$$

$$D\_S\_T_{it} =$$

44.62 27.46

Step 3: In this step, we compute the distance of the goose using the following equation:

$$D\_G_{it} = 0.5 * D\_S\_T_{it} \tag{8}$$

22.31 13.73

Step 4: In the last step, we find the new value of the population using Eq. (11). Alternatively, we find a new X in the new mathematical equation. In Eq.(11), the $Current_Position\ or(Xt)$is added to the speed of the falling object, and the distance of the goose, average of time, and *coe* are multiplied by each other in succession:

$$X_{it} = Xt + F\_F\_S * D\_G_{it} * T\_A^2 * coe \tag{11}$$

$$X_{it} =$$

217.47 33.86

9.3.1.3 These are the Positions for the Third Agent: $x_1 = 6$ and $x_2 = 5$.

Step 1: We set a condition to check the variables *pro* and S_W. If the values of the variables are $pro <= 0.2$ and $S\_W < 12$, then the following equations apply.

We use Eq. (6) to calculate the values of the free fall speed F_F_S. Suppose $S\_W = 5$:

$$F\_F\_S = T\_o\_A\_O_{it} * \left(sqrt\left(S\_W_{it}\right)\right) * g \tag{6}$$

$$F\_F\_S =$$

$$19.08 \quad 4.61$$

Step 2: In this step, we compute the distance of sound travel in the air by multiplying the speed of sound by the time of arrival of the sound.

$$\mathrm{D\_S\_T_{it} = S\_S * T\_o\_A\_S_{it}} \tag{7}$$

$$D\_S\_T_{it} =$$

$$144.14 \quad 267.70$$

Step 3: In this step, we compute the distance of the goose using the following equation:

$$D\_G_{it} = 0.5 * D\_S\_T_{it} \tag{8}$$

$$72.07 \quad 133.85$$

Step 4: In the last step, we find the new value of the population using Eq. (11):

$$X_{it} = Xt + F\_F\_S * D\_G_{it} * T\_A^2 * coe \tag{11}$$

$$X_{it} =$$

$$143.81 \qquad 65.19$$

9.3.1.4 These are the Positions for the Fourth Agent: $x_1 = 4$ and $x_2 = 6$.

Step 1: We set a condition to check the variables *rnd*, *pro*, and *S_W*. If the values of the variables are *rnd* >= 0.5, *pro* > 0.2, and *S_W* >= 12, then the following equations apply.

We use Eq. (6) to calculate the values of the free fall speed F_F_S. Suppose $S\_W = 15$:

$$\mathrm{F\_F\_S} = T\_o\_A\_O_{it} * \left(sqrt\left(S\_W_{it}\right)\right) * 9.81 \tag{6}$$

$$F\_F\_S = \tag{6}$$

33.05 7.98

Step 2: In this step, we compute the distance of sound travel in the air by multiplying the speed of sound by the time of arrival of the sound:

$$D\_S\_T_{it} = S\_S * T\_o\_A\_S_{it} \tag{7}$$

$$D\_S\_T_{it} =$$

44.62 37.75

Step 3: In this step, we compute the distance of the goose using the following equation:

$$D\_G_{it} = 0.5 * D\_S\_T_{it} \tag{8}$$

22.31 18.88

Step 4: In the last step, we find the new value of the population using Eq. (11):

$$X_{it} = Xt + F\_F\_S + D\_G_{it} * T\_A^2 \tag{11}$$

$$X_{it} =$$

50.33 22.06

9.3.1.5 These are the Positions for the Fifth Agent: $x_1 = 9$ and $x_2 = 10$.

Step 1: We set a condition to check the variable *rnd*. If the value of the variable *rnd* < 0.5, then the following equations apply.

The value of the *alpha* variable ranges from 2 to 0. This value decreases dramatically with each iteration of the loop. We use Eq. (12) to improve the result of a new X in the search space. We also use it to calculate the values of *alpha*. Suppose $Loop = 1$, $M\_T = 0.55$, and $Max\_IT = 500$:

$$alpha = \left(2 - \left(\frac{Loop}{\frac{Mx - IT}{2}}\right)\right) \tag{12}$$

$$alpha = 2.$$

Step 2: In the last step, we find the new value of the population using Eq. (13). We ensure that the goose stochastically explores the other individuals in the search space using *randn(1, dim)*. Nevertheless, we use both the *M_T* and *alpha* variables to improve the search ability of GOOSE. In Eq. (13), the minimum time and alpha are multiplied by a random number and then added to the best position (X_{it}) in the search space:

Consider randn(1,dim) = 0.61

$$X_{it} = randn(1, dim) * (M\_T * alpha) + Best\_Pos \tag{13}$$

$$X_{it} =$$

3.67 2.67

In this step, assume for the moment that five agents and a population were generated after the first iteration, as shown in Table 9.2.

9.3.2 Calculating the Second Iteration

Step 1: Assume for the moment that five agents and a population were generated after the first iteration, as shown in Table 9.3.

After the first iteration agent populations are generated, the algorithm reviews the agents that have exceeded the boundary and adjusts them to the limited boundaries of the population. The outcome of the fitness function is sorted from the lowest to

TABLE 9.2
A new *X* was generated

Agent	X_1	X_2
1	-9.00	33.30
2	217.47	33.86
3	143.81	65.19
4	50.33	22.06
5	3.67	2.67

TABLE 9.3
Create an agent population at the first iteration and examine the fitness of each individual

Agent	X_1	X_2	X_1^2	X_2^2	$F(x) = X_1^2 + X_2^2$
1	-9	33.3	81	1108.89	1189.89
2	217.47	33.86	47293.2	1146.5	48439.70
3	143.81	65.19	20681.32	4249.736	24931.05
4	50.33	22.06	2533.109	486.6436	3019.75
5	3.67	2.67	13.4689	7.1289	20.60

the highest value. The Best score = 20.60, Worse score = 48439.70, and Best position = 3.67 2.67. Some of the variables that are randomly assigned between 0 and 1 are *pro*, *rnd*, and *coe*. However, the weights of the stones were randomly assigned between 5 and 25. As a result, we assign some variables at this stage. In the second iteration, for each of the five agents, for instance, $pro = 0.3$, $rnd = 0.7$, $coe = 0.12$, $dim = 2$, $S\_S = 343.2$, gravity (g) = 9.81 and $M\_T = \text{inf}$.

Step 2: To find the weight of the stone, we must evaluate the random function between 5 and 25 using Eq. (1):

$$S\_W_{it} = randi([5,25],1,1) \tag{1}$$

Suppose $S\_W_{it} = 14$.

Step 3: We compute the time of arrival of an object to Earth randomly between (1, dim) using Eq. (2):

$$T\_o\_A\_O_{it} = rand(1, dim) \tag{2}$$

$$T\_o\_A\_O_{it} =$$

$$\begin{matrix} 0.08 & 0.58 \end{matrix}$$

Step 4: We compute the time of arrival of the sound to groups randomly between (1, dim) using Eq. (3):

$$T\_o\_A\_S_{it} = rand(1, dim) \tag{3}$$

$$T\_o\_A\_S_{it} =$$

$$\begin{matrix} 0.40 & 0.68 \\ 0.90 & 0.95 \\ 0.34 & 0.21 \\ 0.87 & 0.01 \\ 0.56 & 0.04 \end{matrix}$$

Step 5: We compute the summation of the time of arrival of the sound to the groups using the following equation:

$$T\_T = sum(T\_o\_A\_S_{it}, T\_o\_A\_O_{it})/dim \qquad t = 10, \tag{4}$$

$$T - T = 0.9$$

Step 6: We compute the average of the time of arrival of the sound to groups using the following equation:

$$T\_A = \frac{T\_T}{2} \tag{5}$$

Suppose $T\_A = 0.5$.

9.3.2.1 These are the Positions for the First Agent: $x_1 = -9.00$ and $x_2 = 33.3$.

Step 7: We set a condition to check the variables *pro*, *rnd*, and *S_W*. If the values of the variables are *rnd* > =0.5, *pro* >0.2, and *S_W* >= 12, then the following equations apply.

We use the following equation to calculate the values of the free fall speed F_F_S:

$$\text{F\_F\_S} = T\_o\_A\_O_{it} * \left(sqrt\left(S\_W_{it}\right)\right) * 9.81 \tag{6}$$

$$\text{F\_F\_S} =$$

26.67 1.58

Step 8: In this step, we compute the distance of sound travel in the air by multiplying speed of sound by the time of arrival of the sound:

$$D\_S\_T_{it} = S\_S * T\_o\_A\_S_{it} \tag{7}$$

$$D\_S\_T_{it} =$$

137.28 233.38

Step 9: In this step, we compute the distance of the goose using the following equation:

$$D\_G_{it} = 0.5 * D\_S\_T_{it} \tag{8}$$

68.64 116.69

Step 10: In the last step, we find the new value of the population using Eq. (9):

$$X_{it} = Xt - F\_F\_S + D\_G_{it} * T\_A^2 \tag{9}$$

$$X_{it} =$$

–5.84 30.26

9.3.2.2 These are the Positions for the Second Agent: $x_1 = 217.47$ and $x_2 = 33.86$.

Step 1: We set a condition to check the variables *pro* and *S_W*. If the values of the variables are *pro* <=0.2 and *S_W* < 12, then the following equations apply.

We use Eq. (10) to calculate the values of the free fall speed F_F_S. Suppose *S_W*=5:

$$F\_F\_S = T\_o\_A\_O_{it} * S\_W_{it} * 9.81 \tag{10}$$

$$F\_F\_S. =$$

3.92 28.45

Step 2: In this step, we compute the distance of sound travel in the air by multiplying the speed of sound by the time of arrival of the sound:

$$D\_S\_T_{it} = S\_S * T\_o\_A\_S_{it} \tag{7}$$

$$D\_S\_T_{it} =$$

308.88 326.04

Step 3: In this step, we compute the distance of the goose using the following equation:

$$D\_G_{it} = 0.5 * D\_S\_T_{it} \tag{8}$$

154.44 163.02

Step 4: In the last step, we find the new value of the population using Eq. (11):

$$X_{it} = Xt + F\_F\_S * D\_G_{it} * T\_A^2 * coe \tag{11}$$

$$X_{it} =$$

21.83 141.81

9.3.2.3 These are the Positions for the Third Agent: $x_1 = 143.81$ and $x_2 = 65.19$.

Step 5: We set a condition to check the variables *pro*, *rnd*, and *S_W*. If the values of the variables are *rnd* >= 0.5, *pro* > 0.2, and *S_W* >= 12, then the following equations apply.

We use Eq. (6) to calculate the values of the free fall speed F_F_S:

$$\mathrm{F\_F\_S} = T\_o\_A\_O_{it} * \left(sqrt\left(S\_W_{it}\right)\right) * 9.81 \tag{6}$$

$$\mathrm{F\_F\_S} =$$

2.72 19.71

Step 6: In this step, we compute the distance of sound travel in the air by multiplying the speed of sound by the time of arrival of the sound.

$$D\_S\_T_{it} = S\_S * T\_o\_A\_S_{it} \tag{7}$$

$$D\_S\_T_{it} =$$

116.69 72.07

Step 7: In this step, we compute the distance of the goose using the following equation:

$$D\_G_{it} = 0.5 * D\_S\_T_{it} \tag{8}$$

58.35 36.04

Step 8: In the last step, we find the new value of the population using Eq. (9):

$$X_{it} = Xt - F\_F\_S + D\_G_{it} * T\_A^2 \tag{9}$$

$$X_{it} =$$

15.54 –8.03

9.3.2.4 These are the Positions for the Fourth Agent: x_1 = 50.33 and x_2 = 22.06.

Step 1: We set a condition to check the variables *pro* and *S_W*. If the values of the variables are *pro* <= 0.2 and *S_W* < 12, then the following equations apply.

We use Eq. (10) to calculate the values of the free fall speed F_F_S. Suppose *S_W* = 5:

$$F\_F\_S = T\_o\_A\_O_{it} * \left(S\_W_{it} * 9.81\right) \tag{10}$$

$$F\_F\_S =$$

3.92 28.45

Step 2: In this step, we compute the distance of sound travel in the air by multiplying the speed of sound by the time of arrival of the sound:

$$D\_S\_T_{it} = S\_S * T\_o\_A\_S_{it} \tag{7}$$

$$D\_S\_T_{it} =$$

298.58 3.43

Step 3: In this step, we compute the distance of the goose using the following equation:

$$D\_G_{it} = 0.5 * D\_S\_T_{it} \tag{8}$$

149.29 1.72

Step 4: In the last step, we find the new value of the population using Eq. (11). Suppose the *coe* value = 0.17:

$$X_{it} = Xt + F\_F\_S * D\_G_{it} * T\_A^2 * coe \tag{11}$$

$$X_{it} =$$

28.54 4.75

9.3.2.5 These are the Positions for the Fifth Agent: x_1=3.67 and x_2=2.67.

Step 1: We set a condition to check the variable *rnd*. If the value of the variable *rnd* < 0.5, then the following equations apply.

We use Eq. (12) to calculate the values of *alpha*. Suppose *Loop* = 450, *M_T* = 0.40, and *Max_IT* = 500:

$$alpha = \left(2 - \left(\frac{Loop}{\frac{Mx\_IT}{2}}\right)\right) \tag{12}$$

$$alpha = 0.2,$$

Step 2: In the last step, we find the new value of the population using Eq. (13).
Consider *randn*(1,*dim*) = 0.71:

$$X_{it} = randn(1, dim) * (M\_T * alpha) + Xt \tag{13}$$

$$X_{it} =$$

0.37 0.20

In this step, assume for the moment that five agents and a population were generated after the second iteration, as shown in Table 9.4.

We repeated the preceding procedures for subsequent iterations.

9.4 CONCLUSION

The GOOSE Algorithm is presented in this study. The purpose of this case study is to clarify any GOOSE Algorithm stages that readers of these algorithms may find

TABLE 9.4
A new X was generated

Agent	X_1	X_2
1	-5.84	30.26
2	21.83	141.81
3	15.54	-8.03
4	28.54	4.75
5	0.37	0.20

unclear. The trial results showed that the GOOSE Algorithms could discover the best solution and enhance and develop quality. By iterative improvement, GOOSE produced superior results.

REFERENCES

Abdullah, J.M. and Ahmed, T. (2019). 'Fitness dependent optimizer: Inspired by the Bee swarming reproductive process', *IEEE Access*, 7, pp. 43473–43486. Available at: https://doi.org/10.1109/ACCESS.2019.2907012.

Gandomi, A.H. and Alavi, A.H. (2012). 'Krill herd: A new bio-inspired optimization algorithm', *Communications in Nonlinear Science and Numerical Simulation*, 17(12), pp. 4831–4845. Available at: https://doi.org/10.1016/J.CNSNS.2012.05.010.

Hamad, R.K. and Rashid, T.A. (2023). 'A systematic study of Krill Herd and FOX algorithms', *Proceedings of the 1st International Conference on Innovation in Information Technology and Business (ICIITB 2022)*, pp. 168–186. Available at: https://doi.org/10.2991/978-94-6463-110-4_13.

Hamad, R.K. and Rashid, T.A. (2024). 'GOOSE algorithm: A powerful optimization tool for real-world engineering challenges and beyond', *Evolving Systems*, pp. 1–26. Available at: https://doi.org/10.1007/S12530-023-09553-6/TABLES/24.

Hassan, B.A. (2021). 'CSCF: A chaotic sine cosine firefly algorithm for practical application problems', *Neural Computing and Applications*, 33(12), pp. 7011–7030. Available at: https://doi.org/10.1007/S00521-020-05474-6/METRICS.

Hassan, B.A. and Rashid, T.A. (2021). 'A multidisciplinary ensemble algorithm for clustering heterogeneous datasets', *Neural Computing and Applications*, 33(17), pp. 10987–11010. Available at: https://doi.org/10.1007/S00521-020-05649-1/METRICS.

Igel, I. (2012). *Hands-on Activity: Measuring Distance with Sound Waves, TeachEngineering*. Available at: www.teachengineering.org/activities/view/nyu_soundwaves_activity1 (Accessed: 1 June 2023).

Karaboga, D. and Basturk, B. (2007). 'A powerful and efficient algorithm for numerical function optimization: Artificial bee colony (ABC) algorithm', *Journal of Global Optimization*, 39(3), pp. 459–471. Available at: https://doi.org/10.1007/S10898-007-9149-X/METRICS.

Kennedy, J. and Eberhart, R. (1995). 'Particle swarm optimization', *Proceedings of ICNN'95 - International Conference on Neural Networks*, 4, pp. 1942–1948. Available at: https://doi.org/10.1109/ICNN.1995.488968.

Maaroof, B.B. *et al.* (2022). 'Current studies and applications of Shuffled Frog Leaping Algorithm: A review', *Archives of Computational Methods in Engineering*, 29(5), pp. 3459–3474. Available at: https://doi.org/10.1007/S11831-021-09707-2/METRICS.

Mirjalili, S. (2016). 'Dragonfly algorithm: A new meta-heuristic optimization technique for solving single-objective, discrete, and multi-objective problems', *Neural Computing and Applications*, 27(4), pp. 1053–1073. Available at: https://doi.org/https://doi.org/10.1007/s00521-015-1920-1.

Mirjalili, S. and Lewis, A. (2016). 'The whale optimization algorithm', *Advances in Engineering Software*, 95, pp. 51–67. Available at: https://doi.org/10.1016/j.advengsoft.2016.01.008.

Rashedi, E., Nezamabadi-pour, H. and Saryazdi, S. (2009). 'GSA: A Gravitational Search Algorithm', *Information Sciences*, 179(13), pp. 2232–2248. Available at: https://doi.org/10.1016/j.ins.2009.03.004.

Xue, J. and Shen, B. (2020). 'A novel swarm intelligence optimization approach: Sparrow search algorithm', *Systems Science & Control Engineering*, 8(1), pp. 22–34. Available at: https://doi.org/10.1080/21642583.2019.1708830.

10 Overcoming Overshooting

The Sigmoid-driven Evolution of the Ant Nesting Algorithm

Mahmood Yashar Hamza and Tarik A. Rashid

10.1 INTRODUCTION

Optimization algorithms are essential in tackling problems across different fields, such as engineering, finance, and machine learning. Essentially, optimization involves the task of identifying the solution from a range of potential options. The determination of what makes a solution "best" is based on criteria or objectives [1]. Optimization algorithms can be classified into two main categories: deterministic algorithms and stochastic algorithms. Deterministic algorithms, including linear programming and descent, adhere to a specific strategy in order to produce an identical solution for a given set of initial conditions. On the other hand, stochastic optimization algorithms, such as simulated annealing, introduce randomness into the optimization process, resulting in different solutions, even with identical inputs. Moreover, optimization techniques can be classified as heuristic or metaheuristic. Heuristic algorithms, like the Hill Climbing and Greedy Algorithms, which focus on nearest neighbor approaches, prioritize efficiency over finding the solution by relying on practical rules of thumb. By contrast, metaheuristic algorithms, such as Particle Swarm Optimization (PSO), Genetic Algorithms, and Ant Colony Optimization (ACO), offer strategies that are problem-independent. They incorporate randomness and global exploration to efficiently navigate through problem landscapes. These distinctions together provide a framework for addressing types of optimization problems [2], also mentioned in Figure 10.1. Furthermore, optimization algorithms can be implemented in various fields. For example, robotics [3] is an important major field in which optimization can be implemented, which helps to provide an optimum solution for the movement of robotic particles. Another important factor that may be optimized for the software development life cycle is automated testing, which aids in selecting and automating the testing of different software applications and locating flaws [4].

There are two ways to categorize algorithms. One is the single solution approach, where it starts with one solution and constantly modifies it, like in the simulated annealing algorithm that uses a single agent and keeps updating that agent. The other

DOI: 10.1201/9781003601555-10

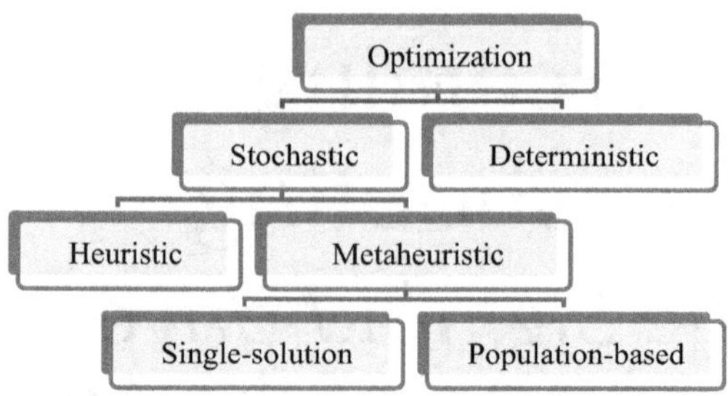

FIGURE 10.1 Classification of optimization algorithms.

is the population-based approach, which utilizes the result of all worker agents, and each agent provides a solution, which produces a collection of solutions and ideas. The population-based approach is used to pick the best one among them [5], which is also presented in Figure 10.1.

To achieve an optimum solution, it is often necessary to go through numerous iterations, which can require a large amount of computational time. To speed up optimization algorithms, a recent study focused on improving the Ant Nesting Algorithm (ANA) by using vectorization techniques. Specifically, they utilized the NumPy library in Python to vectorize ANA. This means that the iterative loop data are processed in parallel, resulting in the much faster execution of the algorithm. By adopting this approach, the study aimed not only to expedite the optimization process but also to effectively handle large datasets and complex solution spaces. This showcases how algorithmic innovation and computational efficiency can work together successfully in optimization fields [6]. In addition, changing Java to Python does not boost the performance of the method since Python is an interpreted language, whereas Java is a compiled language, which makes it slower [7–9]. However, Python includes libraries, like NumPy, which allows algorithms to employ vectorization for speed improvements.

Both PSO and ACO are optimization algorithms that take inspiration from nature. They have become popular for solving problems in various fields including different applications, such as robotic path planning and structural design. PSO mimics the behavior of birds and fish by simulating a group of particles that adjust their positions in the solution space based on their experiences and those of their neighbors. This collaborative approach allows PSO to effectively explore and exploit the solution space, making it highly efficient for optimization problems. ACO is inspired by the foraging behavior of ants and was designed specifically for discrete optimization problems. Ants leave pheromone trails on paths they travel, which guides ants toward routes. Similarly, ACO follows an approach where artificial ants iteratively construct solutions by choosing paths with concentrations of pheromones. This indirect communication and continuous updating of pheromones enable ACO to discover near-optimal solutions in combinatorial optimization scenarios [10, 11].

The Grey Wolf Optimization (GWO) algorithm draws inspiration from the structure of the hunting behavior of grey wolves, which copies the behavior of wolves, such as leadership tactics and collaborative approaches to attacking prey. It mimics the pack structure by representing solutions as individual wolves in a search space. During each iteration, the positions of the wolves are updated, with the alpha wolf representing the first-best solution, the beta wolf representing the second-best solution, and the delta wolf representing the third-best solution. The remaining wolves adjust their positions based on these alpha, beta, and delta wolves imitating the hierarchical nature observed in wolf packs [12].

The Evolutionary Clustering Algorithm (ECA) is an example of a metaheuristic algorithm, which demonstrates that clustering is an effective way of organizing related sets of data into meaningful groupings. The primary difference between this new technique and previous methods is that it ignores a great deal of the data's underlying significance. This issue may be further addressed by grouping more meaningful data. It also works well with complicated data and guarantees that the data after clustering have more meaning. This kind of method is utilized to cluster the data [13, 14].

Formal Context Analysis (FCA) is an approach that creates hierarchies from text, since pairing words and identifying the connection between them to check the process manually requires a significant amount of time. This approach is used to identify pair-related idea words by reconstructing text. This approach can also provide irrelevant pairings of words. For that condition, clustering between FCA and ECA algorithms has been used in order to provide typographical errors, pair ideas, and obtain faster results [15].

Chaotic Sine Cosine Firefly (CSCF) is another type of metaheuristic optimization algorithm that is swarm based, like other optimization algorithms. Most swarm-based algorithms have an issue with speed and complexity, and this problem can be addressed by the CSCF optimization algorithm. This method combines both the firefly and sine cosine algorithms, and is described as a hybrid algorithm. The combination of both optimization algorithms make the CSCF algorithm perform faster [16].

G-HHO is a metaheuristic optimization algorithm that is a combination of the GWO and Harris Hawks Optimization. This algorithm uses a Deep Convolutional Neural Network (DCNN) to detect brain tumors with a faster and more effective approach. This algorithm uses an approach called Otsu thresholding, which identifies the tumor on magnetic resonance imaging (MRI) images. The approach has high accuracy of around 97% [17].

The shuffled frog leaping algorithm is a type of metaheuristic algorithm that is also a hybrid. It combines the particle swarm with memetics and can be implemented to solve various engineering problems [18].

Harmony search is a well-known metaheuristic algorithm that is known to solve complex problems and has the ability to balance between the newly identified solutions and the best solution. This algorithm has been used in a variety of fields, such as healthcare and engineering [19].

iECA is an clustering optimization algorithm that is used to cluster Covid-19 diseases, including patient demographics and diseases. If the optimization technique is not used, the approach requires a significant amount of time for the process of grouping similar data. Different techniques have been implemented for providing the optimum result and solving different real-world problems [20].

In summary, optimization algorithms are used to find the optimum solution from a range of options. They typically aim to minimize or maximize a function, while adhering to given constraints. These algorithms can be classified into methods that provide solutions based on the same initial conditions, and stochastic techniques that introduce randomness for handling complex and uncertain problems. Heuristic algorithms prioritize efficiency to achieve the solution by using rules for problem-solving. On the other hand, metaheuristic algorithms offer strategies that can be adapted to different optimization challenges, regardless of their specific nature. The choice of which approach is most suitable depends on factors such as the problem's characteristics and the need to balance precision, efficiency, and adaptability when navigating diverse solution spaces.

This chapter explores the use of a sigmoid equation to improve ANA. Although the algorithm is effective in optimizing scenarios, it has shown a tendency to overshoot results, which ultimately affects its performance. The proposed improvement incorporates the sigmoid equation to address this issue by aiming for optimal outcomes. By integrating the sigmoid function into ANA, this study aims to enhance its convergence behavior and make it more adaptable to different problem spaces. Through experimentation and analysis, this research aims to demonstrate how the sigmoid function-enhanced ANA can provide accurate and dependable solutions for a wide range of optimization problems. This chapter is structured as follows: introduction to the ANA problem, background of the algorithm and operation, inclination overshooting issue, and comparison of ANA with improved ANA. The primary aim of this chapter is to address the overshooting issue that negatively impacts the performance of ANA. To achieve this, a sigmoid equation is provided to adjust the tendency rate, thereby ensuring a more optimal result that surpasses alternative metaheuristic algorithms.

10.2 ANT NESTING ALGORITHM

Ants have always fascinated humans due to their abilities, with over 10,000 distinct species found worldwide. Despite their size, ants are known for their teamwork, which has been extensively explored in the literature and academic research. One particularly intriguing species is the Leptothorax, which exhibits nesting behaviors in various regions. These ants create colonies that can fit inside a wristwatch casing by building nests in cliffs along the coast of England. Their nests utilize rock cracks, and consist of walls surrounding the queen and her broods. Worker ants diligently arrange building materials to construct a sturdy wall that shields the queen from physical and biological threats. This intricate process involves bulldozing rocks until a packed wall forms around the queen, showcasing remarkable cooperative behavior that continues to inspire scientific inquiry and creative thinking [6, 21].

During the construction of a nest, the Leptothorax ant follows a specific algorithmic model. Each worker ant acts as a search agent, exploring positions around the queen and dropping grains as solutions. The best position discovered by all the ants is known as the potential solution. This position plays a role in nest construction, which is determined by the worker ant's positions, indicating where they prefer to build the nest. The fitness function considers factors like how stones impact wall strength and their ability to bond with stones. To help optimize the algorithm, a random weight called the deposition weight (DW) is introduced for worker ants to decide where to place grains. By utilizing information on previous deposition positions, worker ants can find spots for dropping grains while randomly exploring the colony. Various entities such, as DW and positions, are part of this algorithm [21].

10.3 ANA OPERATIONAL STEPS

The algorithm begins by determining the number of agents, represented as X_i (where $i = 1, 2, 3, ..., n$), which denotes the population. Random values within the range of 100 to -100 are assigned to each dimension of the ant's positions. The number of dimensions is also chosen randomly. Each worker ant has a starting position or aims to find one. If the determined position is worse than the one, the current position is maintained until a superior option is discovered. This iterative process ensures that the optimal position for a deposition is selected. This part explains how the ANA operates. To begin with, the algorithm starts by creating a population of ants and assigning values to their dimensions within a range. Each worker ant then looks for a position to deposit its materials. If the new position is not better than the previously stored one, the ant keeps its position until it finds a superior option. Through this process, the algorithm identifies positions for new depositions. However, there is an issue regarding overshooting caused by a tendency rate problem. In the next section, we discuss this problem in detail and elaborate on potential solutions within ANA. Table 10.1 represents the whole mathematical equation of ANA. All entities are presented in Table 10.2.

10.4 IANA FOR THE TENDENCY OVERSHOOTING PROBLEM

The issue of overshooting in the ANA occurs when a tendency rate is used to minimize results. This tendency rate, which aims to guide the algorithm toward solutions, unintentionally leads to high or low values that significantly deviate from the intended target. As a result, the algorithm's ability to generate a solution is compromised, as the outcomes become unpredictable and vary greatly from what's desired. The image in Figure 10.2 clearly demonstrates the overshooting problem encountered in one of the benchmark test functions, highlighting the importance of understanding and resolving this issue within ANA. Addressing and resolving this overshooting problem is essential for improving the accuracy of the algorithm and ensuring optimization results.

TABLE 10.1
Entities of ANA

Nature	Algorithm
Worker ants	Search agents
Deposition position	Potential solution (current solution)
Deposition position specification	Fitness functions
Worker ants' decision factor	Deposition weight
Stationary stone and/or nestmate	Previous deposition position
Fitness deposition position	Global best

TABLE 10.2
Mathematical equations of ANA

Equations	#
$x_{t,i+1} = x_{t,i} + \Delta x_{t,i+1}$	1
$\Delta x_{t,i+1} = dw \times \left(x_{t,i\ best} - x_{t,i}\right)$	2
$\Delta x_{t,i+1} = r \times x_{t,i}$	3
$\Delta x_{t,i+1} = r \times \left(x_{t,i\ best} - x_{t,i}\right)$	4
$dw = r \times \left(\frac{T}{T_{previous}}\right)$	5
$T = \sqrt{\left(X_{t,ibest} - X_{t,i}\right)^2 + \left(X_{t,ibestfitness} - X_{t,i\ fitness}\right)^2}$	6
$T_{previous} = \sqrt{\left(X_{t,ibest} - X_{t,iprevious}\right)^2 + \left(X_{t,ibestfitness} - X_{t,ipreviousfitness}\right)^2}$	7

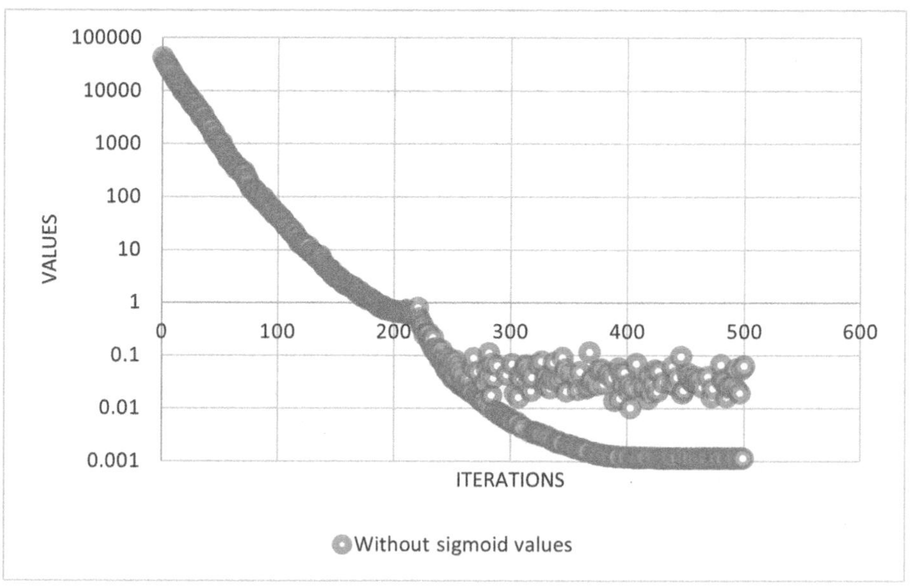

FIGURE 10.2 Tendency overshooting of ANA.

To tackle the issue of the tendency problem in the ANA, it is crucial to emphasize the role played by the sigmoid function in regulating how the algorithm behaves. The tendency problem refers to the algorithm's inclination to lean toward solutions. This problem is addressed by incorporating the sigmoid equation. The sigmoid function, which is frequently denoted by *s*(*x*), is a function that exhibits a curve resembling the letter S. It transforms any number into a range spanning from 0 to 1, which proves valuable for tasks related to binary classification or when used as an activation function in artificial neural networks:

$$S(X) = \frac{1}{1+e^{-x}} \tag{8}$$

Once the sigmoid equation is implemented into the algorithm, it produces a value close to 1 when the input is much higher than the target, and close to 0 when the input is significantly lower. This characteristic allows the sigmoid to effectively limit the output within a range that reduces any impacts from values. As a result, it helps to generate nearly optimal solutions and support the stability of the nesting algorithm, as shown in Figure 10.3 [22].

The tendency issue in optimization algorithms often arises when they tend to lean toward solutions due to uncontrolled deviations from the desired target. To address this problem, the introduction of the sigmoid equation serves as an intervention. The sigmoid function acts as a tool by adjusting the algorithm's outputs to fall within a range. It assigns values to 1 for input values that are attained by the standard test functions and 0 for low inputs, effectively limiting extreme values and curbing the algorithm's inclination toward them. Implementing the sigmoid equation works as a mechanism for controlling the algorithm's behavior and mitigating the tendency issue. This implementation allows for a nuanced approach to generating solutions, leading to outcomes that are not only more stable but also closer to optimal solutions. In summary, utilizing the sigmoid function to normalize outputs plays a role in addressing the tendency problem and enhancing the effectiveness of optimization algorithms.

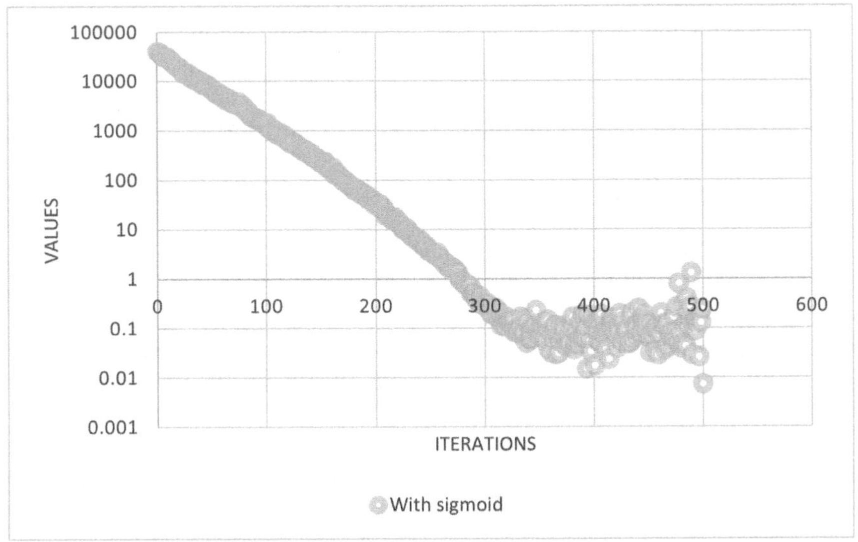

FIGURE 10.3 Implementation of the sigmoid into the tendency rate to output the result.

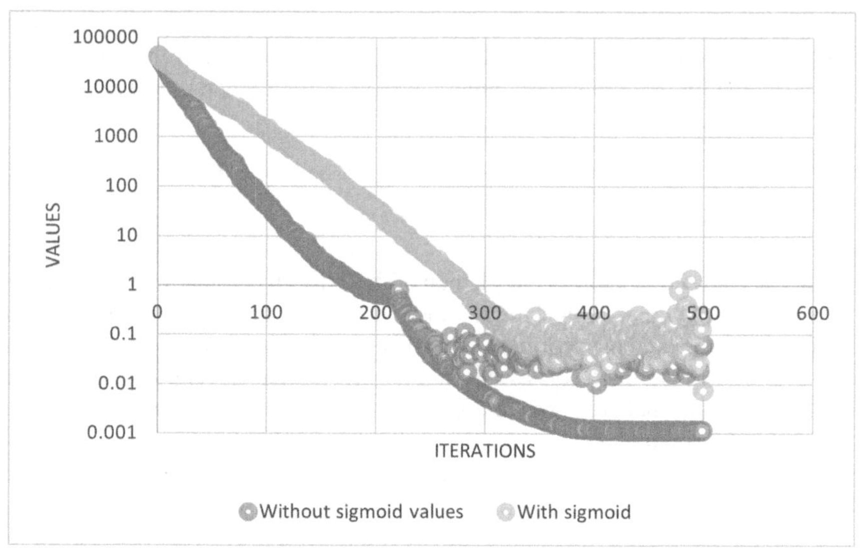

FIGURE 10.4 Tendency output effect with sigmoid.

Figure 10.4 shows how integrating the sigmoid equation into an optimization algorithm can have an impact. Two-line plots are displayed side by side to highlight the difference between using the sigmoid function and not using it. The first plot, labeled "Without Sigmoid", demonstrates that the algorithm behaves erratically, leading to fluctuating results and suboptimal solutions. The second plot, labeled "With Sigmoid", illustrates an improvement. The sigmoid function effectively controls values and stabilizes the algorithm's performance. This implementation does not address the identified issue. It consistently guides the algorithm toward better solutions. Figure 10.5 provides a representation of how incorporating the sigmoid equation enhances optimization outcomes, leading to increased stability and convergence, toward near-optimal solutions. The whole flowchart is presented in the figure, where the green area is the place where the sigmoid equation has been implemented [22].

10.5 ANA VS. IANA

This section discusses how ANA has been improved by applying the sigmoid equation to adjust its tendency rate, resulting in values of 1 and 0. This adjustment has proven to be beneficial in enhancing the algorithm's performance across 26 test functions, which are categorized as unimodal (F1–F7) or multimodal (F9–F18). When comparing the modified version with the original one, it is evident that the modified ANA performs better in functions such as F1, F2, F4, F7, F9, F11, F12, F13, and F15. Detailed results can be found in Tables 10.3 and 10.4, where each data point represents outcomes after running the algorithm for each test function 30 times. Output comparisons are presented in Figure 10.6.

The benchmark functions are known as CEC test functions. Developed by the coding the evolution of computation laboratory, they provide a range of challenges

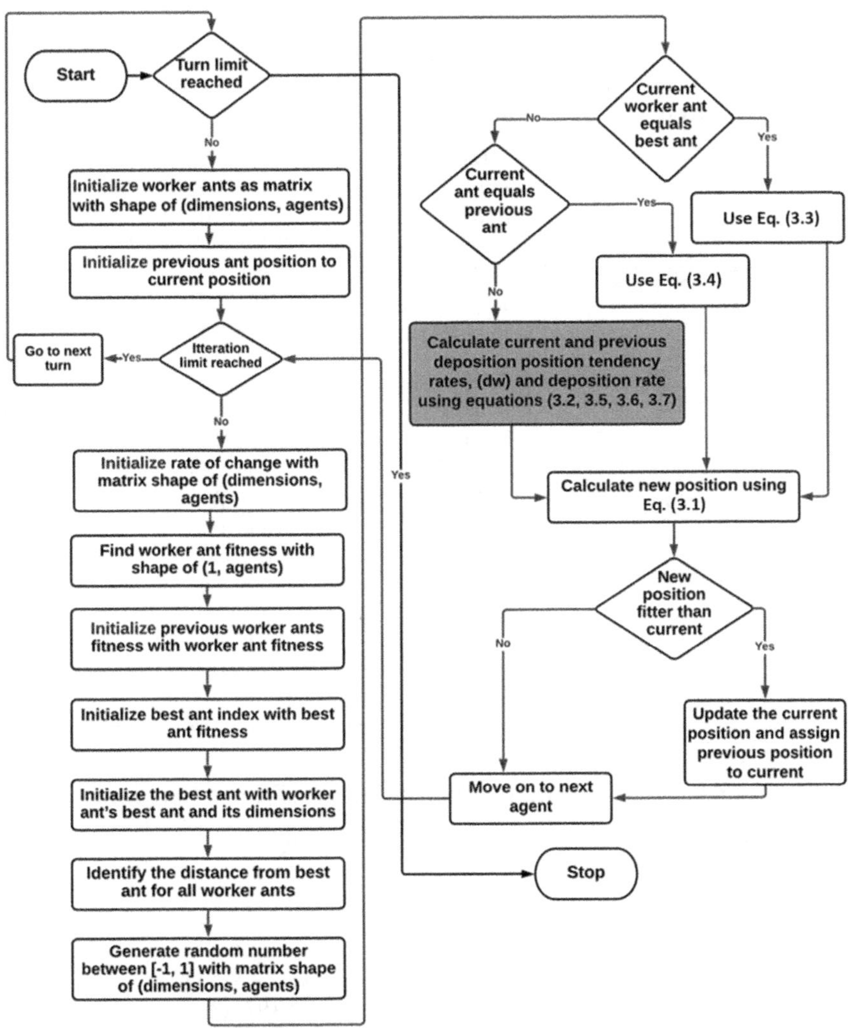

FIGURE 10.5 Flowchart of ANA with sigmoid implementation.

to evaluate optimization algorithms. These functions cover characteristics like dimensionality, multimodality, and nonlinearity, making them an excellent tool to assess the performance of optimization algorithms. After making modifications to ANA, we conducted a comparison study using the CEC test functions (CEC01–CEC10). Notably, according to Table 10.4, due to the slow speed of the algorithm, both CEC01 and CEC06 (which were not handled effectively in ANA) successfully improved the results with the vectorization approach. Clear enhancements in performance were shown for CEC01, CEC05, CEC06, CEC07, CEC08, CEC09, and CEC10 after the implementation of the vectorization technique compared with their counterparts in ANA, and the output representation is presented in Figure 10.7.

TABLE 10.3
ANA vs. IANA output comparison

Function	ANA	VANA
F1	0.016299	4.59E-05
F2	2.36E-05	1.08E-05
F3	1.239172	3.196670112
F4	2.37E-16	2.10885E-16
F5	14.95306	2.83E+18
F7	0.770696	2.92E-07
F9	25.46222	15.74128025
Function	**ANA**	**VANA**
F10	5.54E-15	0.005792241
F11	0.41119	0.410955229
F12	3.219957	1.133507429
F13	1.76E-23	1.07E-23
F14	4.26E-14	3.54E-08
F15	4.89E-06	4.67E-06
F16	23.76092	2.40E+01
F17	223.5622	2.24E+02
F18	31.51015	9.20E+01

TABLE 10.4
ANA vs. IANA output comparison for CEC test functions

Function	ANA	VANA
CEC01	-	388346.8221
CEC02	4	4
CEC03	13.7024	13.70240423
CEC04	38.50888	48.20057349
CEC05	1.224599	1.219721311
CEC06	-	11.98608986
CEC07	116.5962	32.01025539
CEC08	5.472815	5.440928127
CEC09	2.000964	2.000038823
CEC10	2.718282	2.718281828

Implementing the sigmoid equation into the tendency rate demonstrates that the flexibility of the algorithm to hybridize with multiple algorithms for optimization is further proof of its use, which allows a wide range of mixture with other algorithms. This means that it can hybridize with any of the other algorithms, including GOOSE, Fitness Dependent Optimizer (FDO) [23], FOX [24], Donkey and Smuggler Optimization [25], and Child Drawing Development Optimization [26]. By creating hybrid models for the other optimization algorithms, each algorithm can be refined for the use of the tendency rate modulation of the sigmoid equation and one of the strategies can improve the results. More generally, it signals that the importance of algorithmic integration is also just as great in the software testing for optimization context as in the software testing context [4], given the same sort of meticulous comparison and evaluation necessary in the former testing protocol as the latter (Tables 10.5 and 10.6).

10.6 GRAPHICAL REPRESENTATION OF ANA VS. IANA FOR F1–F18

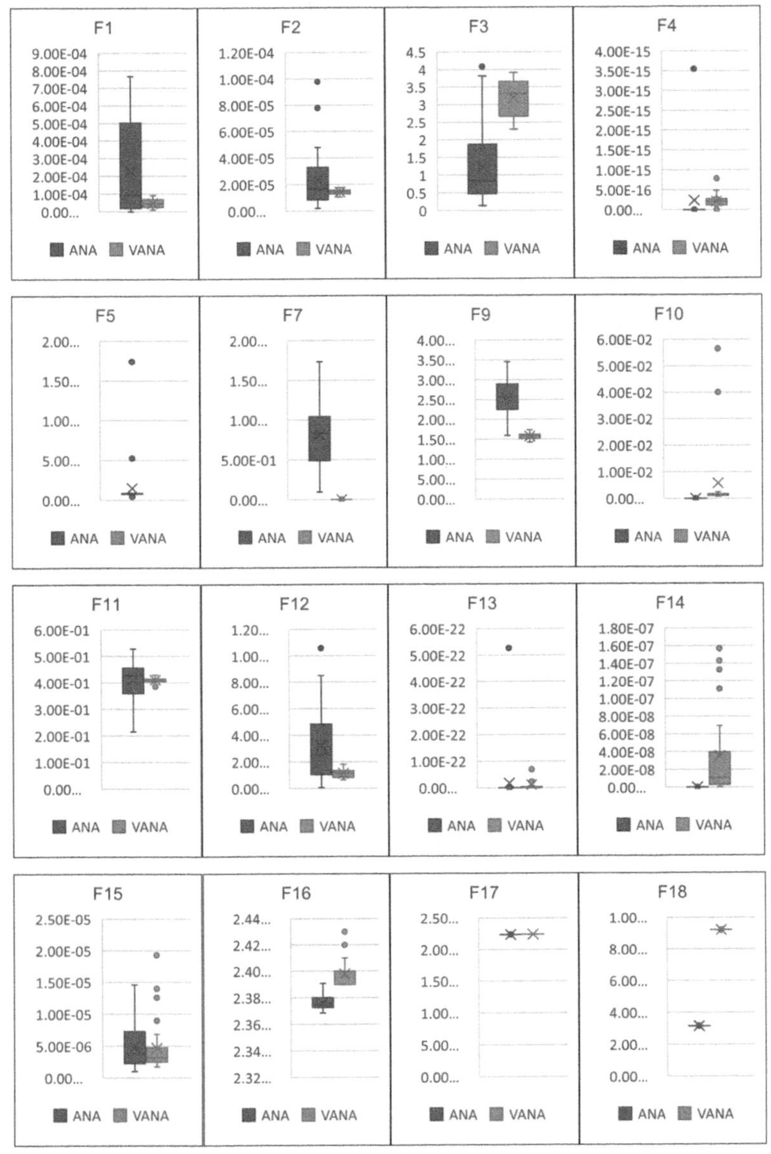

FIGURE 10.6 All standard benchmark comparisons from F01–F18.

10.7 GRAPHICAL REPRESENTATION OF ANA VS. IANA FOR CEC01–CEC10

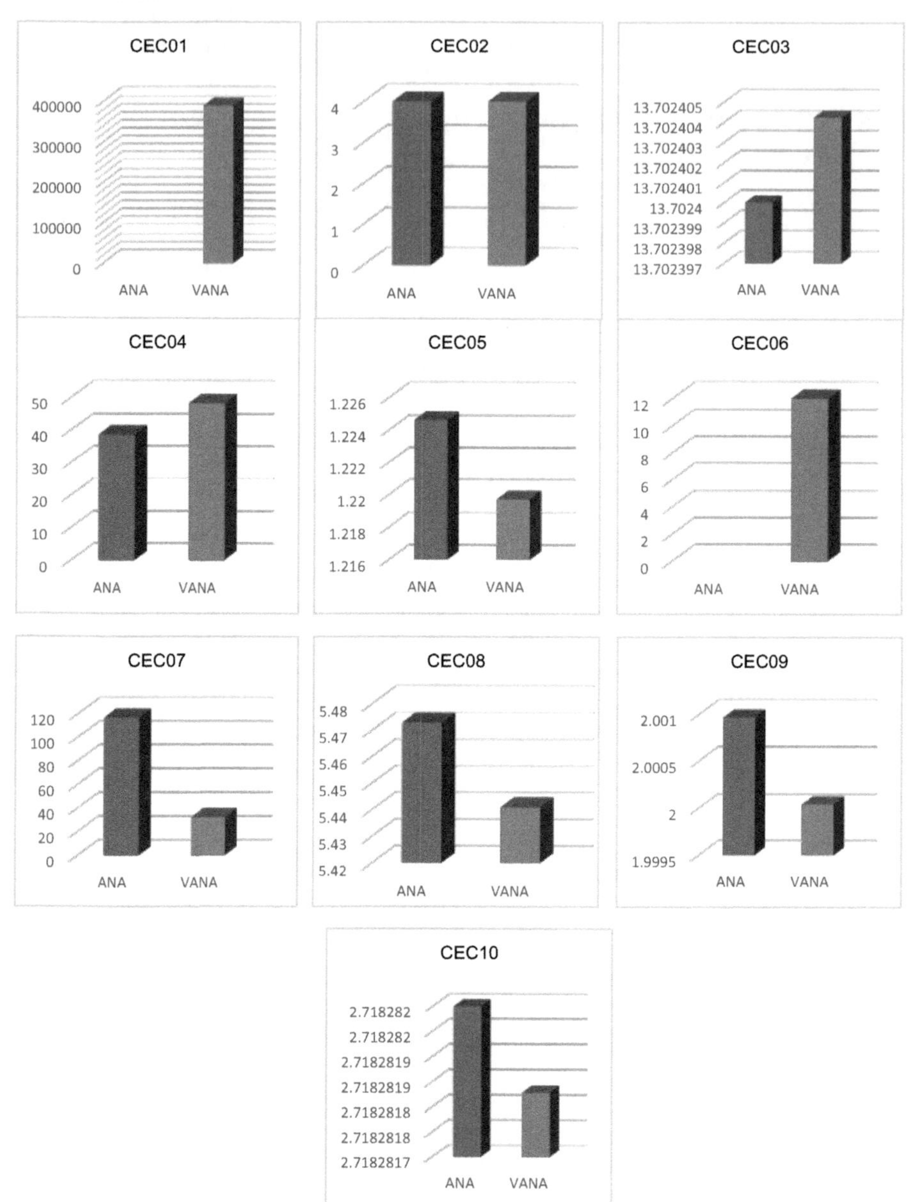

FIGURE 10.7 All CEC test function comparisons.

10.8 ALL STANDARD BENCHMARKS AND CEC TEST FUNCTIONS OUTPUTS

TABLE 10.5
F1–F18 benchmark test function comparison from other metaheuristic algorithms

Functions		VANA	ANA	LPB	DA	PSO	GA
F1	Mean	4.59E-05	0.016299162	0.001877545	2.85E-18	4.20E-18	748.5972
	STD	2.77E-01	0.062457134	0.002093616	7.16E-18	1.31E-17	324.9262
F2	Mean	1.08E-05	2.36E-05	0.005238111	1.49E-05	0.003154	5.971358
	STD	1.08E-05	2.15E-05	0.003652512	3.76E-05	0.009811	1.533102
F3	Mean	3.19667	1.239172335	36.4748883	1.29E-06	0.001891	1949.003
	STD	4.247414324	1.035406371	29.22415523	2.10E-06	0.003311	994.2733
F4	Mean	2.11E-16	2.37E-16	0.393866	0.000988	0.001748	21.16304
	STD	7.799E-16	8.86E-16	0.135818	0.002776	0.002515	2.605406
F5	Mean	2.83E+18	14.95306041	16.76919	7.600558	63.45331	133307.1
	STD	3.97301E+15	30.63521072	22.19251	6.786473	80.12726	85,007.62
F7	Mean	2.92E-07	0.770696384	0.004975	0.010293	0.005973	0.166872
	STD	1.51E-06	0.366042632	0.002965	0.004691	0.003583	0.072571
F9	Mean	15.74128	25.46221873	0.001567	16.01883	10.44724	25.51886
	STD	4.927386119	4.313900987	0.001842	9.479113	7.879807	6.66936
F10	Mean	0.005792	5.54E-15	0.017933	0.23103	0.280137	9.498785
	STD	0.025220399	2.37E-15	0.013532	0.487053	0.601817	1.271393
F11	Mean	0.410955	0.411189712	0.066355	0.193354	0.083463	7.719959
	STD	0.08695936	0.073239096	0.030973	0.073495	0.035067	3.62607
F12	Mean	1.133507	3.219956841	2.78659E-05	0.031101	8.57E-11	1858.502
	STD	1.422485634	2.52657309	3.83626E-05	0.098349	2.71E-10	5820.215

(continued)

TABLE 10.5 (Continued)
F1–F18 benchmark test function comparison from other metaheuristic algorithms

Functions		VANA	ANA	LPB	DA	PSO	GA
F13	Mean	1.07E-23	1.76E-23	0.000309	103.742	150	130.0991
	STD	5.68E-20	9.47E-23	0.000512	91.24364	135.4006	21.32037
F14	Mean	3.54E-08	4.26E-14	0.998004	193.0171	188.1951	116.0554
	STD	1.50E-07	1.54E-13	1.26E-11	80.6332	157.2834	19.19351
F15	Mean	4.67E-06	4.89E-06	0.002358	458.2962	263.0948	383.9184
	STD	3.13E-05	3.31E-06	0.003757	165.3724	187.1352	36.60532
F16	Mean	2.40E+01	23.76092355	-1.03163	596.6629	466.5429	503.0485
	STD	4.64E-01	0.048390796	2.46E-06	171.0631	180.9493	35.79406
F17	Mean	2.24E+02	223.5622125	0.397888	229.9515	136.1759	118.438
	STD	1.05E-01	0.008813889	3.16E-06	184.6095	160.0187	51.00183
F18	Mean	9.20E+01	31.51015225	3.000142	679.588	741.6341	544.1018
	STD	5.33E-02	0.020777872	0.000283	199.4014	206.7296	13.30161

Source: [6, 10, 21, 27].

TABLE 10.6
CEC01–CEC10 benchmark test function comparison from other metaheuristic algorithms

CEC Functions		VANA	ANA	FOX	DA	WOA	SSA	FDO
CEC01	Mean	388346.8221	-	2.58E+04	5.43E+10	4.11E+10	6.05E+09	4585.27
	STD	419228.0172	-	22491.27	6.69E+10	5.42E+10	4.75E+09	20707.627
CEC02	Mean	4.0	4.0	18.3442	78.0368	17.3495	18.3434	4.0
	STD	0	2.87E-14	0.000425	87.7888	0.0045	0.0005	3.22414E-9
CEC03	Mean	13.7024	13.7024	13.7025	13.7026	13.7024	13.7025	13.7024
	STD	5.49661E-08	2.01E-11	7.11E-15	0.0007	0	0.0003	1.6490E-11
CEC04	Mean	48.20057349	38.50887822	1.06E+03	344.356	394.675	41.6936	34.0837
	STD	33.1980625	10.07245727	835.2478	414.098	248.563	22.2191	16.528865
CEC05	Mean	1.219721311	1.224598709	5.3148	2.5572	2.7342	2.2084	2.13924
	STD	0.101222116	0.114632394	0.800019	0.3245	0.2917	0.1064	0.085751
CEC06	Mean	11.98608986	-	5.0325	9.8955	10.7085	6.0798	12.1332
	STD	0.701903068	-	1.499443	1.6404	1.0325	1.4873	0.600237
CEC07	Mean	32.01025539	116.5962143	306.794	578.953	490.684	410.396	120.4858
	STD	23.00293214	8.825046006	141.8588	329.398	194.832	290.556	13.59369
CEC08	Mean	5.440928127	5.472814997	5.4621	6.8734	6.909	6.3723	6.1021
	STD	0.380608936	0.429461877	0.381907	0.5015	0.4269	0.5862	0.756997
CEC09	Mean	2.0	2.000963996	3.7959	6.0467	5.9371	3.6704	2.0
	STD	0.000142271	0.00341781	0.440876	2.871	1.6566	0.2362	1.5916E-10
CEC10	Mean	2.7182	2.7182	20.9818	21.2604	21.2761	21.04	2.7182
	STD	8.88178E-16	4.44E-16	0.011096	0.1715	0.1111	0.078	8.8817E-16

10.9 CONCLUSION

In conclusion, in this chapter, we have deeply looked at the greatest difficulty encountered by ANAs' overshooting issue, including a completely new method of incorporating the sigmoid equation into the tendency rate. This study was chiefly aimed at alleviating the problem of overshooting in ANAs. Moreover, combining the sigmoid equation, in many cases, resulted in increased validity.

The overshooting problem, as the algorithm often goes beyond the limit, has been one of the obstacles in improving the efficiency and accuracy of ACO algorithms. The purpose of this chapter is to introduce the sigmoid equation to modify the tendency coefficient and limit the values within the range 0–1. This chapter sought to confine the values to a range from 0 to a bit more than 1 by introducing the sigmoid equation to adjust the tendency factor of the algorithm. This strategic integration means that the algorithm's propensity to overshoot is kept under control precisely and reliably. The widely used S-shaped formula is of the sigmoid type. As the values that it yields are always between 0 and 1, this device effectively ensures better control and lessens sharp rises. This innovative approach aligns with the broader goal of optimizing algorithmic performance.

In the comparison of the proposed method with other metaheuristic algorithms, the results of the research are well worth considering. As the frontrunner, the ANA was made robust using the sigmoid equation as a continuous modulation function in relation to the propensity quotient. The meticulous attention to overshooting not only distinguishes it but also places it in a better position as a reliable and practical remedy. For future research areas, scholars are encouraged to address further computational intelligence and optimization methodologies. Future work may also address the application of the proposed approach, especially in hybridizing advanced algorithms including GOOSE, FDO, Lagrange Elementary Optimization (Leo), FOX, Donkey and Smuggler Optimization, and Child Drawing Development Optimization, as it represents the emerging frontiers of computational intelligence. These reported algorithms are the most advanced algorithms, so the proposed algorithm can be hybridized with these advanced algorithms to explore more achievements.

REFERENCES

[1] X.-S. Yang, *Nature-inspired optimization algorithms.* Academic Press, 2020.

[2] S. Desale, A. Rasool, S. Andhale, and P. Rane, "Heuristic and meta-heuristic algorithms and their relevance to the real world: A survey," *International Journal of Computer Engineering in Research Trends,* vol. 351, no. 5, pp. 2349–7084, 2015.

[3] M. Yashar and Y. Hashim, "Design and implementation of Arduino controlled Mecanum wheel robot car," in *2024 21st International Multi-Conference on Systems, Signals & Devices (SSD),* 2024, pp. 474–479: IEEE.

[4] H. M. Ali, M. Y. Hamza, and T. A. Rashid, "A comprehensive study on automated testing with the software lifecycle," *Journal of Duhok University,* vol. 26, no. 2, pp. 613–620, 2023.

[5] S. Bandaru and K. Deb, "Metaheuristic techniques," in *Decision sciences,* 2016, pp. 709–766: CRC Press.

[6] M. Yashar and T. A. Rashid, "VAPI: Vectorization of algorithm for performance improvement," *arXiv preprint arXiv:2308.01269,* 2023.

[7] S. K. Omer *et al.*, "Comparative analysis of encapsulation in Java and Python: Syntax and implementation differences," *Journal of Duhok University,* vol. 26, no. 2, pp. 379–389, 2023.

[8] H. M. Ali, M. Y. Hamza, and T. A. Rashid, "Exploring polymorphism: Flexibility and code reusability in object-oriented programming," *Passer Journal of Basic and Applied Sciences,* vol. 6, no. Special Issue, pp. 502–512, 2024.

[9] S. D. Mirkhan, S. K. Omer, T. A. Rashid, H. M. Ali, M. Y. Hamza, and P. Nedunchezhian, "Effective delegation and leadership in software management," *Journal of Duhok University,* vol. 26, no. 2, pp. 407–415, 2023.

[10] J. Kennedy and R. Eberhart, "Particle swarm optimization," in *Proceedings of ICNN'95-International Conference on Neural Networks*, 1995, vol. 4, pp. 1942–1948: IEEE.

[11] M. Dorigo, M. Birattari, and T. Stutzle, "Ant colony optimization," *IEEE Computational Intelligence Magazine,* vol. 1, no. 4, pp. 28–39, 2006.

[12] S. Mirjalili, S. M. Mirjalili, and A. Lewis, "Grey wolf optimizer," *Advances in Engineering Software,* vol. 69, pp. 46–61, 2014.

[13] B. A. Hassan and T. A. Rashid, "A multidisciplinary ensemble algorithm for clustering heterogeneous datasets," *Neural Computing and Applications,* vol. 33, no. 17, pp. 10987–11010, 2021.

[14] B. A. Hassan, T. A. Rashid, and S. Mirjalili, "Performance evaluation results of evolutionary clustering algorithm star for clustering heterogeneous datasets," *Data in Brief,* vol. 36, p. 107044, 2021.

[15] B. A. Hassan, T. A. Rashid, and S. Mirjalili, "Formal context reduction in deriving concept hierarchies from corpora using adaptive evolutionary clustering algorithm star," *Complex & Intelligent Systems,* vol. 7, no. 5, pp. 2383–2398, 2021.

[16] B. A. Hassan, "CSCF: A chaotic sine cosine firefly algorithm for practical application problems," *Neural Computing and Applications,* vol. 33, no. 12, pp. 7011–7030, 2021.

[17] S. M. Qader, B. A. Hassan, and T. A. Rashid, "An improved deep convolutional neural network by using hybrid optimization algorithms to detect and classify brain tumor using augmented MRI images," *Multimedia Tools and Applications,* vol. 81, no. 30, pp. 44059–44086, 2022.

[18] B. B. Maaroof *et al.*, "Current studies and applications of shuffled Frog Leaping Algorithm: A review," *Archives of Computational Methods in Engineering,* vol. 29, no. 5, pp. 3459–3474, 2022.

[19] M. T. Abdulkhaleq *et al.*, "Harmony search: Current studies and uses on healthcare systems," *Artificial Intelligence in Medicine,* vol. 131, p. 102348, 2022.

[20] B. A. Hassan, T. A. Rashid, and H. K. Hamarashid, "A novel cluster detection of COVID-19 patients and medical disease conditions using improved evolutionary clustering algorithm star," *Computers in Biology and Medicine,* vol. 138, p. 104866, 2021.

[21] D. N. Hama Rashid, T. A. Rashid, and S. Mirjalili, "ANA: Ant nesting algorithm for optimizing real-world problems," *Mathematics,* vol. 9, no. 23, p. 3111, 2021.

[22] S. Narayan, "The generalized sigmoid activation function: Competitive supervised learning," *Information Sciences,* vol. 99, no. 1–2, pp. 69–82, 1997.

[23] J. M. Abdullah and T. Ahmed, "Fitness dependent optimizer: Inspired by the bee swarming reproductive process," *IEEE Access,* vol. 7, pp. 43473–43486, 2019.

[24] H. Mohammed and T. Rashid, "FOX: A FOX-inspired optimization algorithm," *Applied Intelligence,* vol. 53, no. 1, pp. 1030–1050, 2023.

[25] A. S. Shamsaldin, T. A. Rashid, R. A. Al-Rashid Agha, N. K. Al-Salihi, and M. Mohammadi, "Donkey and smuggler optimization algorithm: A collaborative working approach to path finding," *Journal of Computational Design and Engineering,* vol. 6, no. 4, pp. 562–583, 2019.

[26] S. Abdulhameed and T. A. Rashid, "Child drawing development optimization algorithm based on child's cognitive development," *Arabian Journal for Science and Engineering,* vol. 47, no. 2, pp. 1337–1351, 2022.

[27] C. M. Rahman and T. A. Rashid, "A new evolutionary algorithm: Learner performance based behavior algorithm," *Egyptian Informatics Journal,* vol. 22, no. 2, pp. 213–223, 2021.

11 VAPI

Vectorization of Algorithms for Performance Improvement

Mahmood Yashar, Tarik A. Rashid and Aso M. Aladdin

11.1 INTRODUCTION

Most metaheuristic algorithms used to solve different real-world problems in the traditional way, that are used for small-scaled problems and are not impractical for complex problems, cannot be implemented due to the large amount of time and resources required for the algorithm's execution. All metaheuristic algorithms find the optimum solution through an iterative approach. When the problem becomes complex and the number of worker agents increases, optimization algorithms lack speed. Standard benchmark test functions are implemented to measure algorithms' performance. Most optimization algorithms provide the optimum solution, but not in a fast period of time. This approach can be enhanced by implementing the vectorization technique. Vectorization is a technique that can be implemented for different optimization algorithms to modify their behavior and enable them to execute multiple values at once. The vectorization-induced change in algorithm behavior results in parallel processing, which reduces the number of iterations, improves algorithm performance in terms of speed, and performs test functions that cannot be executed in a short amount of time. Regarding time complexity, running large algorithms or complex equations in optimal time in real-world applications is very important [1]. The vectorization technique can be implemented in various ways, one of which is through the NumPy library, which can be found in the Python language. To speed up processing, vectorization is crucial in a variety of scientific areas, such as machine learning, image processing, and audio processing [2]. Numerous real-world problems can be solved by optimization algorithms; nevertheless, in some situations, metaheuristic algorithms run slowly or fail to get the desired results from test functions. This work focuses on using the vectorization approach to increase the performance speed of a metaheuristic algorithm called the Ant Nesting Algorithm (ANA).

This study presents a vectorized algorithm to minimize computing time, enhance algorithm speed, make the algorithm clearer to understand, and reduce iterations in the algorithm, which are the major reasons for algorithm speed reduction. Python is a programming language that offers the NumPy library, which can be used to vectorize

DOI: 10.1201/9781003601555-11

TABLE 11.1
Difference between the vectorized and non-vectorized approaches

Non-Vectorized Approach	Vectorized Approach

`array1 = [1, 2, 3, 4]`
`array2 = [5, 6, 7, 8]`
`result = []`
`for i in range(len(array1)):`
`result.append(array1[i] +`
`  array2[i])`
`print(result)`
	`import numpy as np`
	`array1 = np.array([1,`
	`  2, 3, 4])`
	`array2 = np.array([5,`
	`  6, 7, 8])`
	`result = array1 + array2`
	`print(result)`

mathematical calculations, save memory copying, and reduce operation counts. The NumPy array is a multidimensional array, represented as (n-dimensional array), that is represented as a block of memory to easily manipulate numbers. It employs a central processing unit to work in memory efficiently, storing and accessing a multidimensional array [3]. Most programming languages misguidedly use iterations, which results in poor algorithm performance speed, particularly when applied to large datasets. When performing an arithmetic operation on an array, for instance, iterations are required. However, when using vector representation, the algorithm may do the same action on every item without requiring iterations, leading to a faster method overall. NumPy arrays provide a lightweight and efficient method of working with and manipulating data in matrices or vectors [3, 4]. An example is represented in the code in Table 11.1 between vectorized and non-vectorized approaches. As shown, the vectorized approach does not require any iterations directly implemented in the code.

The objective of this research is to apply vectorization approaches to the ANA metaheuristic algorithm and to enhance efficiency, reduce iterations, and improve a test function that does not generate output. Section 11.2 presents related vectorization work, Section 11.3 provides an experiment on the ANA metaheuristic algorithm on how agents move during each iteration, Section 11.4 discusses the mathematical modeling of ANA, Section 11.5 presents the conversion of ANA to the vectorized version, Section 11.6 is based on a comparison of speed between vectorized and non-vectorized ANA, and finally, Section 11.7 concludes the chapter.

11.2 RELATED WORK

The NumPy library has been utilized in several domains, such as engineering, economics, finance, chemistry, physics, and astronomy. For example, the NumPy library was used in software for gravitational wave finding [5] and the first imaging of a black hole was performed using the NumPy library. The NumPy library is a strong programming concept that has been used in much scientific research to improve the performance speed of algorithms. The NumPy library also improves the speed of algorithms by improving the collaboration reducing error and making the code simpler to understand. Here are a few examples of how NumPy's vectorization is used to speed up algorithms in real-world applications.

Image Processing: NumPy makes it easy to handle images quickly. The basic idea is to use functions that work with collections of coordinates instead of for loops that go through pixel coordinates. Because of this, the solutions are ten times faster to obtain than easy solutions In conclusion, NumPy and SciPy make it easy to handle images quickly. By making the code "vectorized", which means that it doesn't use for loops over pixel coordinates but instead uses built-in functions for handling arrays, image data can be processed quickly in every manner [1].

The vectorization approach can be used in a variety of fields. For example, one study used a clustering algorithm known as the Evolutionary Clustering Algorithm (ECA) which is used to cluster datasets for better outputs. The study was conducted on 32 heterogeneous datasets, and overall, ECA outperformed other algorithms to produce the best results. Vectorization can be used in this field to increase the run-speed of the algorithm [6]. The improved version of the algorithm is called IECA, which was designed to include Covid-19 by implementing three approaches for the data analysis elbow method, cleansing the data, and analyzing Covid-19 data [7].

Another research area called Formal Context Analysis is for the automation and improvement of parsing text and extracting pairs of data. This approach helps to automate the process to reduce time consumption. In addition, in a study, the procedure was implemented into 385 samples and the accuracy rate was 89% throughout the automation of the process. Vectorization can be implemented to improve algorithms' speed and avoid iterative approaches [8].

Another approach is the ECA*, which is used to compare five different algorithms. Overall, this approach can be used for multi-feature datasets and improves time consumption over manual processing. The data vectorization approach can also improve the significance of the speed [9].

Another kind of swarm-based metaheuristic optimization method is the Chaotic Sine Cosine Firefly (CSCF), similar to previous optimization techniques. The CSCF optimization technique can solve the performance and complexity issues that plague the majority of swarm-based algorithms. This approach, which is referred to as a hybrid algorithm, combines the Sine Cosine and Firefly Algorithms. The algorithm operates more quickly when both optimization algorithms are combined, like other optimization algorithms [10].

Grey Wolf and Harris Hawks Optimizations are combined to create G-HHO, a metaheuristic optimization method. This technique employs a quicker and more efficient method of brain tumor detection by using a Deep Convolutional Neural Network. This program employs an Otsu thresholding technique, which has a high accuracy rate of 97% in identifying tumors for magnetic resonance imaging [11]. The Shuffled Frog Leaping Algorithm is a hybrid metaheuristic algorithm that integrates memetics with particle swarms and it may be used for a range of engineering issues [12].

A well-known metaheuristic method for solving complicated issues that might strike a balance between the best answer and newly discovered solutions is the Harmony Search Algorithm. This algorithm has been used in a number of industries, including engineering, healthcare, and architecture [13]. The vectorization approach can be implemented in various fields, such as robotics [14], software testing to parallelly test the data for speed improvement [15], changing the Java and Python structure to the vectorized approach [16, 17], and finally, in software management tools to enhance the software's runtime speed [18].

11.3 EXPERIMENT

To experimentally assess vectorization outcomes, ANA, which is a metaheuristic algorithm, was modified to the vectorization method. In this section, we examine how ANA works, as well as how to vectorize it, and compare its results to non-vectorized versions of the algorithm. This research examines the vectorization of ANA and checks the performance of the algorithm after modification to vectorization.

ANA is a swarm intelligence metaheuristic algorithm that simulates ant behavior while looking for grains to construct a new nest. ANA analyzes its current and previous positions together with the fitness value to determine the optimal location to lay the grains. It also generates deposition weights using the Pythagorean theorem [19]. ANA finds the minimum number among other ants. As it is a metaheuristic algorithm, the number of agents is initialized to 30, with an iteration of 500, and the algorithm finds the optimal solution among 30 agents. The aim is to find the minimum number of agents that reach the destination, as shown in Figure 11.1 and 11.2. All agents try to reach the destination. In each iteration, each ant moves closer to the destination.

11.4 MATHEMATICAL MODELING OF ANA

ANA simulates ant behavior to construct a new nest. Its significant role is to discover the best solution among other ants. Ants' natural behavior is to gather little grains and transport them to their destinations, as demonstrated in Figures 11.1 and 11.2. ANA works to find the minimal distance, which is used to put the grains close together to construct a strong wall around the queen ant. First, the algorithm starts by initializing ten random numbers in the range [lower boundary to upper boundary] as [-100, 100] for each ant to determine which position to put the grains in and find the best position. In this case, the new position is compared with the old position to determine which one is the optimal position. The worker ant starts searching for a new position randomly, where the new position is $X_{t',i+1}$, which represents the new deposition position, $X_{t',i}$ represents the current position, and the deposition rate of change is denoted by $\Delta X_{t',i+1}$:

$$x_{t,i+1} = x_{t,i} + \Delta x_{t,i+1} \tag{1}$$

$\Delta X_{t',i+1}$ represents the rate of change of the deposition position, which depends on the *dw* deposition position multiplied by the difference between the best worker ant, which is represented as $X_{t',i\ best'}$, and the current deposition position, represented as $X_{t',i}$:

$$\Delta x_{t,i+1} = dw \times \left(x_{t,i\ best} - x_{t,i}\right) \tag{2}$$

As the algorithm starts, there is no best ant, so it chooses the first ant as the best ant and then the comparison begins. The fitness function for each ant is determined for all 30 ants, then the best fitness is selected, and two new arrays are created for

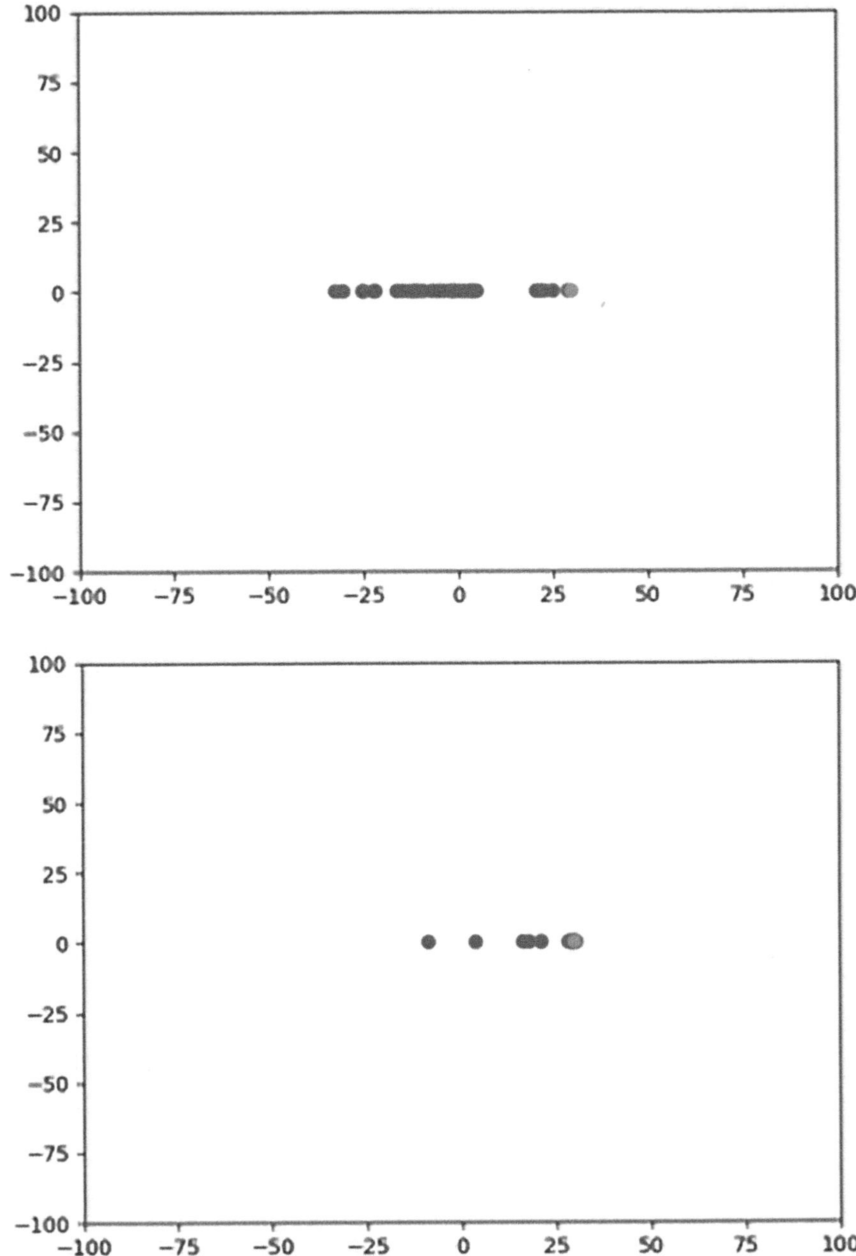

FIGURE 11.1 Agents move to the destination from iterations 1–250.

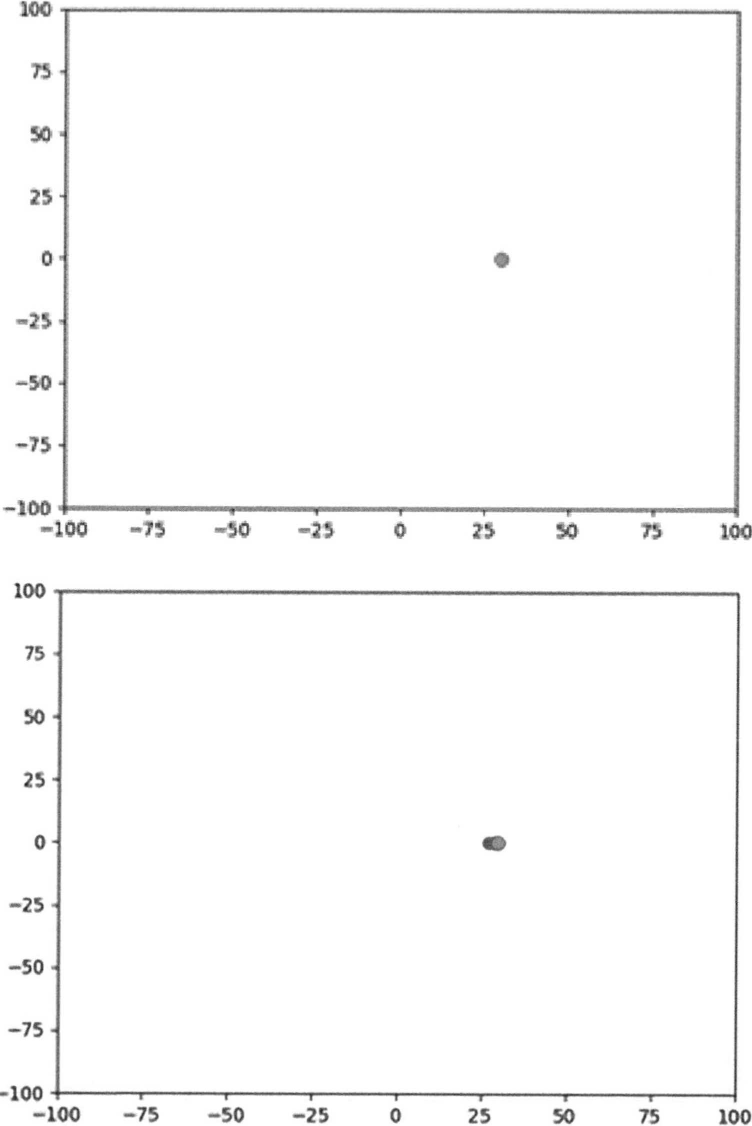

FIGURE 11.2 Agents move to the destination from iterations 250–499.

a new position and the previous position. After this, the algorithm compares every 10 random numbers for each ant between the first and last ant to find the distance from the best ant. The distance is calculated by subtracting the position of the best ant from the position of the first ant, which creates the previous variable and creates a simple random walk in the range [-1,1]. Then the algorithm passes through three conditions.

1. If the current ant equals the best ant, the following is used:

$$\Delta x_{t,i+1} = r \times x_{t,i} \quad (3)$$

2. If the current ant equals the previous ant, the following is used:

$$\Delta x_{t,i+1} = r \times \left(x_{t,i\ best} - x_{t,i}\right) \quad (4)$$

3. Else, the following is used:

$$dw = r \times \left(\frac{T}{T_{previous}}\right) \quad (5)$$

where *dw* represents the deposition weight and *r* is a random number in the range [-1,1]. The algorithm uses Levy flight to make the algorithm more suitable. In addition, *dw* depends on *T*, which is the known tendency rate using the Pythagorean theorem and *Tprevious* is the previous tendency rate:

$$T = \sqrt{\left(X_{t,i\ best} - X_{t,i\ previous}\right)^2 + \left(X_{t,i\ best\ fitness} - X_{t,i\ previous\ fitness}\right)^2} \quad (6)$$

$$T_{previous} = \sqrt{\left(X_{t,i\ best} - X_{t,i\ previous}\right)^2 + \left(X_{t,i\ best\ fitness} - X_{t,i\ previous\ fitness}\right)^2} \quad (7)$$

Eqs. (6) and (7) in ANA have errors, contrary to what is given in the Pythagorean theorem equation. The errors, along with the proper calculations, may be found in Eqs. (6) and (7). The preceding piece of work had a typographical error. The incorrectly calculated equation would not provide the right output for the algorithm. The error in the equation was that it used subtraction instead of addition [19]. The corrected equation for the Pythagorean theorem has been applied to ANA. In addition, the corrected equation has been implemented in this research. The Vectorized Ant Nesting Algorithm (VANA) uses the correct version of the equation, and all of the output of VANA uses the correct equation of the Pythagorean theorem. This is also stated in the pseudocode of the algorithm.

According to the Pythagorean theorem, the square of the length of the hypotenuse in a triangle with a right angle is equal to the sum of the squares of the lengths of the other two sides of the triangle. The hypotenuse is the side that is opposite the right angle. It is possible to describe it mathematically using the formula $a^2 + b^2 = c^2$, where a and b refer to the lengths of the two legs (the sides next to the right angle), and c refers to the length of the hypotenuse. Pythagoras, an ancient Greek mathematician, is given credit for both the discovery of this theorem and its proof. Consequently, this theorem bears his name. It finds use in a wide variety of disciplines, including geometry, trigonometry, and physics. The optimization algorithm

Algorithm 1: ANA Algorithm Pseudocode

```
Initialize worker ant population x_i(i = 1, 2, 3, ..., n)
Initialize worker ant previous position X_iprevious
while iteration (t) limit not reached do
    for each artificial worker ant x_t,i do
        Find best artificial worker ant x_t,ibest
        Generate random walk r in [-1, 1] range
        if (x_t,i == X_t,ibest) then
            calculate Δx_t,i+1 using equation (3)
        else if (x_t,i == X_t,iprevious) then
            calculate Δx_t,i+1 using equation (4)
        else
            calculate T using equation (6)
            calculate T_previous using equation (7)
            calculate dw using equation (5) (for minimization)
            calculate Δx_t,i+1 using equation (2)
        calculate x_t,i using equation (1)
        if (x_t,i+1 fitness < X_t,ifitness (for minimization)) then
            move and accept x_t,i assigned to X_t,iprevious
        else
            maintain the current position
```

FIGURE 11.3 ANA pseudocode.

also uses the Pythagorean theorem to identify the best position with the help of this equation. ANA uses this equation to help find the optimum solution by referring to the tendency rate and previous tendency rate, as stated in Eqs. (6) and (7). Finally, the Pythagorean theorem is implemented in the algorithm to find the optimum solution. Improving the speed of an algorithm is important in a variety of scientific fields. This chapter discusses how ANA may be vectorized to increase its processing speed. This allows the algorithm to do several operations on the data being processed simultaneously, rather than just one.

Some of the test functions were significantly slow when ANA was performed, as demonstrated in Tables 11.2 and 11.3. This suggests that (F9, F11, F12, F15, F17, and F18) and primarily CEC01 and CEC06 are likewise very slow [19] Because of the slowness of CEC01 and CEC06, their output was not used in ANA, which prioritizes the slow performance of the algorithm, as shown in Table 11.4. After the vectorization approach is used, it increases the speed of the algorithm and receives CEC01 and CEC06 output. The explanation for the algorithm's slow performance is clarified in the next section.

11.5 VECTORIZATION OF ANA

As previously mentioned, ANA is extremely slow because of the large number of iterations since it starts by initializing the ants with 10 random numbers in the range [lower boundary to upper boundary] as [-100, 100] for each ant, for which it cycles 300 times; however, the NumPy function can handle this looping issue with its library, as shown in Figure 11.4.

TABLE 11.2
Time in seconds for F1–F18 test functions

Test Function	Execution Time (seconds) ANA
F1	32
F2	57
F3	52
F4	28
F5	26
F7	26
Test Function	**Execution Time (seconds) ANA**
F9	106
F10	16
F11	168
F12	140
F13	6
F14	39
F15	151
F16	79
F17	81
F18	96

TABLE 11.3
Time in seconds for CEC01–CEC10 test functions

Test Function	Execution Time ANA
CEC01	23705 seconds
CEC02	58 seconds
CEC03	511 seconds
CEC04	208 seconds
CEC05	337 seconds
CEC06	10606 seconds
CEC07	341 seconds
CEC08	266 seconds
CEC09	148 seconds
CEC10	17 seconds

Instead of dealing with loops to repeat 300 times to initialize 10 random numbers in the range [lower boundary, upper boundary] as [-100, 100] for each ant, the NumPy module generates the size of a (dimensions, agents) matrix so that the process can be vectorized to enhance the performance of the algorithm, as shown in Figure 11.5.

TABLE 11.4
CEC01 and CEC06 haven't been recorded

	ANA	
Test Function	**Mean**	**Standard Deviation**
CEC01	-	-
CEC02	4	2.87 x 10-14
CEC03	13.70240422	2.01 x 10-11
CEC04	38.50887822	10.07245727
CEC05	1.224598709	0.114632394
CEC06	-	-
CEC07	116.5962143	8.825046006
CEC08	5.472814997	0.429461877
CEC09	2.000963996	0.00341781
CEC10	2.718281828	4.4 x 10-16

```
for continue to loop in range of agents do
 | create array for initializing dimensions
 | for continue to loop in range of dimensions do
 |  | append random numbers [-100, 100] to array of dimensions
 | end
 | give the ant function dimensions, test function, previous dimensions,
 |   previous test function
 | store the new generated value to the worker ants
end
```

FIGURE 11.4 Initializing a random number in the range [lower boundary, upper boundary] with for loops.

```
Self . worker_ants = np . random . uniform (self . lower_
                     bound, self . upper_bound, (self .
                     dimension, self . number_of_agents))
```

FIGURE 11.5 NumPy array for creating a matrix of (dimensions, agents) random numbers in the range [lower boundary, upper boundary].

Following the execution of the np.random function, as shown in Figure 11.5, a matrix of (dimensions, agents) size is created. As shown in Figure 11.6, the rows represent the dimensions of ants and the column represents the number of agents.

The algorithm requires a previous position after initializing the ants. If there is no previous ant in the first step, the current worker ant will be initialized as the previous one. To transfer all current worker ants to previous worker ants, no iteration is required in the NumPy array, otherwise the speed of the algorithm would reduce. The previous position is used to keep track of the current position and newly generated

$$\textbf{Worker Ants} = \begin{bmatrix} \text{Ant}_1 & \text{Ant}_2 & \cdots & \cdots & \text{Ant}_{n-1} & \text{Ant}_n \\ A_1, D_1 & A_2, D_1 & \cdots & \cdots & A_{n-1}, D_1 & A_n, D_1 \\ A_1, D_2 & A_2, D_2 & \cdots & \cdots & A_{n-1}, D_2 & A_n, D_2 \\ \cdots & \cdots & \cdots & \cdots & \cdots & \cdots \\ \cdots & \cdots & \cdots & \cdots & \cdots & \cdots \\ A_1, D_{n-1} & A_2, D_{n-1} & \cdots & \cdots & A_{n-1}, D_{n-1} & A_{n-1}, D_{n-1} \\ A_1, D_n & A_2, D_n & \cdots & \cdots & A_{n-1}, D_n & A_n, D_n \end{bmatrix} \begin{matrix} \text{Dimention}_1 \\ \text{Dimention}_2 \\ \\ \\ \\ \text{Dimention}_{n-1} \\ \text{Dimention}_n \end{matrix}$$

FIGURE 11.6 Current view of worker ants of the (dimensions, agents) matrix.

$$\begin{matrix} \text{Copy of the wor ker ant} \end{matrix}$$

$$\text{Pr evious Ant} = \begin{bmatrix} Ant_1 & Ant_2 & \cdots & \cdots & Ant_{n-1} & Ant_n \\ A_1, D_1 & A_2, D_1 & \cdots & \cdots & A_{n-1}, D_1 & A_n, D_1 \\ A_1, D_2 & A_2, D_2 & \cdots & \cdots & A_{n-1}, D_2 & A_n, D_2 \\ \cdots & \cdots & \cdots & \cdots & \cdots & \cdots \\ \cdots & \cdots & \cdots & \cdots & \cdots & \cdots \\ A_1, D_{n-1} & A_2, D_{n-1} & \cdots & \cdots & A_{n-1}, D_{n-1} & A_{n-1}, D_{n-1} \\ A_1, D_n & A_2, D_n & \cdots & \cdots & A_{n-1}, D_n & A_n, D_n \end{bmatrix} \begin{matrix} Dimention_1 \\ Dimention_2 \\ \\ \\ \\ Dimention_{n-1} \\ Dimention_n \end{matrix}$$

FIGURE 11.7 Current view of previous ants.

$$\text{Rate Of Change} = \begin{bmatrix} \text{Ant}_1 & \text{Ant}_2 & \cdots & \cdots & \text{Ant}_{n-1} & \text{Ant}_n \\ 0 & 0 & \cdots & \cdots & 0 & A_0, D_1 \\ 0 & 0 & \cdots & \cdots & 0 & 0 \\ \cdots & \cdots & \cdots & \cdots & \cdots & \cdots \\ \cdots & \cdots & \cdots & \cdots & \cdots & \cdots \\ 0 & 0 & \cdots & \cdots & 0 & 0 \\ 0 & 0 & \cdots & \cdots & 0 & 0 \end{bmatrix} \begin{matrix} Dimention_1 \\ Dimention_2 \\ \\ \\ \\ Dimention_{n-1} \\ Dimention_n \end{matrix}$$

FIGURE 11.8 Current view of the (dimensions, agents) matrix of zeros.

random value to know which value is more optimum. NumPy arrays contain a copy library that eliminates the need for iterative processes, which copies all current ants to the previous ants, as shown in Figure 11.7.

The rate of change will create a matrix with the shape of (dimensions, agents) with zero values as shown in Figure 11.8.

Worker ant fitness, as stated in Figure 11.9, finds the fitness of all ants according to the test function and stores them in a (1, 30) shape matrix.

$$\text{Wor ker AntFit} = \begin{bmatrix} \text{Ant}_1 & \text{Ant}_2 & \cdots\ \cdots & \text{Ant}_{n-1} & \text{Ant}_n \\ Fitnesso\ fant_1 & Fitnesso\ fant_2 & & Fitnesso\ fant_{n-1} & Fitnesso\ fant_n \end{bmatrix} \text{Dimention}_1$$

FIGURE 11.9 Worker ant fitness contains the fitness of all ants with the shape of the (1,30) matrix.

$$\text{PrevAntFit} = \begin{bmatrix} \text{Ant}_1 & \text{Ant}_2 & \cdots\ \cdots & \text{Ant}_{n-1} & \text{Ant}_n \\ PrevAntFit_1 & PrevAntFit_2 & & PrevAntFit_{n-1} & PrevAntFit_n \end{bmatrix} \text{Dimention}_1$$

FIGURE 11.10 Current view of previous ant fitness.

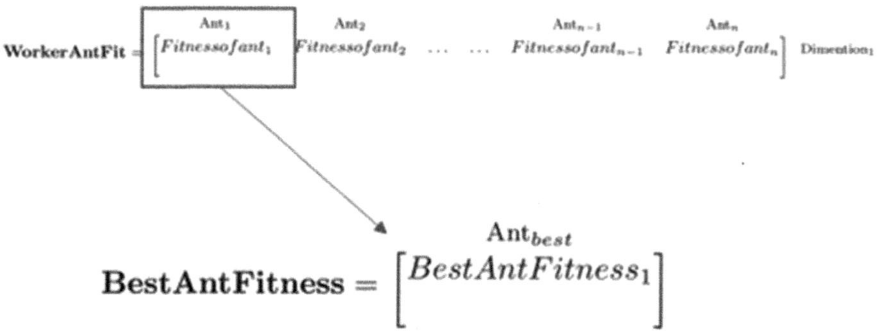

FIGURE 11.11 Current view of best ant fitness with index.

ANA also finds the previous ant fitness to keep track of which ant has the optimal fitness value, the same as worker ant fitness with the shape of (1, 30), as shown in Figure 11.10.

The best ant index stores the index of the best ant fitness case. If the best fitness is the first ant, it will store the index of 0 and the best ant fitness will store the fitness of the best ant, as shown in Figure 11.11.

Best ant stores the worker ant best ant, which indicates the selection of all columns of the best ant index, and reshapes it according to the dimension of the fitness function. Index is mainly used to indicate the position of the best ant, and the best ant index is the index of the best ant, as shown in Figure 11.12.

The distance from the best ant stores all matrices of worker ants (dimensions, agents) then subtracts the best ant by all other ants. This process of subtracting the ants from the best ant is to find the distance between the best ant and other ants, as shown in Figure 11.13.

Temp_xs, with the help of the NumPy library, copies all worker ants to temp_xs without using iteration, as shown in Figures 11.14 and 11.15.

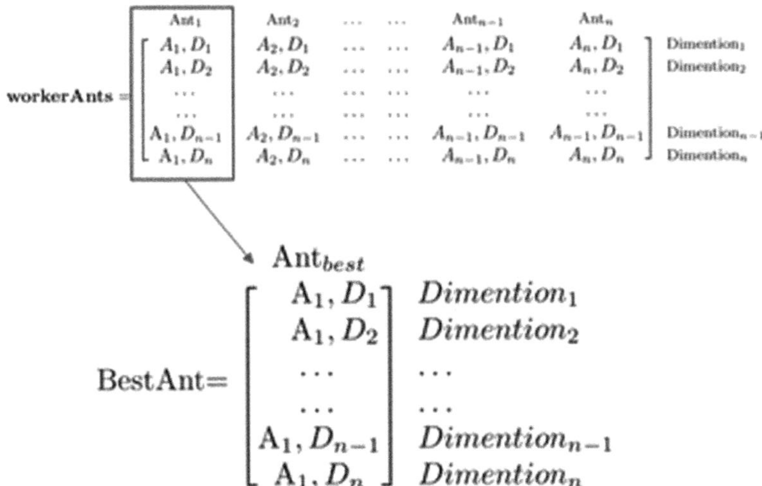

FIGURE 11.12 Current view for identifying the best ant.

The NumPy library also provides matrix initialization of ones with the same dimension as the worker ant, as stated in Figure 11.16.

The (10,30) matrix array is initialized to calculate a simple random walk for the ants, as shown in Figure 11.17.

After all these steps are applied, the algorithm passes through three conditions:

- If the current ant is equal to the best ant, use Eq. (3).
- If the current ant is equal to the previous ant, use Eq. (4).
- If the above points aren't satisfied, use Eq. (5).

These steps are stated in Section 11.4 in a more detailed way.

Then the process passes through three conditions to calculate all the processes, as shown in Figure 11.18.

As shown in Figure 11.3, ANA performed slowly due to the large number of iterations. After the algorithm was modified to vectorization, fewer iterations were used, as shown in Figs. 11.4 and 11.5. The initialization of a random number of ants reduced from 300 iterations to a matrix of size (dimensions, agents). In addition, for copying the current worker ant to the previous worker ant, iteration was also avoided, as shown in Figure 11.7. In addition, for creating the rate of change, 300 iterations were also avoided. All the functionality of ANA that was performed with iteration was converted to matrices, which helped to improve the algorithm's performance and speed. The full pseudocode of VANA is presented in Figure 11.19 and the flowchart of VANA is presented in Figure 11.20.

Modifying ANA to a vectorized method improved the speed of the algorithm, as shown in Table 11.5, and Table 11.6 indicates that the algorithm performed much faster, mainly in test functions F9, F12, F15, F17, and F18. In addition, for CEC01,

$$\text{Distance From Best Ant} = \left[\begin{array}{cccccc} \text{Ant}_1 & \text{Ant}_2 & & & \text{Ant}_{n-1} & \text{Ant}_n \\ A_1, D_1 - \textit{Best Ant} & A_2, D_1 - \textit{Best Ant} & \cdots & \cdots & A_{n-1}, D_1 - \textit{Best Ant} & A_n, D_1 - \textit{Best Ant} \\ A_1, D_2 - \textit{Best Ant} & A_2, D_2 - \textit{Best Ant} & \cdots & \cdots & A_{n-1}, D_2 - \textit{Best Ant} & A_n, D_2 - \textit{Best Ant} \\ \cdots & \cdots & \cdots & \cdots & \cdots & \cdots \\ \cdots & \cdots & \cdots & \cdots & \cdots & \cdots \\ A_1, D_{n-1} - \textit{Best Ant} & A_2, D_{n-1} - \textit{Best Ant} & \cdots & \cdots & A_{n-1}, D_{n-1} - \textit{Best Ant} & A_{n-1}, D_{n-1} - \textit{Best Ant} \\ A_1, D_n - \textit{Best Ant} & A_2, D_n - \textit{Best Ant} & \cdots & \cdots & A_{n-1}, D_n - \textit{Best Ant} & A_n, D_n - \textit{Best Ant} \end{array}\right] \begin{array}{l} \textit{Dimention}_1 \\ \textit{Dimention}_2 \\ \\ \\ \\ \textit{Dimention}_{n-1} \\ \textit{dimention}_n \end{array}$$

FIGURE 11.13 Current view of subtracting the best ant from all other ants.

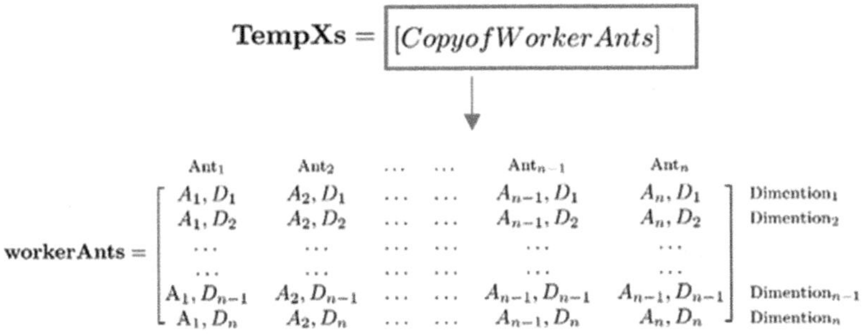

$$\mathbf{TempXs} = \boxed{[CopyofWorkerAnts]}$$

$$\downarrow$$

$$\mathbf{workerAnts} = \begin{bmatrix} \text{Ant}_1 & \text{Ant}_2 & \cdots & \cdots & \text{Ant}_{n-1} & \text{Ant}_n \\ A_1, D_1 & A_2, D_1 & \cdots & \cdots & A_{n-1}, D_1 & A_n, D_1 \\ A_1, D_2 & A_2, D_2 & \cdots & \cdots & A_{n-1}, D_2 & A_n, D_2 \\ \cdots & \cdots & \cdots & \cdots & \cdots & \cdots \\ \cdots & \cdots & \cdots & \cdots & \cdots & \cdots \\ A_1, D_{n-1} & A_2, D_{n-1} & \cdots & \cdots & A_{n-1}, D_{n-1} & A_{n-1}, D_{n-1} \\ A_1, D_n & A_2, D_n & \cdots & \cdots & A_{n-1}, D_n & A_n, D_n \end{bmatrix} \begin{matrix} \\ \text{Dimention}_1 \\ \text{Dimention}_2 \\ \\ \\ \text{Dimention}_{n-1} \\ \text{Dimention}_n \end{matrix}$$

FIGURE 11.14 Current view of copying worker ants to temp_xs.

$$\text{TempXs} = \begin{bmatrix} Ant_1 & Ant_2 & \cdots & \cdots & Ant_{n-1} & Ant_n \\ A_1, D_1 & A_2, D_1 & \cdots & \cdots & A_{n-1}, D_1 & A_n, D_1 \\ A_1, D_2 & A_2, D_2 & \cdots & \cdots & A_{n-1}, D_2 & A_n, D_2 \\ \cdots & \cdots & \cdots & \cdots & \cdots & \cdots \\ \cdots & \cdots & \cdots & \cdots & \cdots & \cdots \\ A_1, D_{n-1} & A_2, D_{n-1} & \cdots & \cdots & A_{n-1}, D_{n-1} & A_{n-1}, D_{n-1} \\ A_1, D_n & A_2, D_n & \cdots & \cdots & A_{n-1}, D_n & A_n, D_n \end{bmatrix} \begin{matrix} Dimention_1 \\ Dimention_2 \\ \\ \\ \\ Dimention_{n-1} \\ Dimention_n \end{matrix}$$

FIGURE 11.15 Current view of worker ants to temp_xs.

$$\text{r} = \begin{bmatrix} \text{Ant}_1 & \text{Ant}_2 & \cdots & \cdots & \text{Ant}_{n-1} & \text{Ant}_n \\ 1 & 1 & \cdots & \cdots & 1 & 1 \\ 1 & 1 & \cdots & \cdots & 1 & 1 \\ \cdots & \cdots & \cdots & \cdots & \cdots & \cdots \\ \cdots & \cdots & \cdots & \cdots & \cdots & \cdots \\ 1 & 1 & \cdots & \cdots & 1 & 1 \\ 1 & 1 & \cdots & \cdots & 1 & 1 \end{bmatrix} \begin{matrix} Dimention_1 \\ Dimention_2 \\ \\ \\ \\ Dimention_{n-1} \\ Dimention_n \end{matrix}$$

FIGURE 11.16 Current view of the (10,30) matrix of ones.

the non-vectorized algorithm took around 6 hours to execute the equation, but the vectorized algorithm took only 1.5 minutes. Moreover, for CEC06, the former took around 3 hours, whereas the latter took 30 seconds, which leads to the conclusion that the speed of the algorithm was mainly from iteration, which can be improved by vectorization. All the comparisons of ANA and VANA stated in Section 11.5 are shown in Figures 11.21–11.24 and the process of VANA is illustrated in Figure 11.19.

$$
r = \begin{bmatrix} \mathrm{Ant}_1 & Ant_2 & \cdots & \cdots & Ant_{n-1} & Ant_n \\ [-1,1] & [-1,1] & \cdots & \cdots & [-1,1] & [-1,1] \\ & [-1,1] & \cdots & \cdots & [-1,1] & [-1,1] \\ \cdots & \cdots & \cdots & \cdots & \cdots & \cdots \\ \cdots & \cdots & \cdots & \cdots & \cdots & \cdots \\ [-1,1] & [-1,1] & \cdots & \cdots & [-1,1] & [-1,1] \\ [-1,1] & [-1,1] & \cdots & \cdots & [-1,1] & [-1,1] \end{bmatrix} \begin{matrix} Dimention_1 \\ Dimention_2 \\ \\ \\ \\ Dimention_{n-1} \\ Dimention_n \end{matrix}
$$

FIGURE 11.17 Current view of random walk [-1,1] with the shape of (10,30).

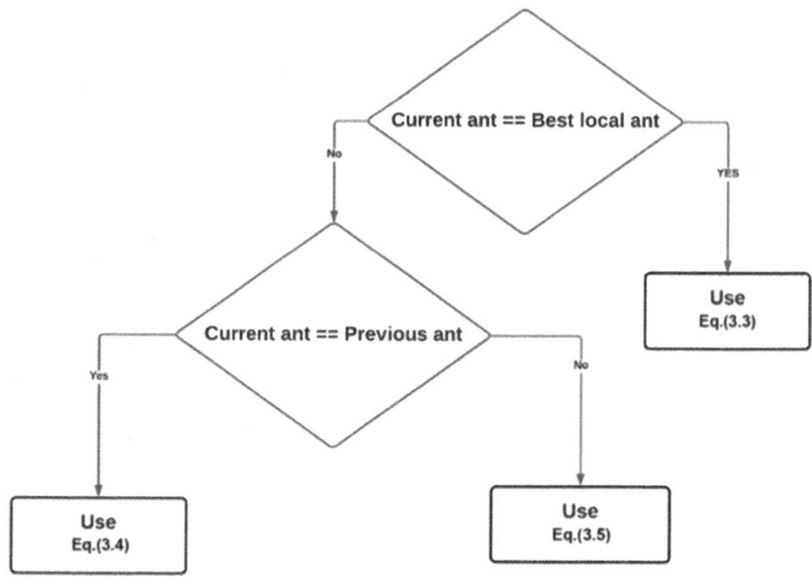

FIGURE 11.18 Three conditions depending on the result.

11.6. QUALITATIVE ANALYSIS BETWEEN VECTORIZED AND NON-VECTORIZED TECHNIQUES

Following the implementation of the vectorized technique on the metaheuristic algorithm and neglecting the iterative approach, as illustrated in the flowchart in Figure 11.21, the speed is compared between the vectorized and non-vectorized approaches. Table 11.5 represents the speed improvement between the vectorized and non-vectorized approaches. Additionally, ANA was vectorized using Python 3.10, and all benchmark test functions were converted to vectorized Python versions in order to improve the method's speed performance. A personal computer with the

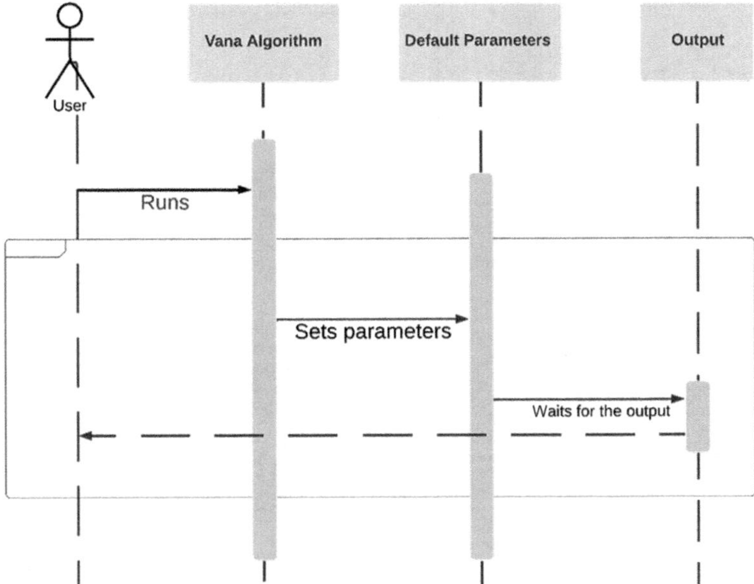

FIGURE 11.19 Sequence diagram of VANA.

for *(continue till number of turns reach limit)* **do**
  worker-ants = np. random.uniform(lb, ub,(dim,no-of-agents))
  $\text{previous}_x = np.copy(workerants)$
  Initalize best-ant-index, best-ant-fitness = to empty array
  for *(continue till number of iterations reach limit)* **do**
    rate-of-change = create (dims,agents) size of the matrix of zeros
    worker-ant-fitness = find the fitness of worker ant
    previous-fitness = find the fitness of the previous ant
    best-ant-index, best-ant-fitness = stores index of best fitness and stores best ant fitness
    best-ant = store best ant by ant index
    distance-from-best-ant = calculate the distance of all ants from a best ant
    temp-xs = store worker ant in a temporary ant
    r = simple random walk between [-1,1]
    if $(x_{t,i} == X_{t,ibest})$ **then**
      calculate $\Delta x_{t,i+1}$ using Equation (3.3)
    else if $(x_{t,i} == X_{t,iprevious})$ **then**
      calculate $\Delta x_{t,i+1}$ using equation (3.4)
    else
      calculate T using equation (3.6)
      calculate $T_{previous}$ using equation (3.7)
      calculate dw using equation (3.5) (for minimization)
      calculate $\Delta x_{t,i+1}$ using equation (3.2)
  calculate $x_{t,i}$ using equation (3.1)
  temp-fitness = store fitness ant to temporary fitness **if** $(x_{t,i+1}\ fitness < X_{t,ifitness}$ *(for minimization)* **then**
    move and accept $x_{t,i}$ assigned to $X_{t,iprevious}$
  else
    maintain the current position

FIGURE 11.20 VANA pseudocode.

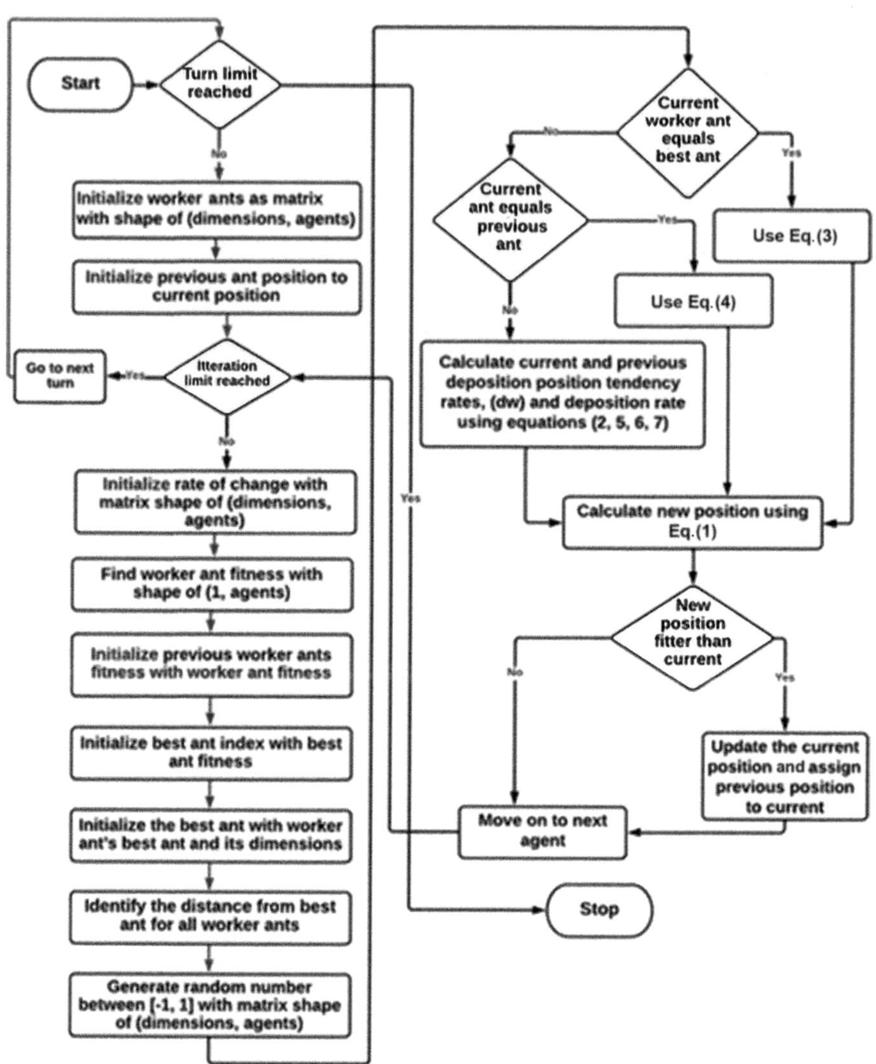

FIGURE 11.21 Flowchart of the VANA.

specifications of Core i7 5500U 2.40 GHz, RAM 16 GB DDR3, and GPU Intel(R) HD Graphics 5500 was used to implement the algorithms. One trade off of the vectorization technique is that it requires a robust GPU. The results can vary between PCs because computers with higher specs can provide faster results.

TABLE 11.5
Time in seconds for F1–F18 test functions

	Execution Time (seconds)		
Test Function	**ANA**	**VANA**	**Improvement (%)**
F1	32	9.5	70.25
F2	57	9.7	82.9
F3	52	9.6	81.35
F4	28	25.9	7.32
F5	26	12.4	52.27
F7	26	11	57.15
F9	**106**	**11.1**	89.6
F10	16	11.7	26.56
F11	**168**	**12**	92.83
F12	**140**	**18**	87.1
F13	6	4.9	18.17
F14	39	6.7	82.77
F15	**151**	**11.6**	92.27
F16	79	15.5	80.38
F17	**81**	**15**	81.39
F18	**96**	**17.5**	81.73

TABLE 11.6
Time in seconds for CEC01–CEC10 test functions

	Execution Time (seconds)		
Test Function	**ANA**	**VANA**	**Improvement (%)**
CEC01	**23705**	**94**	99.60
CEC02	58	18.4	68.13
CEC03	511	458	10.36
CEC04	208	12.9	93.77
CEC05	337	12.7	96.20
CEC06	**10606**	**22.9**	99.78
CEC07	341	22.9	93.27
CEC08	266	17.8	93.30
CEC09	148	12.9	91.24
CEC10	17	12.5	26.29

11.7 ANA VS. VANA SPEED COMPARISON

COMPARISONS OF ANA AND VANA ARE SHOWN IN FIGURES 11.22–11.25.

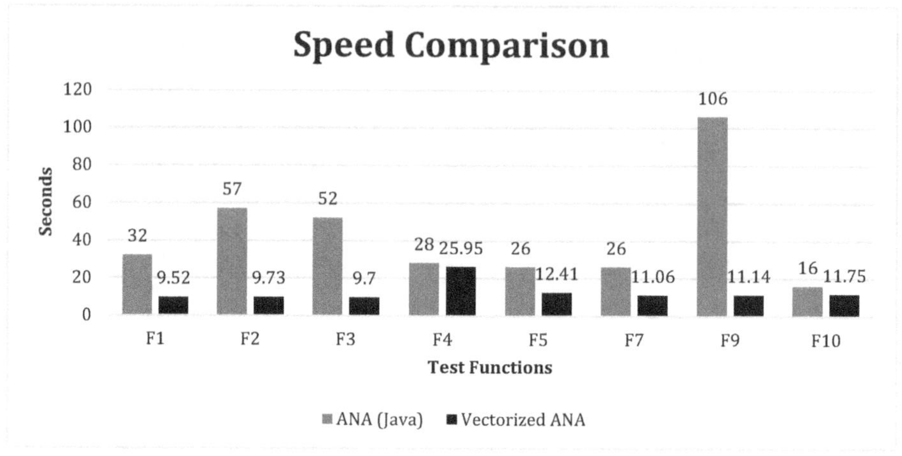

FIGURE 11.22 Comparison of VANA and ANA from F1–F10 in seconds.

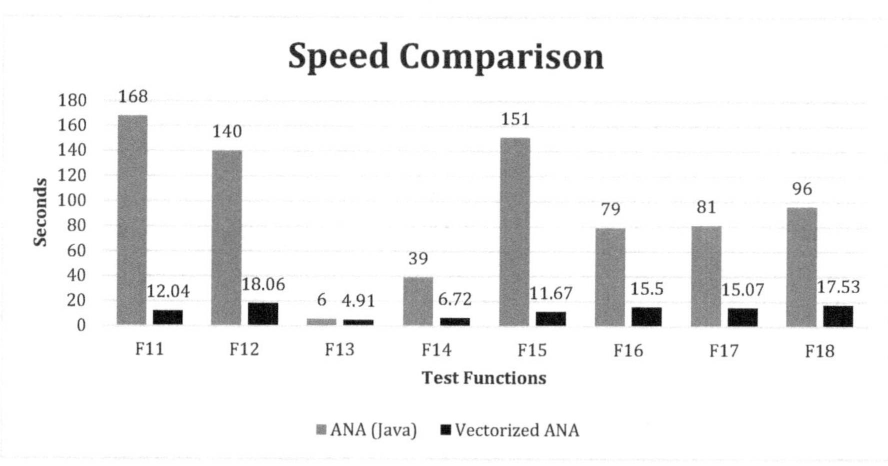

FIGURE 11.23 Comparison of ANA and VANA from F11–F18 in seconds.

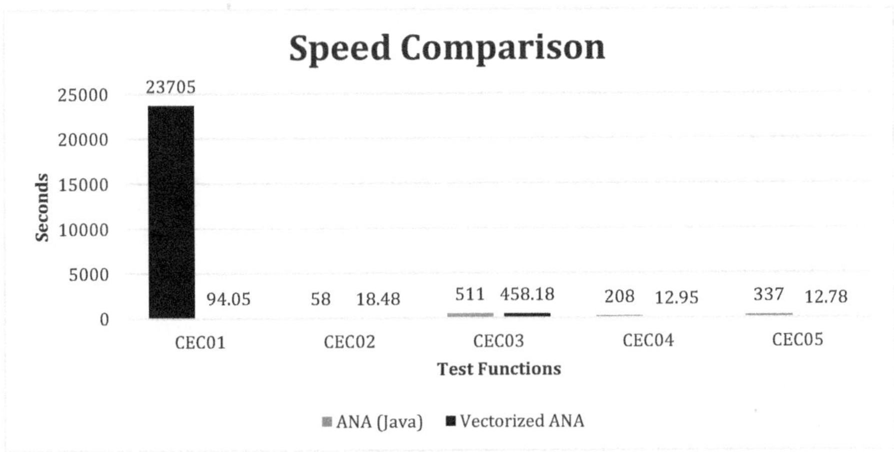

FIGURE 11.24 Comparison of ANA and VANA from CEC01–CEC05 in seconds.

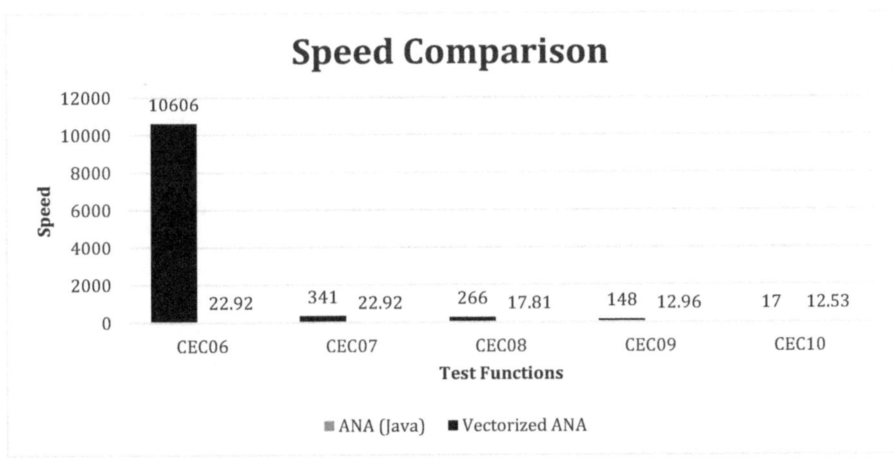

FIGURE 11.25 Comparison of ANA and VANA frosm CEC06–CEC10 in seconds.

11.8 CONCLUSION

This chapter introduced a new method for using optimization techniques to improve the algorithm's speed and performance, since the metaheuristic algorithm uses test functions that perform slowly in execution time. The vectorized method improved the speed of the algorithm. Vectorization is a process of executing multiple statements at a time, which enhances the performance speed of the algorithm. This method was used straightforwardly to change the algorithm to vectorization to implement slow test functions, reduce processing time and processing load, and reduce delay. Overall, the vectorization of algorithms is strong and well suited for metaheuristic algorithms to solve the problems of optimization in a shorter period. Vectorization can be applied to other algorithms, especially those with real-world applications that require rapid speed.

REFERENCES

[1] C. Bauckhage, *NumPy / SciPy recipes for image processing: Avoiding for loops over pixel coordinates,* 2018.

[2] J. Patterson and A. Gibson, *Deep learning: A practitioner's approach.* O'Reilly Media, Inc., 2017.

[3] S. V. D. Walt, S. C. Colbert, and G. Varoquaux, "The NumPy array: A structure for efficient numerical computation," *Computing in Science & Engineering,* vol. 13, no. 2, pp. 22–30, 2011.

[4] C. R. Harris *et al.*, "Array programming with NumPy," *Nature,* vol. 585, no. 7825, pp. 357–362, 2020.

[5] S. Bandaru and K. Deb, "Metaheuristic techniques," in *Decision sciences*: CRC Press, 2016, pp. 709–766.

[6] B. A. Hassan and T. A. Rashid, "A multidisciplinary ensemble algorithm for clustering heterogeneous datasets," *Neural Computing and Applications,* vol. 33, no. 17, pp. 10987–11010, 2021.

[7] B. A. Hassan, T. A. Rashid, and H. K. Hamarashid, "A novel cluster detection of COVID-19 patients and medical disease conditions using improved evolutionary clustering algorithm star," *Computers in Biology and Medicine,* vol. 138, p. 104866, 2021.

[8] B. A. Hassan, T. A. Rashid, and S. Mirjalili, "Formal context reduction in deriving concept hierarchies from corpora using adaptive evolutionary clustering algorithm star," *Complex & Intelligent Systems,* vol. 7, no. 5, pp. 2383–2398, 2021.

[9] B. A. Hassan, T. A. Rashid, and S. Mirjalili, "Performance evaluation results of evolutionary clustering algorithm star for clustering heterogeneous datasets," *Data in Brief,* vol. 36, p. 107044, 2021.

[10] B. A. Hassan, "CSCF: A chaotic sine cosine firefly algorithm for practical application problems," *Neural Computing and Applications,* vol. 33, no. 12, pp. 7011–7030, 2021.

[11] S. M. Qader, B. A. Hassan, and T. A. Rashid, "An improved deep convolutional neural network by using hybrid optimization algorithms to detect and classify brain tumor using augmented MRI images," *Multimedia Tools and Applications,* vol. 81, no. 30, pp. 44059–44086, 2022.

[12] B. B. Maaroof *et al.*, "Current studies and applications of shuffled Frog Leaping Algorithm: A review," *Archives of Computational Methods in Engineering,* vol. 29, no. 5, pp. 3459–3474, 2022.

[13] M. T. Abdulkhaleq *et al.*, "Harmony search: Current studies and uses on healthcare systems," *Artificial Intelligence in Medicine,* vol. 131, p. 102348, 2022.

[14] M. Yashar and Y. Hashim, "Design and implementation of Arduino controlled Mecanum wheel robot car," in *2024 21st International Multi-Conference on Systems, Signals & Devices (SSD)*, 2024, pp. 474–479: IEEE.

[15] H. M. Ali, M. Y. Hamza, and T. A. Rashid, "A comprehensive study on automated testing with the software lifecycle," *Journal of Duhok University,* vol. 26, no. 2, pp. 613–620, 12/24 2023.

[16] S. K. Omer *et al.*, "Comparative analysis of encapsulation in Java and Python: Syntax and implementation differences," *Journal of Duhok University,* vol. 26, no. 2, pp. 379–389, 2023.

[17] H. M. Ali, M. Y. Hamza, and T. A. Rashid, "Exploring polymorphism: Flexibility and code reusability in object-oriented programming," *Passer Journal of Basic and Applied Sciences,* vol. 6, no. Special Issue, pp. 502–512, 2024.
[18] S. D. Mirkhan, S. K. Omer, T. A. Rashid, H. M. Ali, M. Y. Hamza, and P. Nedunchezhian, "Effective delegation and leadership in software management," *Journal of Duhok University,* vol. 26, no. 2, pp. 407–415, 2023.
[19] D. N. Hama Rashid, T. A. Rashid, and S. Mirjalili, "ANA: Ant nesting algorithm for optimizing real-world problems," *Mathematics,* vol. 9, no. 23, p. 3111, 2021.

12 MOANA
Multi-objective Ant Nesting Algorithm

Yasameen Sami, Tarik A. Rashid and Aso M. Aladdin

12.1 INTRODUCTION

The field of multiple-criteria decision-making encompasses Multi-Objective Optimization Problems (MOPs), also known as multi-objective programming, multi-criteria optimization, and Pareto optimization. MOPs lie under the umbrella of multi-objective programming. Many real-world optimization challenges include the simultaneous optimization of many elements or goals. These issues may be redefined as MOPs to illustrate their complexity. The objective functions of the issues may conflict with one another (Deb, et al., 2002). For instance, the goals are to achieve maximum system performance while simultaneously achieving a minimum total system cost. As a result of the competing objective functions (performance and cost), the optimal solution includes a set trade-off between the objectives. Taking this into consideration, the value of one objective is reduced, while the value of another target is increased in exchange. Therefore, choosing a single optimal solution might be challenging for a decision-maker. In other words, it becomes difficult to solve the issue since there are many trade-offs and optimal solutions, but only one will be implemented. As a direct result of this, heuristic and metaheuristic techniques, including Multi-objective Evolutionary Algorithms (MOEAs), have arisen as one of the most popular strategies for addressing MOPs. However, MOEAs consist of three separate approaches: Dominance-Based, Decomposition-Based, and Indicator-Based Evolutionary Algorithms (IBEAs) (Brockhoff, et al., 2008) (Deb & Jain, 2014).

Dominance-based approaches, like the Non-Dominated Sorting Genetic Algorithm (NSGA) and its iterations, such as NSGA-II and NSGA-III, as well as Multi-objective Particle Swarm Optimization (MOPSO) and the Strength Pareto Evolutionary Algorithm, are built on the basis of Pareto dominance. This idea includes contrasting several solutions in terms of the values their objectives aim to achieve and determining if one solution dominates another, which would indicate that it excels in at least one target without compromising its performance in any of the others. This idea of dominance is used by these algorithms to direct the process of searching toward the Pareto front.

Decomposition-based approaches involve breaking down the multi-objective problem into multiple individual single-objective subproblems and solving them concurrently or in succession. Instances of such approaches encompass the weighted sum

DOI: 10.1201/9781003601555-12

technique, goal programming, and MOEAs grounded in decomposition, like MOEA/D.

Finally, IBEAs have garnered a great deal of attention because of the strong theoretical foundation and background they provide (Brockhoff, Friedrich & Neumann, 2008). One of the most significant aspects of MOEAs is the selection guide method. The method for selecting individuals who are going to stay is also in charge of the solution set's diversity and convergence. IBEAs prove valuable within the framework of multi-objective evolutionary optimization as they assess both the diversity and convergence of unmatched solutions in the objective space. Since IBEA indicator functions inherently address the diversity challenge within their population solutions, there is no requirement for any form of diversity-preserving approach.

An example would be the hypervolume indicator-based algorithm introduced by Zitzler, Deb, and Thiele (2000). As a consequence, various IBEAs have been developed employing preference-based information and local search optimizers (Chugh, et al., 2015). IBEA's main drawback is that it takes longer to calculate hypervolumes while solving many objectives' problems.

Two aspects must be taken into consideration to successfully solve MOPs. First, the accuracy with which estimated Pareto-optimal solutions may be found. The second consideration is the diversity of predicted Pareto-optimal solutions. These two, often competing criteria, need to be taken into consideration by multi-objective metaheuristics. In most cases, they begin with an initial solution that is chosen randomly and continue to refine the solution through iterations until the end condition is met (Abdullah, et al., 2023).

12.2 RELATED WORK

This section covers Multi-Objective Optimization (MOO) and its approach in detail. In addition, it examines convergence and diversity, two key components of MOO. The chapter also presents the most important MOO principles, methodologies, and improvements to offer a summary. It highlights major works and their contributions to provide a full grasp of MOO's cutting-edge features. This chapter seeks to summarize MOO's recent developments.

12.2.1 Brief History of Multi-objective Optimization

Zakas, Betts, and Bellman were early MOO researchers in the 1960s. Steuer introduced the Pareto efficient border in the 1970s, when Pareto dominance gained popularity. Evolutionary algorithms gained popularity in the 1980s with Goldberg's Genetic Algorithm (GA) and Goldberg and Holland's NPGA. Both advances occurred in the 1980s. MOEAs were developed in the 1990s by Deb, Coello Coello, and Zitzler. These academics greatly advanced MOEAs. In the early 2000s, biodiversity conservation and metrics-based initiatives were priorities. The combination of machine learning, hybrid methods, and metaheuristic algorithms is among recent improvements. The history of MOO shows the development of theoretical ideas, computational advances, and unique optimization paradigms to solve problems with many competing goals (Olivier, 2004).

Several researchers within the field are responsible for the creation of Multi-objective GAs. Goldberg, who introduced the concept of Pareto-based optimization using GAs in the late 1980s, was one of the early explorers in this field (Murtada & Ishibuchi, 1995). MOPSO was proposed by Coello and Lechuge in 2002 and motivated by the choreography of a bird flock (Dasheng, 2008). MOBA was developed by Yang in 2011 and motivated by using the echolocation behavior of bats and their hunting strategies to solve optimization problems (Yang, 2011).

The MOABC algorithm was developed by various researchers, and various research groups have proposed modifications and enhancements inspired by honeybees' foraging behavior, in which they investigate the environment, communicate their findings, and make collective decisions (Akbari & Hedaytzadeh, et al., 2012). MOFA was developed by a group of researchers led by Mirjalili and motivated by the blinking behavior of fireflies. It initializes a population of fireflies, each representing a solution. Multi-objective functions determine firefly fitness. Based on fitness, fireflies migrate toward brighter ones (Yang, et al., 2012). MOCSA was introduced by (Yang & Deb, 2013). It takes inspiration from the brood parasitism behaviors shown by cuckoos and seeks to replicate the natural sequence of nest selection, egg laying, and reproductive processes.

The MODF Algorithm was developed by Mirjalili simultaneously with a single objective and derived from the observable patterns of fixed and changing group behavior shown by dragonflies (Mirjalili, 2015). MOGWO was proposed by Mirjalili et al. (2016). It's based on how grey wolves search and hunt and is made up of three steps: initialization, hunting, and updating. MOSSA was developed by Mirjalili simultaneously with a single objective. It was inspired by the navigational and feeding habits of salp colonies in the ocean (Mirjalili, et al., 2017).

Both Mirjalili and Lewis were responsible for the creation of MOWOA. The hunting behavior of humpback whales served as inspiration for this strategy, which is separated into three stages: the first stage is referred to as "encircling prey", the second stage is termed "bubble-net attacking also known as the hunting process", and the third stage is referred to as "search for prey" (Mirjalili & Siddiqi, et al., 2019). MOACO was developed by a team of researchers led by Marco Dorigo. The concept is derived from foraging activity shown by social ants (Liu & Liu, et al., 2019). MOLPBA was developed by Rahman et al. (2022). The inspiration for this method was student's transition from secondary education to college, that is, high school graduates begin the college transfer process.

MOFDO was developed by Jaza (Abdullah, et al., 2023). It was motivated by the bee-swarming reproductive process. The process starts with an initial population of scout populations that serve as possible options inside the search area. GMOCSO was developed by Ahmed & Rashid et al. (2023). It is an optimization technique that draws inspiration from nature and is modeled after the hunting behavior of cats. Both internal and exterior populations are accounted for by the algorithm.

12.2.2 Multi-objective Optimization Approach

MOO simplifies complex problems using two non-mathematical methods: Pareto and scalarization. The Pareto approach generates dominated and non-dominated solutions.

A constantly updated algorithm generates these solutions. However, scalarization uses weighting variables to combine many objective functions into one solution. Scalarization uses equal, rank-order centroid, and rank-sum weights. Next, the Pareto technique's solution is used to define MOO's different aspects and provide a compromise solution called a Pareto-optimal front. The scalarization approach yields an effectiveness indicator that integrates a scalar function into the fitness function. Gunantara (2018) stated that each solution is a Pareto-optimal front.

12.2.3 Enhancing Multi-objective Algorithms

Addressing convergence and diversity in multi-objective algorithms helps solve optimization problems with numerous conflicting objectives. Convergence is an algorithm's ability to find solutions around the true Pareto front (the least dominated possibilities). Diversifying solutions over the Pareto front guarantees a range of trade-offs. These two essential aspects of optimization have been addressed by many methods and metrics by researchers and practitioners. Using a range of problem-solving methods, these strategies guide optimization algorithms toward well-converged and diverse solutions. Techniques like Pareto dominance are common. Solution performance on many objectives is compared in this manner. Using random initialization or space-filling designs for population initialization helps diversify solution space search. Population initialization methods can do this. Tournament and elitism (archiving) selection techniques maintain convergence and variety, preventing early convergence to specific regions.

Crossover and mutation operator designs are also crucial. Operating systems that discourage overexploitation let the algorithm explore more trade-offs. Ecology-based niching creates separate solution niches, thereby avoiding Pareto front overconcentration. By dynamically adjusting algorithm parameters, adaptive parameter control techniques make it simpler to balance the exploration and exploitation stages of optimization. Several more methods are used to quantify convergence and diversity: the Generational Distance (GD), Inverse GD (IGD), and the Hyper-volume metric. The Hyper-volume metric measures the volume of the dominated space, whereas GD and IGD measure the algorithmically generated solutions' Pareto front fit. Spacing and spread quantify the solution distribution, and the epsilon indicator measures how closely approximated responses match ideal solutions for a reference front.

Handling convergence and diversity in multi-objective algorithms is difficult and involves several methods and measurements. By strategizing using multiple methods, researchers and practitioners maximize across numerous aims. This gives scientists significant tools to handle complex optimization problems in many fields (Dasheng, 2008).

12.3 METHODOLOGY

The biological and social behavior of ants, which served as inspiration for the Ant Nesting Algorithm (ANA) suggested by (Hama, et al., 2021), is covered first in this section, followed by an overview of the concept of MOO. The modeling procedure of the algorithm, together with the multi-objective field, is then provided.

Ants are small insects with over 10,000 species worldwide that have long fascinated humans. Many novels and study articles have been written on their lives, from pure literary (Maeterlinck, 1939) to scientific studies (Hölldobler & Wilson, 1990, 1994). Their remarkable social structure, despite their limited individual capacities, is especially astonishing. They have inspired scientists and researchers for years. This section summarizes Leptothorax's ant nest-building behavior. Leptothorax ants build nests, as shown in Figure 12.1.

Leptothorax ant colonies make nests in flat rock fissures to protect themselves from physical and biological threats. They surround themselves with dirt or grains—earth or stone particles—to seal themselves within their niche. The wall can accommodate a queen, broods, and up to 500 worker ants. The nest is built by 2–3-mm worker ants. Ants established a Leptothorax ant nest in the lab Figure 12.2.

Leptothorax ants that work avoid constructing a roofing or flooring structure inside their nests. Their nest solely consists of a surrounding wall that encloses the queen and the various stages of eggs, larvae, and pupae. The construction of this nest is initiated when worker ants leave their cluster, pick up building material, and randomly deposit it. Subsequently, the ants clear stones from the area with the highest accumulation. The selection of a suitable location for clearing is essential for consolidating the wall. This method was inspired by the stigmergic interaction of worker

FIGURE 12.1 Shows Leptothorax ants constructing a wall to enclose the core brood cluster and the queen.

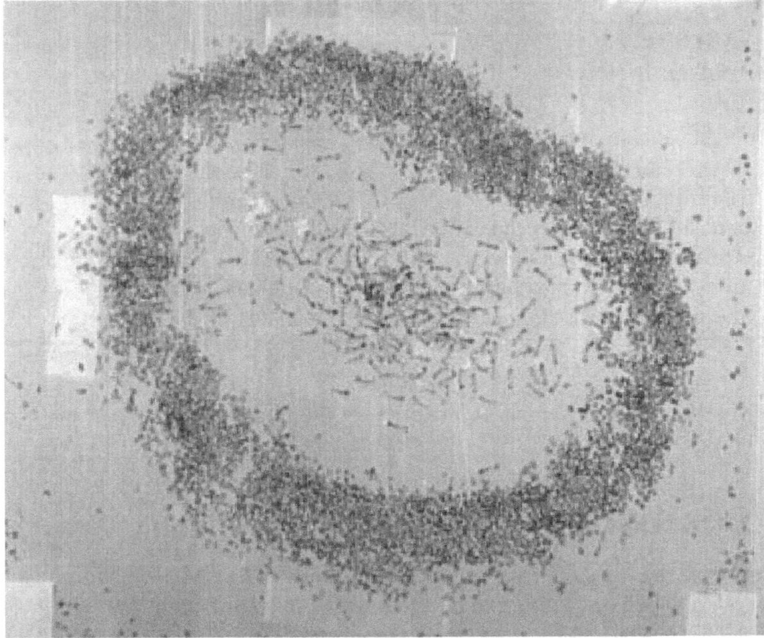

FIGURE 12.2 Laboratory Leptothorax ant nest.

ants depositing particles at a specific distance from the nest's center, guided by the presence of previous placements.

Wall formation is a result of each worker ant's individual choices. When forming a new nest, worker ants perform bulldozing activities from the region with the highest grain concentration around the queen (Hama, et al., 2021).

The choice of each worker ant to deposit grain at a particular spot around the queen ant represents a possible solution that could be exploited by an artificial search agent in an algorithmic model of the behavior of the Leptothorax ant species during nest construction. In other words, the queen ant is the target of an artificial search agent. The best location that can be taken advantage of by all worker ants is the global optimal solution. Therefore, the placement of worker ants determines the deposition. The fitness functions of the algorithm could include the influence that the deposition location has on the consolidation of the wall and the distance between individual stones. The deposition weight, abbreviated *dw*, is a variable that indicates the grain deposit decision factor for each worker ant in the algorithm. *dw* represents a random weight of worker ants that will be used to deposit grain at a certain location, and is calculated by making use of the solution and fitness information at the grain's previous and current deposition sites. The key components of the method are outlined in Table 12.1.

The prior deposition position of worker ants, denoted by $X_{t,i\ previous}$, is used to represent the fixed stones and nestmates that the worker ants encounter as they wander randomly around the nest. Therefore, the algorithm's stationary stone and/or

TABLE 12.1
ANA entities

Nature	Algorithm
Worker ant	Search agent
Deposition position	Potential solution
Deposition position specification	Fitness function
Worker ant's decision factor	Deposition weight
Fittest deposition position	Optimum solution
Stationary stone and/or nestmate	Previous deposition position

nestmate is the previous deposition location of the worker ant that is now being used (Hama, et al., 2021).

12.4 MODELING OF MOANA

In this section, we discuss fundamental concepts and essential terminology related to MOO. MOANA is explained in depth in terms of mathematics and programming. The amount of information is set up in a way that makes it easy for other experts to perform the same task.

12.4.1 Pareto-Optimal Solutions Set

In a mathematical sense, MOPs may be expressed in the following way:

$$\text{Minimize: } F(\vec{x}) = \{f_1(\vec{x}), f_2(\vec{x}), \ldots, f_n(\vec{x})\} \tag{1}$$

Subject to:

$$g_{i(\vec{x})} \leq 0, \qquad i = 1, 2, \ldots, m$$

$$h_{i(\vec{x})} = 0, \qquad i = 1, 2, \ldots, p$$

In this context, n is the number of separate goals, whereas g and h each refer to a different set of constraints. m represents an inequality constraint, whereas p represents an equality constraint. Due to the multi-objective character of the issue, as well as the fact that there are competing goals inside the same optimization problem, this sort of problem cannot be optimally solved by the traditional single-objective method. The optimum trade-offs between goals are represented by a collection of optimal solutions called the Pareto-optimality solution set. Pareto optimality is briefly covered in this subsection for readability reasons. The definitions listed below may be used to describe Pareto-optimality solutions (Coello, 2009):

Definition 1: Think about the vectors (solutions) $\vec{a}$ and $\vec{b}$ within the optimization space. For $i = 1,2,\ldots,m$, $\vec{a}$ should be less than or equal to $\vec{b}$. This implies that the objectives of vector $\vec{a}$ should be smaller than the objectives of vector $\vec{b}$, with at least one instance where $\vec{a_i} < \vec{b_i}$.

Definition 2: If $\vec{a} \leq \vec{b}$, then $\vec{a}$ dominates $\vec{b}$, denoted by $\vec{a} \prec \vec{b}$.

Definition 3: When Definition 1 is not applied, it's possible that two solutions might not have dominance over each other. In this case, solutions $\vec{a}$ and $\vec{b}$ do not possess mutual dominance and are represented as $\vec{a} \nprec \vec{b}$. The collection of all solutions that are not dominated is referred to as the Pareto-optimal solution set P_s and defined as

$$P_s := \{\vec{a}, b \in A | \exists F(a) \succ F(b)\} \tag{2}$$

Definition 4: Pareto-optimal front P_f is the collection containing all object values of Pareto-optimal solutions in P_s and is represented by

$$P_f := \{F(\vec{a}) | \vec{a} \in | P_s\} \tag{3}$$

12.4.2 Mathematical Modeling of the Multi-objective Ant Nesting Algorithm

The most recent work on the single-objective ANA (Hama, et al., 2021) served as the foundation for MOANA. ANA draws inspiration from Leptothorax ants and emulates their actions as they search for suitable spots to place grains during nest construction. ANA modifies an individual's position by calculating the change in rate to the current position, as demonstrated in

$$X_{i,t+1} = X_{i,t} + \Delta X_{i,t+1} \tag{4}$$

where i denotes the present individual index, t represents the current iteration, x signifies the deposition position of the artificial worker ant, and $\Delta X_{i,t+1}$ represents the rate of change. MOANA uses the same mechanism. This method makes use of a variety of mathematical symbols; Table 12.2 provides an overview of all of these symbols. To calculate the rate of change, however, it is necessary to consider the three conditions presented as follows:

- The current placement of the worker ant's deposition corresponds to the local best deposition position of the worker ant:

$$\Delta X_{i,t+1} = r * X_{i,t} \tag{5}$$

- The current deposition position of the worker ant it is equal to the previous deposition position:

$$\Delta X_{i,t+1} = r * (X_{i,tbest} + X_{i,t}) \tag{6}$$

TABLE 12.2
Notation used in mathematics by ANA

Nature	Algorithm
t	Current iteration
i	Worker ant index
m	Iteration number
n	Total number of worker ants
$X_{i,t}$	Worker ant's current deposition position
$X_{i,tbest}$	Worker ant's local best deposition position
$X_{i,t\ previous}$	Worker ant's previous deposition position
$X_{i,tfitness}$	Worker ant's current deposition position fitness
$x_{t,i\ bestfitness}$	Worker ant's local best deposition position fitness
$x_{t,i\ previousfitness}$	Worker ant's previous deposition position fitness
$X_{i,t+1}$	Worker ant new deposition position
$\Delta X_{i,t+1}$	Worker ant new deposition position rate of change
T	Worker ant's current deposition tendency rate
$T_{previous}$	Worker ant's previous deposition tendency rate
dw	Deposition weight
r	Random number in the [-1, 1] range

- However, the rate of change is dependent on the deposition weight, also known as dw, as well as the distance that separates the deposition site of the current worker ant, $X_{i,t}$, from that of the best worker ant in the area, $X_{i,tbest}$:

$$\Delta X_{i,t+1} = dw * (X_{i,tbest} + X_{i,t}) \tag{7}$$

The concept of deposition weight (dw) relates to the mathematical representation of the random movement pattern shown by worker ants. It is dependent on the historical behavior of an artificial worker ant, denoted by $T_{previous}$, as well as its present tendency, denoted by T, to deposit grain at a certain position. The variables T and $T_{previous}$ represent the slopes of the sides in the Pythagorean theorem, pertaining to the present and prior deposition positions of the worker ant, along with the most favorable deposition position identified thus far. The difference in their fitness values corresponds to the lengths of the other sides. Thus, minimization problems can calculate dw as follows:

$$dw = r \times \left(\frac{T}{T_{previous}} \right) \tag{8}$$

for worker ants as a deposition factor for managing dw, where r is a random value in the range [-1 1]. Random numbers may be generated using a variety of methods.

Because of the superior distribution curve that it offers, Levy flight was chosen as the optimal option since it results in more consistent movement.

The ant worker's tendency rate of deposition (T) is calculated as

$$T = \sqrt{\left(x_{i,best} - x_{i,t}\right)^2 + \left(\sum x_{i,t\ best\ fitness} - \sum x_{i,t\ fitness}\right)^2} \quad (9)$$

The MOANA equation has been modified from the single-objective ANA, as well as for $T_{previous}$.The ant worker's tendency rate of deposition ($T_{privous}$) is calculated as

$$T_{previous} = \sqrt{\left(x_{i,tbest} - x_{i,t\ previous}\right)^2 + \left(\sum x_{i,t\ best\ fitness} - \sum x_{i,t\ previous\ fitness}\right)^2} \quad (10)$$

It is necessary to make slight modifications to use ANA for solving maximizing issues. The modification is to replace Eq. (5) with

$$dw = r \times \left(\frac{Tprevious}{T}\right) \quad (11)$$

given that Eq. (11) is only an inverse version of Eq. (5).

Although the algorithm structure is similar to that of an ANA with a single objective, to some extent, MOANA includes the following enhancements:

1. In the optimization process, an archive (also known as a repository) is used to save Pareto front solutions since this kind of storage has seen extensive application in relevant research (Deb & Jain, 2014). Archiving helps preserve variety within the collection of non-dominated solutions, which is a benefit of archiving. A broad variety of trade-offs and possible solutions may be investigated if several solutions are proposed. This results in a deeper comprehension of the problem's context, as well as the options that are available to address it.
2. Since there is more than one goal to consider in MOPs, the optimal solution cannot be taken at its simplest as a global guide, as could be the case in single-objective optimization. In most cases, these goals are incompatible. This means you should think more carefully before settling on a global guide. Similar to the work on MOPSO done in (Coello Coello & Lechuge, 2002), the global guide individual, or x^*, is a global best non-dominated solution chosen from the least populous area using artificial worker ants. The archive is divided into several grids (sub-hyperspheres in multidimensional problems) of equal size using a technique called the archive controller; these grids are hypercubes. Because each grid in a hypercube has a different number of solutions, the algorithm can easily identify the least densely packed region. See Figure 12.3 for an illustration of how the least densely inhabited region provides the world's finest answer.

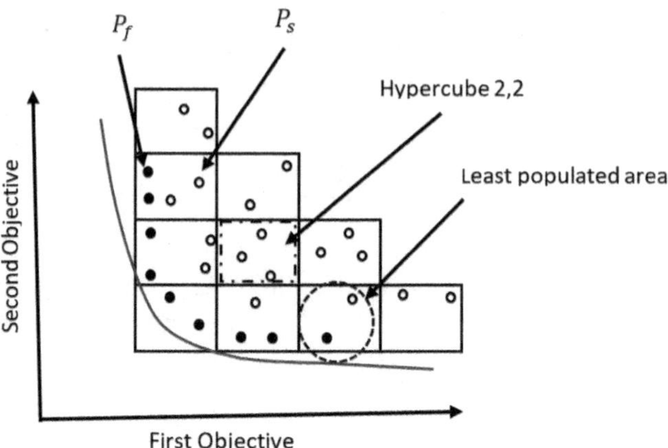

FIGURE 12.3 Illustrates the Pareto solution, Pareto front, and hypercube grids, all of which aid in choosing between global and local guides.

Figure 12.4 shows the pseudocode that demonstrates the concept of generating a hypercube grid. The grid generation function requires three parameters to be input as shown on line 1. In line 2, it gathers the fitness values of all the ants and stores them in a matrix denoted by *c*. Lines 4 and 5 determine the lowest and highest possible values of *c* along the second dimension, which is the dimension that corresponds to the fitness values. Therefore, the value of each ant's minimal fitness is in the *cmin* column, whereas the value of each ant's maximum fitness is stored in the *cmax* column. On lines 7, 8, and 9, *cd* determines the range of possible fitness values for each ant by subtracting the minimum from the maximum value for that ant's fitness. In MOO, the inflation rate can be used to balance the exploration of the Pareto front (maintaining the diversity of solutions) and the exploitation of promising regions. A higher inflation rate can encourage more diversity, whereas a lower rate can focus the algorithm on improving the solutions in the current Pareto front. A frequent strategy is to begin with an inflation rate that is slightly below 1 (e.g., 0.9) and another that is slightly above 1 (e.g., 1.1). To make this range larger, *cmin* and *cmax* are modified by multiplying it by the inflation rate and then either adding or removing the result from the values of the original *cmin* and *cmax*, depending on the equation. Because of this modification, the grid is now a little bit bigger than the actual fitness values. Following that, on line 11, the number of rows in the c array is determined and this information is used to determine the number of goals, also known as fitness values. Lines 13 and 14 show that a grid structure with the name empty grid is given to the empty grid that is built. These two fields, known as the Lower Bound (LB) and Upper Bound (UB), establish the bounds of the grid for each individual target. The Grid variable is initiated as an array of empty grid structures with the specified number of objective elements, one element for each goal. Lines 17–19 demonstrate that a linearly spaced vector *cj* is generated between the given values of $cmin(j)$ and $cmax(j)$ using grid number + 1 points that are equally spaced. *cj* denotes the limits

```
1  Generate grid inputs(ants, grid number, inflation rate)
2  Store and extract the fitness value of all ants c = [ants. fitness]
3  Calculate the minimum and maximum fitness value for each objective (column-wise) and
   store them in cmin and cmax respectively.
4          Cmin = min (c, [], 2).
5          Cmax = max (c, [], 2).
6  Calculate dc, cmin, cmax
7          dc = cmax- cmin;
8          cmin = cmin – inflation_rate *dc.
9          cmax = cmax + inflation_rate *dc.
10 Determine number of objectives
11         nObj=size(c,1)
12 Initialize Empty Grid Structure
13         empty _grid LB = [].
14         empty _grid UB = [].
15         Grid=repmat(empty_grid,nObj,1).
16                 for each objective
17                         cj = linspace(cmin(j),cmax(j), Grid_Number +1);
18                         Grid (j).LB = [-inf,cj].
19                         Grid (j).UB = [cj, +inf].
20                 end
```

FIGURE 12.4 Generation hypercube grid pseudocode.

of the grid in accordance with the jth objective. Then the jth grid's LB and UB are configured. The bottom limit has been specified as [-inf, *cj*], which indicates that it ranges from negative infinity all the way to the initial value of *cj*. The upper limit has been specified as [*cj*, +inf], which indicates that it ranges from the most recent value of *cj* to a number that is positive infinity.

After grids are generated, indexes should be initialized for each grid. The pseudocode in Figure 12.5 explains the notion of setting grid indexes for grids. It examines each fitness value in Ant. Fitness and compares it to the UB value in the delete factor structure that corresponds to that fitness value. It searches for the first grid level in which the fitness value is lower than the UB value and then saves the grid sub-index that it discovers in the Ant.GridSubIndex(j) variable, as demonstrated on Lines 5 and 6. After all the sub-indices have been determined, the code proceeds to compute the overall ant's grid index, then provides the sub-index of the first fitness result as the starting point for the ant's grid index variable, as shown on line 8. Then it proceeds to loop over all the fitness values starting at 2 to the number of objectives, and changes the ant's grid index in the manner that is shown on lines 10–12. After 1 is subtracted from the ant's grid index, that result is multiplied by the entire number of grid levels, and then the sub-index of the most recent fitness value is added.

The same method of creating a hypercube grid has been used to split the search area into cells of equivalent size, and the best individual solution has been chosen to serve as the local guide inside each cell.

```
1   Find grid index (ant, delete Factor)
2   Determine the number of fitness values.
3   Initialize grid sub-index to zeros.
4         for number of objectives
5               Ant Grid Sub Index(j)=
6               find (Ant.Fitness(j)<Delete_Factor(j). UB,1, 'first').
7         end
8               Ant.GridIndex=Ant.GridSubIndex(1).
9         for number of objectives
10              Ant.GridIndex= Ant.GridIndex-1.
11              Ant.GridIndex= nGrid*Ant.GridIndex.
12              Ant.GridIndex= Ant.GridIndex+ Ant.GridSubIndex(j).
13        end
```

FIGURE 12.5 Pseudocode for assigning the index to a hypercube grid.

Due to the varying number of solutions in each grid of a hypercube, the algorithm is capable of effectively discerning the area with the lowest density. The process of picking the best ant is shown as pseudocode in Figure 12.6. This function goes over each element in the archive array and gets the grid index field, then puts it in the array grid index *GI*, as demonstrated on line 3. It finds the one-of-a-kind grid indices included within the *GI* array and saves them in the occupied cell OC. These indices are representative of the occupied cells found inside the archive on line 5. The number of ant individuals that are stored in each occupied cell is recorded in an array called N, which is initialized with zeros on lines 7. On line 8, for the loop covering the index of each occupied cell, inside the loop, the number of ants in *GI* that correspond to the current occupied cell index $\mathrm{OC}(\mathrm{K})$ are counted and the result of that count is saved in $N(K)$. This provides the number of particles that can be found in every occupied cell. As shown on line 12, a selection probability, denoted by P, is determined for each occupied cell using the number of particles currently housed in that cell as the basis. It is possible to alter how much the total number of individuals has an effect by adjusting the value of the parameter known as best ant selection factor. Cells that contain a greater number of particles do better in environments with lower levels of the best ant selection factor. Line 13 adjusts the selection probabilities such that their combined total is 1, and normalizes them. As indicated on line 15 and in Figure 12.6, selection is performed using a roulette wheel to choose an occupied cell depending on the probability associated with that pick. As illustrated in Figure 12.7, Roulette Wheel Selection is a method that is often used in GAs and other types of optimization algorithms for the purpose of picking individuals from a population depending on the fitness values they possess. Roulette Wheel Selection helps create a balance between exploring the solution space and exploiting potential regions by preferring fitter people, while still allowing less fit individuals to be picked. This favoring of fitter individuals is accomplished by allowing less fit individuals to be selected. The maintenance of this equilibrium is essential to the viability of optimization techniques.

Following line 17 in Figure 12.5, the index of the cell that was chosen by the Roulette Wheel Selection method is retrieved and line 19 identifies the indices of

```
1   Select best ant (archive, best ant selection factor)

2   Grid Index of All Repository Members
3           GI= [Archive.GridIndex];

4   Occupied Cells
5           OC=unique (GI).

6   Number of Particles in occupied cells
7           N=zeros (size (OC)).
8               for K=1 :numel(OC)
9                       N(K)=numel (find (GI==OC(K))).
10              end

11  Selection Probabilities
12          P=exp(-Best_Ant_Selection_factor*N).
13          P=P/sum(P);

14  Selected Cell Index
15          sci=Roulette Wheel Selection(P).

16  Selected Cell
17          sc=OC (sci).

18  Selected Cell Members
19          SCM=find (GI==sc).

20  Selected Member Index
21          smi=randi ([1 numel (SCM)]).

22  Selected Member
23          sm=SCM (smi).

24  Leader
25          Best_Ant=Archive(sm).
```

FIGURE 12.6 Pseudocode for selecting the best ant.

```
1 Initialize variable as fitness score for sum of all fitness value.
2 Generate random numbers between 0 and sum of fitness value.
3 Initialize variable as selected individual.
4 For each Individual in population.
5     Calculate each fitness value and store in fitness score.
6             if fitness score>=random number
7                     Selected individual = individual
8                     break
```

FIGURE 12.7 Pseudocode for the Roulette Wheel Selection method for selecting individuals.s

every element in GI that is associated with the cell that has been chosen. On line 21, an index is chosen at random from those contained within the selected cell's elements. Finally, on lines 23 and 25, the index of the chosen member inside the selected cell is retrieved and then the best ant is returned by indexing the archive with the selected member index *sm*.

After polynomial mutation is applied to the Pareto front solutions, the non-dominated solution is then saved to the archive (Liagkouras & Metaxiotis, 2013). Polynomial mutation is an example of a variation operator that has been used in MOEAs and its definition is as follows (Li, et al., 2016):

$$S_i = \left(x_{1,}x_{2,} \ldots, x_n\right)$$

$$S_{Ni}\left(x_j\right) = S_i\left(x_j\right) + \alpha\,\beta_{max}\left(x_j\right), \quad i = 1,2,\ldots,NP, \quad j = 1,2,\ldots,n \tag{12}$$

$$\alpha = \begin{Bmatrix} (2\upsilon)^{\frac{1}{(q+1)}} - 1, & \upsilon < 0.5 \\ 1 - \left(2(1-\upsilon)\right)^{\frac{1}{(q+1)}}, & otherwise \end{Bmatrix} \tag{13}$$

$$\beta_{max}\left(x_j\right) = Max\left[S_i\left(x_j\right) - l_j,\ u_j - S_i\left(x_j\right)\right],\ i = 1,2,\ldots,NP, j = 1,2,\ldots n \tag{14}$$

where S_{Ni} represents a newly proposed solution and $S_i\left(x_j\right)$ denotes the existing solution. The symbol $\beta_{max}\left(x_j\right)$ represents the maximum perturbation acceptable between the initial solution and the modified solution. In this scenario, *NP* signifies the population's magnitude, q represents a positive real value, v is a random number distributed evenly within the specified range [0, 1], l is the lower limit of the decision variable x, u is the upper limit of the decision variable x, and n is the count of decision variables, also referred to as the problem's dimensions. The perturbance factor, which is often referred to as α, is used to determine the mutation strength and controls the degree of mutation. Polynomial mutation is a technique that makes minute adjustments to solutions, which helps solutions avoid prematurely converging toward a specific region within the solution space. Polynomial mutation helps in the discovery of a broad variety of trade-off solutions along the Pareto front. This is accomplished by maintaining the diversity of the population. The pseudocode that is shown in Figure 12.8 provides an overview of this technique.

The goal of picking the global guide from the region with the fewest people is to ensure that the Pareto front solutions have enough variety to meet the needs of all individuals. Because of this, those in charge of making decisions have a wider variety of options (solutions) to take into consideration. In spite of this, the size of the archive is restricted. If an additional non-dominated solution is discovered when the archive is already at its maximum capacity, the controller of the archive removes the least desirable solution from the grid that contains the most populated grid. With this step, we want to provide room for the introduction of a new solution that is not dominated

```
1   Mutation operation inputs (x_j, q, l_j, u_j)
2   For each vector in an ant individual.
3   Specify the mutation rage using equation ( 14).
4.  Calculate the maximum perturbation acceptable for selected decision variable β_max
    (x_j) = Max [S_i (x_j) – l_j]
5        if β_max (x_j) < l_j
6             β_max (x_j) = l_j
7        end
8   Calculate the maximum perturbation acceptable for selected decision variable β_max
    (x_j) = Max [u_i – S_i (x_j)]
9        if β_max (x_j) > u_j
10            β_max (x_j) = u_j
11       End
12  Generate random number υ
13  Apply mutation for each vector using equation ( 13)
14       if υ < 0.5
15            Calculate perturbance factor α = (2υ)^(1/(q+1)) – 1
16            Calculate new solution using equation ( 12).
17       else
18            Calculate perturbance factor α = 1 – (2(1 – υ))^(1/(q+1))
19            Calculate new solution using equation ( 12).
20       end
```

FIGURE 12.8 Polynomial mutation pseudocode.

by the existing one. Therefore, the newly found solution may be implemented as long as it is superior to the solution that has the dubious distinction of being the worst in the archive.

12.4.3 Pareto-Optimal Solutions Set

The first step of MOANA is the random distribution of artificial worker ants over the exploration space. This procedure is outlined in both the pseudocode depicted in Figure 12.9 and the flowchart illustrated in Figure 12.10.

The procedure determines whether one member of a pair of individuals in the population dominates the other member of that pair as demonstrated on line 3. After that, hypercube grids are built and an archive with a predetermined size that stores non-dominated solutions is created. The primary iteration of the algorithm starts on line 6 and continues either until a predefined number of times have passed or a certain set of criteria have been satisfied.

In accordance with the count of individuals, the actions taken by each individual (artificial worker ant) from line 7 to line 34 is repeated in line 6 for each and every search. To calculate the rate of change, these steps include determining the global

```
1  Initialize the artificial worker ant X_{i,t} I =1,2,..., n, and t=1,2,... .,m
2  Check ant workers dominance.
3              if all (x. fitness <=y. fitness) && any (x. fitness < y. fitness)
4  Creating archive for nondominated solutions
5  Generate Hypercube grid.
6  While (t) iteration limit not reached (m)
7           for each artificial worker ant X_{i,t}
8                   find best artificial worker ant X*_{i,t}
9                   generating random walk r in [-1, 1] rang
10                  if (X_{i,t} == X_{i,tbest})
11                     calculate ΔX_{i,t+1} using equation ( 5)
12                  else if (X_{i,t} == X_{i,tprevious})
13                     calculate ΔX_{i,t+1} using equation (6)
14                  else
15                     calculate T using equation (9)
16                     calculate T_previous using equation ( .10)
17                     calculate dw using equation ( 8) (for minimization)
18                     calculate ΔX_{i,t+1} using equation ( .7)
19                  end if
20                  calculate X_{i,t+1} using equation ( 4)
21                  if (X_{i,t+1} fitness dominates ΔX_{i,t} fitness)
22                     move accepted and ΔX_{i,t} assigned to X_{i,tprevious}
23                  else
24                       calculate ΔX_{i,t+1} with previous rate of change
25                       if (X_{i,t+1} fitness dominates X_{i,t} fitness)
26                            move accepted and X_{i,t} assigned to X_{i,tprevious}
27                       else
28                            maintain current position.
29                  end if
30                  end if
31                  Apply polynomial mutation.
32                  Add non-dominated Ants (solution) to archive.
33                  Keep only non-dominated members in the archive.
34                  Update Hypercube Grid Indices.
35          end for
36 end while
```

FIGURE 12.9 MOANA pseudocode.

optimal deposition site, generating a random number using Levy flight, and applying criteria derived from Eqs. (5), (6), and (7) on lines 11, 13, and 18. Following this, line 20 uses Eq. (7) to determine a new deposition site for the deposit.

When a new deposition point is found, the algorithm verifies constantly to see whether the newly acquired solution (cost function) is more important than the prior outcome (line 21). In the event that this criterion is satisfied, the new position is authorized, and the rate of change is saved for possible use in the future, as shown on line 22. On the other hand, if this is not the case, a previously stored rate of change is used instead, supposing that it is available. This strategy is implemented with the

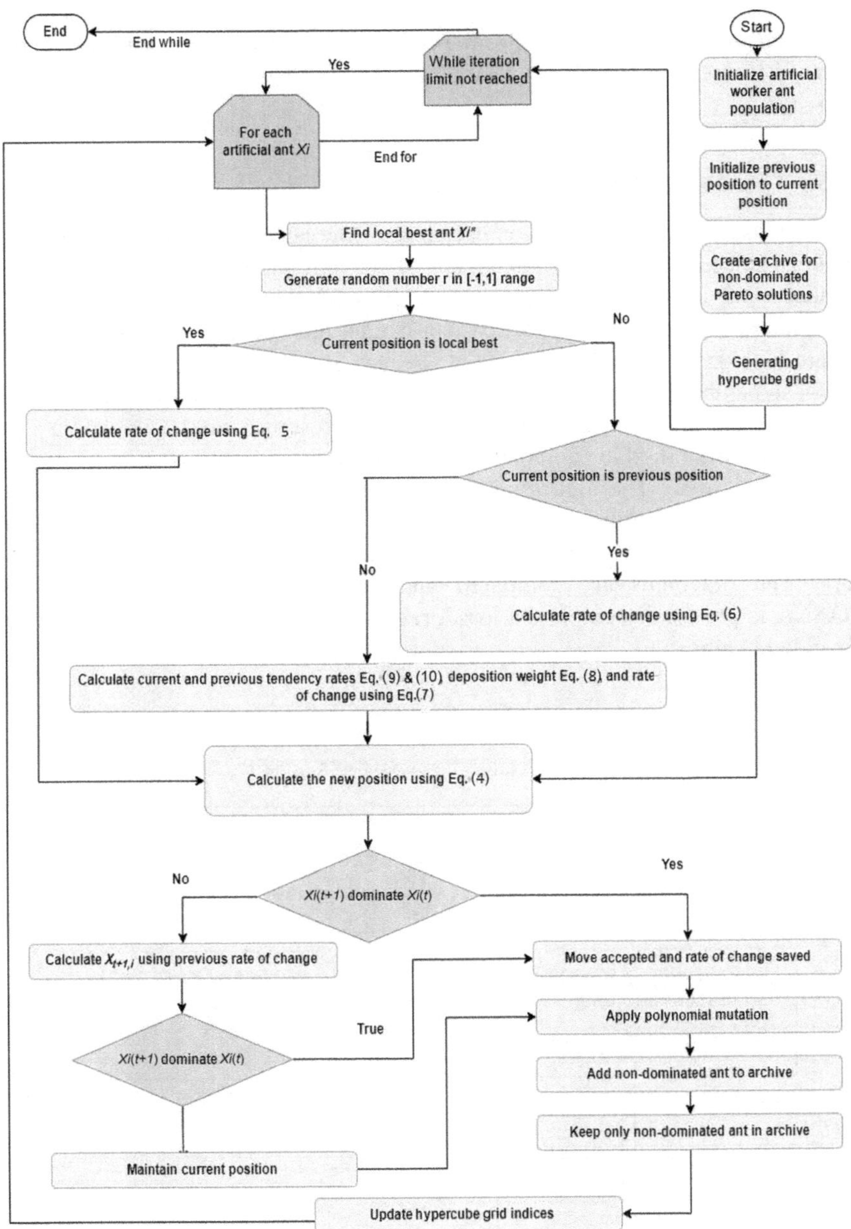

FIGURE 12.10 Flowchart illustrates MOANA's logical operation.

hope that it will lead to a more favorable conclusion than the previous one, unless the search agent continues to operate from its present location, lines 20–29.

On line 31, polynomial mutation is used to produce a greater variety of alternative solutions. This helps to ensure that all possible outcomes are considered. The next lines, lines 32 and 33, investigate whether the result may be stored inside the archive. According to the result on line 34, the indices of the hypercube grid are modified so that they are in agreement with the changes that have been made to the search landscape.

12.5 TESTING AND EVALUATION

As previously mentioned, ANA is extremely slow because of the large number of iterations since it starts by initializing ants with ten random numbers between [lower boundary and upper boundary] as [-100, 100] for each ant, which it cycles 300 times; however, the NumPy function can handle this looping issue with its library (Figure 12.11).

As can be observed from the data illustrated in Figure 12.12, the academic realm offers a wide array of methodologies to assess and scrutinize the algorithm's practicality. These methods encompass a diverse range of testing functions and real-world applications. The algorithm's efficacy is evaluated in comparison to other state-of-the-art algorithms, such as MOFDO, MOLPB, Multi-objective Dragonfly Algorithm (MODA), MOPSO, and NSGA-III, utilizing the Classical ZDT benchmark. The outcomes are compared and analyzed for disparities. Moreover, MOANA is practically employed to address a problem involving the optimization of welded beams.

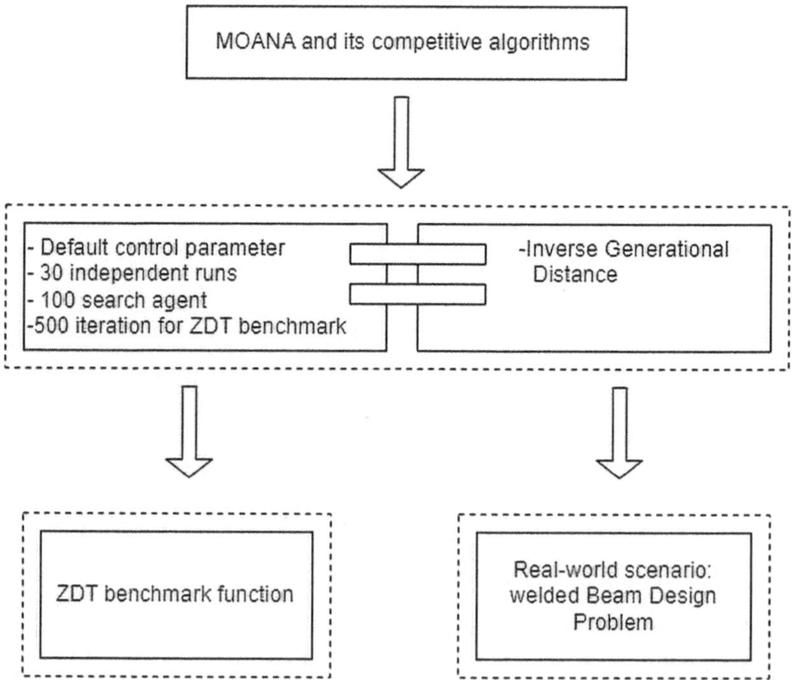

FIGURE 12.11 Fundamental structure for measuring MOANA's performance.

TABLE 12.3
Mathematical equations of the ZDT functions

Functions	Mathematical Equation
ZDT1	$g(x) = 1 + 9\left(\sum_{i=2}^{n} x_i\right)/(n-1)$ $F_1(x) = x_1$ $F_2(x) = g(x)[1 - \sqrt{x_1/g(x)}]x \in [0,1]$.
ZDT2	$g(x) = 1 + 9(\sum_{i=2}^{n} x_i)/(n-1)$ $F_1(x) = x_1$ $F_2(x) = g(x)[1 - (x_1/g(x))^2]x \in [0,1]$.
ZDT3	$g(x) = 1 + 9\left(\sum_{i=2}^{n} x_i\right)/(n-1)$ $F_1(x) = x_1$ $F_2(x) = g(x)[1 - \sqrt{x_1/g(x)} - x_1/g(x)\sin(10\pi x_1)]x \in [0,1]$.
ZDT4	$g(x) = 91 + \sum_{i=2}^{n}[x_i^2 - 10\cos(4\pi x_i)]$ $F_1(x) = x_1$ $F_2(x) = g(x)[1 - \sqrt{x_1/g(x)}]x_1 \in [0,1],\ x_i \in [-5,5]\ i = 2,\ldots,10$.
ZDT5	$g(x) = 1 + 9\left[(\sum_{i=2}^{n} x_i)/(n-1)\right]^{0.25}$ $F_1(x) = 1 - \exp(-4x_i)\sin^6(6\pi x_1)$ $F_2(x) = g(x)[1 - (f_1(x)/g(x))^2]x \in [0,1]$.

12.5.1 Classical ZDT Benchmarks

The field ZDT benchmarks, which are often referred to as the Zitzler–Deb–Thiele benchmarks, are a collection of MOO test functions. These benchmarks are frequently employed to assess the efficacy of evolutionary algorithms and other optimization approaches. It is believed that Zitzler, Deb, and Thiele were responsible for developing these standard functions. As can be seen in the article that Zitzler, Deb, and Thiele wrote together and then published (Zitzler, Deb & Thiele, 2000), the formation of these benchmarks was the result of the combined efforts of Zitzler, Deb, and Thiele. These benchmark challenges include a total of five different tasks, and each one is designated by its own corresponding number, which are as follows: ZDT1, ZDT2, ZDT3, ZDT4, and ZDT5. Table 12.3 contains the mathematical definitions of these concepts. Every ZDT test function has a unique collection of characteristics, some of which may include linearity, convexity, non-convexity, disconnectedness, and scalability. Other qualities may also be included, such as scalability and sphericity. These benchmarks provide a diverse set of optimization tasks that may be used in the process of evaluating and comparing different MOO strategies.

The findings of MOANA are evaluated against those of five well-known algorithms: MOFDO, MOLPB, MODA, MOPSO, and NSGA-III. Table 12.4 displays the result. The algorithm is permitted to execute up to 500 iterations. Each

TABLE 12.4
Parameter configurations for the chosen multi-objective algorithms

Parameters	Parameter Value
Polynomial Mutation Rate	5
Number of Grids per Dimension	7
Best Ant Selection Factor	2
Delete Factor	2
Inflation Rate	1

approach is furnished with an initial group of 100 search entities and an archive of size 100. The parameter configurations for each algorithm remain consistent with those documented in their original sources (Mirjalili, 2015), as outlined in Table 12.3. For every method, the identical initial set of 100 search particles and an archive of size 100 are allocated.

Then one of the most important measures is used to assess the results that were achieved; the resulting equation may be seen below. The Inverted Generational Distance (IGD) is an assessment method that uses the true Pareto front of the issue as a benchmark. Using the methodology described in (Sierra & Coello Coello, 2005), the measurement based on IGD involves contrasting each aspect of the problem with the P_f that was determined using that algorithm. In a nutshell, the computation of IGD calls for two distinct groups of answers: the genuine Pareto-optimal front, which is indicated by the notation P_f, and the variety of answers that are produced by the method that is being evaluated:

$$IGD = \frac{\sqrt{\sum_{i=1}^{n} d_i^2}}{n} \tag{15}$$

where d_i denotes the Euclidean distance between the closest achieved Pareto-optimal solutions and the ith actual Pareto-optimal solution that is included inside the reference set. Given that n represents the number of really optimum Pareto solutions, it is abundantly evident that an IGD value of 0 indicates that all of the components that were created are located on the genuine Pareto front of the problem. After accumulating the IGD findings from 30 separate iterations of each method, the data are analyzed to determine the average (mean) and standard deviation (STD), shown in Table 12.5. The IGD score indicates the distance, in terms of the true Pareto-optimal front, that separates the solutions found by the algorithm from the ideal solution. Lower values of IGD are desired since they suggest greater convergence toward the genuine Pareto front and, therefore, higher performance of the optimization process. This is the case because lower values of IGD have a smaller standard deviation.

In this section, the rankings of each algorithm are shown through table comparisons, as seen in Table 12.6. For instance, if MOANA came in first place in ZDT1, its position

TABLE 12.5
Classical benchmark results

Functions	Algorithms	IGD AVG.	IGD STD.
ZDT 1	MOANA	0.04913	0.0120
	MOLPB	0.1213	0.046
	MOFDO	0.06758	0.030911
	MODA	0.07653	0.012071
	MOPSO	0.07843	0.008848
	NSGA-III	0.52599	0.509184
ZDT 2	MOANA	0.05739	0.01416
	MOLPB	0.0732	0.0458
	MOFDO	0.03511	0.00404
	MODA	0.00292	0.00026
	MOPSO	0.03243	0.00093
	NSGA-III	0.13972	0.02626
ZDT 3	MOANA	0.07508	0.01623
	MOLPB	0.0732	0.0194
	MOFDO	0.06676	0.023913
	MODA	0.07653	0.014411
	MOPSO	0.07758	0.005755
	NSGA-III	0.19474	0.080043
ZDT 4	MOANA	0.4965	0.4828
	MOLPB	0.4778	0.209
	MOFDO	0.6802	0.352945
	MODA	64.9628	2.847807
	MOPSO	0.46175	0.047785
	NSGA-III	0.73731	0.307518
ZDT 5	MOANA	0.31536	0.28196
	MOLPB	0.035	0.061
	MOFDO	0.35853	0.161795
	MODA	0.11349	0.01827
	MOPSO	0.26862	0.136598
	NSGA-III	0.66397	0.235754

TABLE 12.6
Ranking table displays the performance of the algorithms in Table (4.2)

Function	MOANA Rank	MOLPB Rank	MOFDO Rank	MODA Rank	MOPSO Rank	NSGA-III Rank
ZDT 1	1	5	2	3	4	6
ZDT 2	4	5	3	1	2	6
ZDT 3	3	2	1	4	5	6
ZDT 4	3	2	4	6	1	5
ZDT 5	4	1	5	2	3	6
Total Ranking	15	15	15	16	15	29

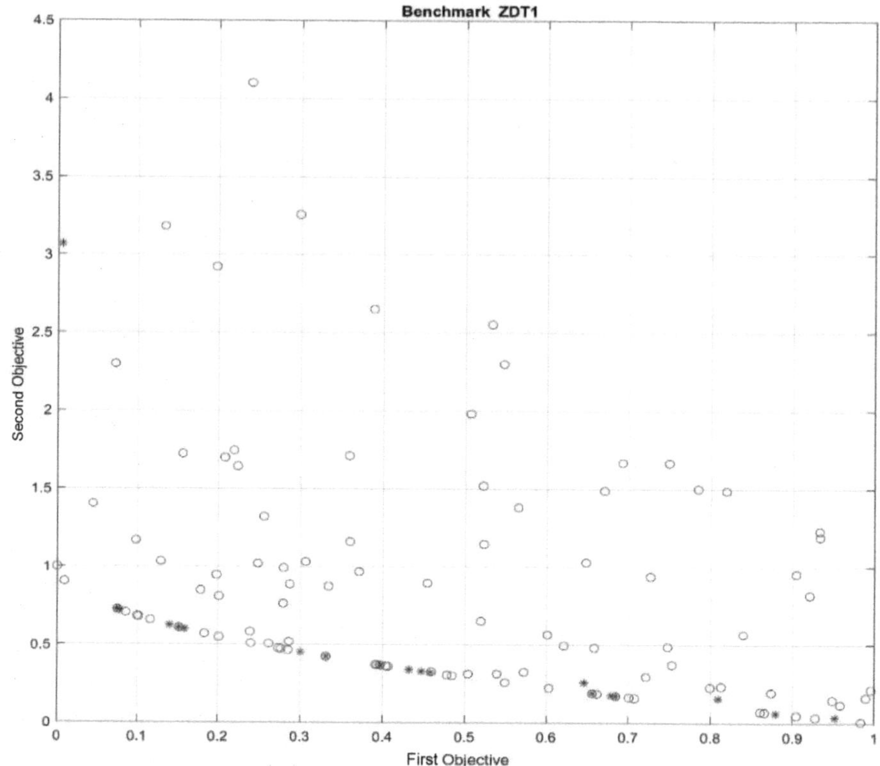

FIGURE 12.12a In iteration 14 of MOANA, 19 P_f solutions were discovered.

is equal to 1, and if MOANA came in fourth place in ZDT2, its position is equal to 4, and so forth. The overall rank is the total of all rankings that a certain algorithm has collected throughout the time of its operation. The ranking table is a straightforward method for demonstrating the superiority of a particular algorithm in comparison to the group of competing algorithms.

MOANA outperforms MODA in the ZDT1, ZDT3, and ZDT4 test functions. However, MOANA underperforms MODA in ZDT2 and ZDT 5. Moreover, MOANA outperforms MOFDO in ZDT1, ZDT4, and ZDT5. As seen in the ranking table shown in Table 4.4, finally, MOANA contributes better outcomes compared to MOFDO, MODA, MOPSO, and NSGA-III.

As an indication of MOANA's performance, the ZDT1 solution landscape may be seen in Figure 12.12. The process of locating solutions that were originally randomly dispersed is shown in Figure 12.12a, which uses a total of just 19 P_{fs}. After that, MOANA was able to successfully enhance the number of P_f solutions that were evenly distributed, as shown in Figure 12.12b.

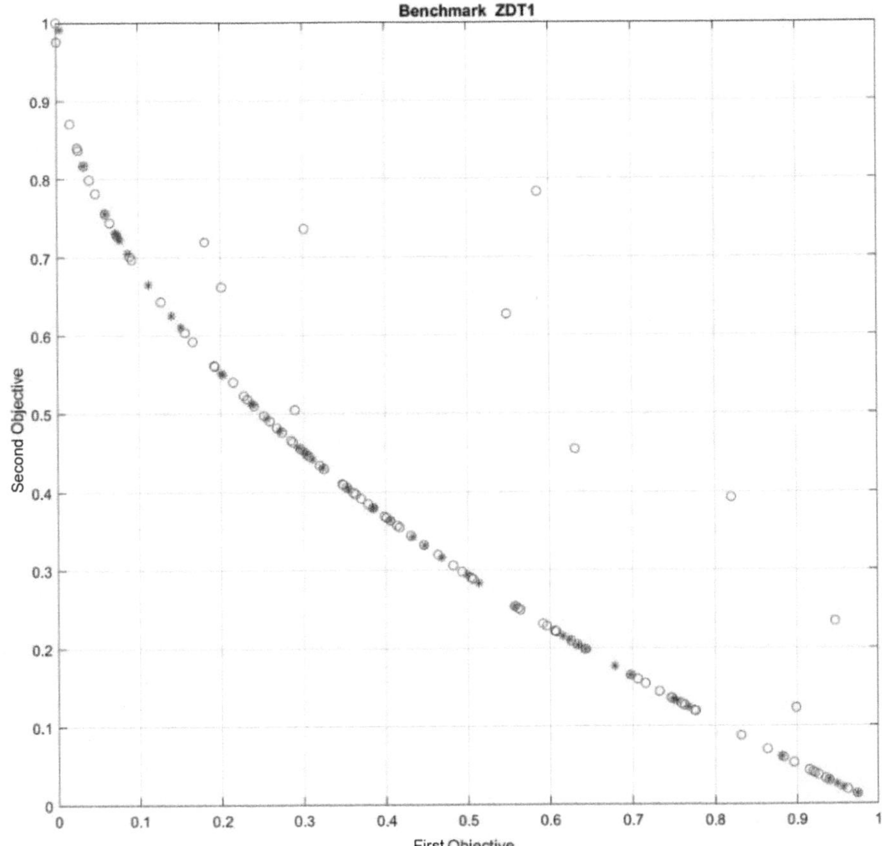

FIGURE 12.12b In iteration 48 of MOANA, 37 P_f solutions were discovered.

12.6 CONCLUSION

A multi-objective algorithm called MOANA was used to solve a novel single-objective ant nesting method inspired by Leptothorax ants by mimicking their movement to deposit grains while building a new nest. MOANA archives all non-dominant solutions and removes dominant solutions. MOANA selects global guides using a hypercube guide and local guides using a roulette wheel. No convergence to the local optimum or parameter modification issues are related to MOANA.

Additionally, ZDT benchmark test functions and the Welded Beam Design Challenge were used to evaluate the proposed technique. Then, the results were compared to six other popular algorithms. Based on the findings, MOANA virtually competed with the best and broke even. Because MOANA provided well-distributed answers, it was easily employed to tackle real-world problems. This may provide decision-makers with more options. The recommended method may successfully advance toward Pareto front solutions while keeping its variety in this region.

REFERENCES

Abdullah, J.M., Rashid, T.A., Maaroof, B.B., Mirjalili, S., 2023. Multi-objective fitness-dependent optimizer algorithm. *Neural Computing and Applications* 35, 11969–11987.

Ahmed, A.M., Rashid, T.A., Saeed, S.Ab.M., Noori, K.A., Hassan, B.A., Rahman, C.M., Ahmed, O.H., Umar, S.U., Yaseen, Z.M., 2023. GMOCSO: Multi-objective Cat Swarm Optimization Algorithm based on a Grid System.

Akbari, R., Hedayatzadeh, R., Ziarati, K., Hassanizadeh, B., 2012. A multi-objective artificial bee colony algorithm. *Swarm and Evolutionary Computation* 2, 39–52.

Brockhoff, D., Friedrich, T., Neumann, F., 2008. *Analyzing Hypervolume Indicator Based Algorithms*. Springer-Verlag, 651–660.

Chugh, T., Sindhya, K., Hakanen, J., Miettinen, K., 2015. An Interactive Simple Indicator-Based Evolutionary Algorithm (I-SIBEA) for multiobjective optimization problems. *Lecture Notes in Computer Science* 277–291.

Coello Coello, C.A., 2009. Evolutionary multi-objective optimization: Some current research trends and topics that remain to be explored. *Frontiers of Computer Science in China* 3, 18–30.

Coello Coello, C.A., Lechuga, M.S., 2002. MOPSO: A proposal for multiple objective particle swarm optimization. *Proceedings of the 2002 Congress on Evolutionary Computation.* CEC'02 (Cat. No.02TH8600).

Coello Coello, C.A., Pulido, G.T., Lechuga, M.S., 2004. Handling multiple objectives with particle swarm optimization. *IEEE Transactions on Evolutionary Computation* 8, 256–279.

Dasheng L., 2008. Multi-objective particle swarm optimization: Algorithm and application, a thesis for the degree of doctor, in department of electrical & computer engineering.

Deb, K., Jain, H., 2014. An evolutionary many-objective optimization algorithm using reference-point-based nondominated sorting approach, part I: Solving problems with box constraints. *IEEE Transactions on Evolutionary Computation* 18, 577–601.

Deb, K., Pratap, A., Agarwal, S., Meyarivan, T., 2002. A fast and elitist multiobjective genetic algorithm: NSGA-II. *IEEE Transactions on Evolutionary Computation* 6, 182–197.

Deb, K., Sundar, J., Udaya Bhaskara, R.N., Chaudhuri, S., 2006. Reference point based multi-objective optimization using evolutionary algorithms. *International Journal of Computational Intelligence Research* 2.

Gunantara, N., 2018. A review of multi-objective optimization: Methods and its applications. *Cogent Engineering* 5, 1502242.

Hama Rashid, D.N., Rashid, T.A., Mirjalili, S., 2021. ANA: Ant nesting algorithm for optimizing real-world problems. *Mathematics* 9, 3111.

Li, L.-M., Lu, K.-D., Zeng, G.-Q., Wu, L., Chen, M.-R., 2016. A novel real-coded population-based extremal optimization algorithm with polynomial mutation: A non-parametric statistical study on continuous optimization problems. *Neurocomputing* 174, 577–587.

Liagkouras, K., Metaxiotis, K., 2013. An elitist polynomial mutation operator for improved performance of MOEAs in computer networks. *2013 22nd International Conference on Computer Communication and Networks (ICCCN).*

Liu, J., Liu, J., 2019. Applying multi-objective ant colony optimization algorithm for solving the unequal area facility layout problems. *Applied Soft Computing* 74, 167–189.

Long, Q., Wu, C., Wang, X., Jiang, L., Li, J., 2015. A Multi-objective genetic algorithm based on a discrete selection procedure. *Mathematical Problems in Engineering* 2015, 1–17.

Mirjalili, S., 2015. Dragonfly algorithm: A new meta-heuristic optimization technique for solving single-objective, discrete, and multi-objective problems. *Neural Computing and Applications* 27, 1053–1073.

Mirjalili, S., Gandomi, A.H., Mirjalili, S.Z., Saremi, S., Faris, H., Mirjalili, S.M., 2017. Salp Swarm Algorithm: A bio-inspired optimizer for engineering design problems. *Advances in Engineering Software* 114, 163–191.

Mirjalili, S., Saremi, S., Mirjalili, S.M., Coelho, L. dos S., 2016. Multi-objective grey wolf optimizer: A novel algorithm for multi-criterion optimization. *Expert Systems with Applications* 47, 106–119.

Mirjalili, S., Siddiqi, F.A., Mofizur Rahman, C., 2019. Evolutionary multi-objective Whale Optimization Algorithm. *Advances in Intelligent Systems and Computing* 431–446.

Olivier L., 2004. *Multi-Objective Optimization: History and Promise.* System Division Engineering.

Patil, D.D. Dangewar, B., 2014. Multi-Objective Particle Swarm Optimization (MOPSO) based on Pareto Dominance Approach. *International Journal of Computer Applications* 107, 13–15

Rahman, C.M., Rashid, T.A., Ahmed, A.M., Mirjalili, S., 2022. Multi-objective learner performance-based behavior algorithm with five multi-objective real-world engineering problems. *Neural Computing and Applications* 34, 6307–6329.

Ray, T., Liew, K.M., 2002. A Swarm metaphor for multiobjective design optimization. *Engineering Optimization* 34, 141–153.

Sierra, M.R., Coello Coello, C.A., 2005. Improving PSO-based multi-objective optimization using crowding, mutation and ∈-Dominance. *Lecture Notes in Computer Science* 505–519.

Xiujuan, L., Zhongke, S., 2004. Overview of multi-objective optimization methods. *Journal of Systems Engineering and Electronics*, 15(2), 142–146.

Yang, X.S., 2011. Bat algorithm for multi-objective optimisation. *International Journal of Bio-Inspired Computation* 3, 267.

Yang, X.-S., 2012. Multiobjective firefly algorithm for continuous optimization. *Engineering with Computers* 29, 175–184.

Yang, X.-S., Deb, S., 2013. Multiobjective cuckoo search for design optimization. *Computers & Operations Research* 40, 1616–1624.

Zitzler, E., Deb, K., Thiele, L., 2000. Comparison of multiobjective evolutionary algorithms: Empirical results. *Evolutionary Computation* 8, 173–195.

13 Multi-objective Optimization Vectors

Mathematical Benchmark Overview

Mzhda S. Abdulkarim, Araz Ibrahim Mustafa, Soran R. Mohammed-Taha, Aso M. Aladdin, Dler O. Hasan and Tarik A. Rashid

13.1 INTRODUCTION

Various methods can be employed to optimize systems, with the three main techniques being classical, numerical, and evolutionary optimization [1]. Classical optimization methods, which are analytical techniques, are used to find optimal solutions for problems involving continuous and differentiable functions. These methods can determine maximum and minimum points for both unconstrained and constrained continuous objective functions. They are applicable to optimization problems involving single-variable functions, multivariable functions without constraints, and multivariable functions with inequality and equality constraints [2]. Based on these principles, various algorithms have been developed, categorized by behavior, such as heuristic, evolutionary, and gradient-based [3]. Consequently, to assess the effectiveness of the solutions generated by these algorithms, a set of basic benchmarks is necessary. These benchmarks are derived from fundamental mathematical principles. The functions used in these benchmarks vary in complexity; some are relatively simple, whereas others are more complex. This range of benchmarks ensures a thorough evaluation of the algorithms' performance across different types of problems [4].

Clearly, this chapter focuses on functions derived for multi-objective optimization, utilizing a set of multi-objective benchmarks [5]. The main problem of the research is the absence of a comprehensive library of benchmark functions for systematically evaluating multi-objective optimization algorithms. To tackle this, the research objectives are as follows: aiming to compile a detailed collection of benchmark functions, developing an evaluation framework with performance metrics and statistical analysis, ensuring that benchmarks reflect real-world problem characteristics like modality, separability, and scalability, and improving the accuracy and effectiveness of optimization algorithms in identifying optimal solutions [6].

DOI: 10.1201/9781003601555-13

This chapter compiles the fundamental functions, and highlights their specific characteristics and solution points. As a result, the following contributions of this chapter are outlined clearly and precisely:

- It curates a comprehensive library of benchmark functions, meticulously categorized by their objectives and inherent characteristics.
- Each function is systematically analyzed based on its purpose (minimization/maximization) and key features like modality, separability, and scalability.
- This focus on real-world applicability ensures that the functions are practical for evaluating algorithm performance in diverse scenarios.
- The library includes functions of varying dimensionality and modality, such as the Zitzler–Deb–Thiele (ZDT) Series, alongside both constrained and unconstrained test problems. Additionally, the chapter provides detailed mathematical formulations for each function, promoting understanding and the replication of tests.
- This comprehensive approach empowers researchers to select the most suitable functions for their specific needs, ultimately leading to the development of more effective and efficient optimization solutions for real-world problems.

The remainder of this chapter includes the review methodology in Section 13.2, and a detailed categorization and overview of benchmarking in multi-objective optimization in Section 13.3. Section 13.4 presents a mathematical classification of test functions for categorized benchmarks. Section 13.5 is dedicated to explaining the diverse collection of general test functions. Section 13.6 discusses methods for enhancing accuracy at locating points. Finally, the chapter concludes with a clear and concise summary of the key points.

13.2 REVIEW METHODOLOGY

The methodology for this systematic review consists of several key steps, as shown in Figure 13.1. First, a comprehensive literature search and selection process was conducted using databases like PUBMED, IEEE Xplore, Google Scholar, and Springer, with keywords such as "multi-objective optimization" and "benchmark test functions". Inclusion criteria focused on multi-objective optimization benchmark functions, their applications, and mathematical characteristics. Studies outside this scope were excluded.

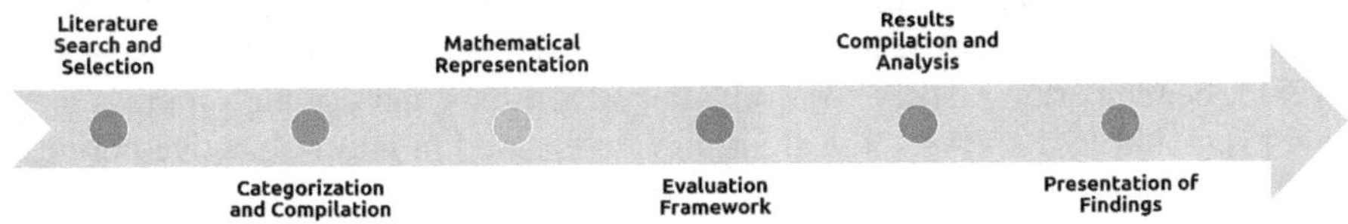

FIGURE 13.1 Flowchart of the review methodology used.

Next, a diverse set of benchmark functions used in multi-objective optimization was collected and categorized based on characteristics such as modality, separability, scalability, and the nature of their objective functions (e.g., ZDT, Deb–Thiele–Laumanns–Zitzler (DTLZ), Walking Fish Group (WFG), and Unconstrained Functions (UF) Series). Detailed mathematical formulations for each function, including objective functions, constraints, and other relevant properties, were provided. The characteristics of each function, such as Pareto optimality, dimensionality, nonlinearity, and complexity, were also analyzed and documented.

The evaluation framework involved identifying and describing performance metrics to assess the effectiveness of optimization algorithms on the benchmark functions. Various optimization algorithms (e.g., Non-Dominated Sorting Genetic Algorithm 2 (NSGA-II), Multi-objective Evolutionary Algorithm based on Decomposition (MOEA/D), Strength Pareto Evolutionary Algorithms (SPEA2), Multi-objective Simultaneous Hybrid Exploration that is Robust, Progressive and Adaptive (MO-SHERPA), and Multi-objective Particle Swarm Optimization (MOPSO)) were compared using these benchmarks to evaluate their strengths and weaknesses. Results were collected from the literature, detailing algorithm performance, and subjected to rigorous statistical analysis to assess factors like convergence speed, solution diversity, and computational efficiency.

Finally, findings were presented through summary tables and graphs, clearly highlighting algorithm performance on various benchmark functions. The discussion addressed the implications of the findings, potential applications, and limitations of the study.

13.3 BENCHMARKING IN MULTI-OBJECTIVE OPTIMIZATION AS A FOUNDATIONAL STRATEGY

Benchmarking in multi-objective optimization compares the performance of different algorithms using standardized test problems. It helps researchers to evaluate algorithm efficiency, robustness, and scalability, guiding the development of new techniques and identifying state-of-the-art methods for solving complex optimization problems.

The reasons for conducting benchmark studies on optimization algorithms vary widely, just like the algorithms and problems involved. Besides scientific goals, benchmarking can also be employed to promote a specific algorithmic method or highlight a particular problem [7].

Benchmarking is important for assessing algorithms because it gives a fair way to compare them. It helps us to see how well algorithms work, what they're good at, and where they need improvement. Benchmarking also helps researchers to find the best techniques and check if new algorithms are reliable compared to existing ones. Overall, benchmarking is key for evaluating algorithms and driving progress in the field [8]. Table 13.1 shows some algorithms often used in multi-objective optimization that utilize ZDT functions for testing [6, 9].

Benchmark functions in Multi-objective Optimization (MOO) deal with multiple goals simultaneously, which might not agree with each other. These goals could be different things we're trying to accomplish. The goal is to find solutions that balance these goals effectively.

TABLE 13.1
Selected algorithms are examined with a focus on their description as multi-objective optimization techniques

Algorithm	Description	Background
NSGA-II (Non-dominated Sorting Genetic Algorithm II)	An efficient method for solving problems with multiple objectives by sorting and combining solutions to find the best ones. It's widely used due to its speed and effectiveness.	NSGA-II is an advanced version of NSGA that manages multiple conflicting objectives by sorting solutions based on dominance and promoting diversity, efficiently finding a diverse set of Pareto-optimal solutions [6, 10].
MOEA/D (Multi-Objective Evolutionary Algorithm based on Decomposition)	Splits complex problems into simpler parts and solves them concurrently. It's effective for real-world problems with many objectives.	MOEA/D handles complex multi-objective problems by breaking them into simpler subproblems and solving them together, making it efficient for diverse real-world scenarios [10, 11].
SPEA2 (Strength Pareto Evolutionary Algorithm 2)	Balances finding the best solutions with maintaining diversity. Useful for complex problems with many objectives.	SPEA2 is an enhanced version of the SPEA algorithm, aiming to find a diverse set of solutions along the Pareto front while maintaining a balance between convergence and diversity. It achieves this by combining Pareto dominance with a density estimation technique to guide the search towards the Pareto front.
MO-SHERPA	It is a method for finding the best hyperparameters for a model in multi-objective optimization. It explores different setups to balance multiple goals and efficiently handles large solution spaces to find optimal configurations.	Multi-Objective SHERPA (MO-SHERPA) is an adaptation of the SHERPA algorithm for multi-objective problems. It manages multiple objectives separately and generates solutions optimal for each objective. MO-SHERPA uses non-dominated sorting to prioritize solutions, differing from NSGA-II and NSGA in several ways [6].
MOPSO (Multi-Objective Particle Swarm Optimization)	Simulates birds' behavior to find good solutions for problems with multiple objectives. It adapts solutions based on what it learns over time.	MOPSO is inspired by the behavior of birds flying in a group. It maintains a population of solutions and updates them based on their performance and interaction with others. This approach enables MOPSO to adaptively search for good solutions to problems with multiple objectives over time [12].

These benchmark functions in MOO are made to test algorithms that aim to find solutions that aren't worse than any other possible solution in terms of the goals we're trying to achieve [5]. They usually have these features:

- **Pareto Optimality:** In MOO, Pareto optimality is key. A solution is Pareto optimal if no other choice improves one thing without making another thing worse. Benchmark functions often display sets of Pareto-optimal solutions and how they manage different goals that might clash.
- **Performance Metrics:** In addition to assessing proximity to the true Pareto front, MOO performance metrics evaluate solution diversity along this front. They measure an algorithm's ability to discover diverse solutions that address multiple goals effectively.
- **Dimensionality:** Benchmark functions in MOO often involve numerous variables and goals, similar to real-world situations. These functions present a challenge for algorithms to navigate the problem space efficiently.
- **Nonlinearity and Complexity:** Just like in single-objective optimization, MOO benchmark functions can be tricky. They might have curves, several high points, and other difficult features to imitate real-world problems.

Multi-objective optimization often tests algorithms using standard functions like ZDT, DTLZ, and WFG. These functions mimic various problem types and help assess how well algorithms perform. Table 13.2 focuses on a comparison of four commonly used artificial benchmark functions for multi-objective optimization [13].

The cited works in this chapter provide a solid foundation for multi-objective optimization through widely recognized benchmark functions like ZDT, DTLZ, WFG, and UF, which offer standardized comparisons for algorithms. However, these functions often lack real-world relevance and primarily focus on theoretical evaluations, missing the complexity of practical problems. Similarly, optimization algorithms such as NSGA-II, MOEA/D, SPEA2, MO-SHERPA, and MOPSO are well-established and extensively tested, yet their evaluations are usually conducted under static conditions. This approach fails to capture the dynamic nature of real-world scenarios, and current metrics may not fully assess the performance of these algorithms in diverse and evolving contexts.

Existing research on multi-objective optimization has several gaps that this study aims to fill. Current benchmark functions often lack real-world relevance and are mostly static, failing to capture dynamic changes in objectives and constraints typical of real-world problems. Additionally, there is insufficient focus on high-dimensional objective spaces, which limits the assessment of algorithm scalability. Evaluation metrics currently used may not fully capture the robustness and practical applicability of algorithms. This research addresses these gaps by developing benchmark functions that better mimic real-world scenarios, including dynamic and high-dimensional aspects, and by creating a comprehensive evaluation framework with advanced performance metrics. It also aims to design benchmarks inspired by practical problems from various domains, ensuring relevance and applicability in real-world contexts.

TABLE 13.2
Comparison of four widely used artificial benchmark functions for multi-objective optimization

Benchmark Function	Description	Characteristics	Optimal Solutions	Usage
Zitzler–Deb–Thiele (ZDT)	The ZDT benchmark suite consists of a series of test problems with two objectives. These problems are designed to capture different aspects of multi-objective optimization challenges, such as nonlinearity, multimodality, and Pareto front shapes. The functions are scalable, meaning the number of decision variables can vary.	Nonlinearity - Multimodality - Pareto front shapes	Known Pareto-optimal frontiers	Widely used in benchmarking and comparing the performance of multi-objective optimization algorithms.
Deb–Thiele–Laumanns–Zitzler (DTLZ)	The DTLZ benchmark suite extends the ZDT benchmark suite by including problems with more than two objectives. These problems are designed to evaluate the scalability and performance of algorithms in higher-dimensional objective spaces. Like the ZDT functions, the DTLZ functions are also scalable.	Scalability to higher-dimensional objective spaces	Known Pareto-optimal frontiers	Widely used for benchmarking algorithms in multi-objective optimization, especially in studies focusing on scalability and higher-dimensional optimization problems.
Walking Fish Group (WFG)	The WFG benchmark suite consists of a set of scalable test problems with varying degrees of difficulty and characteristics, including convex, concave, disconnected, and multimodal Pareto fronts. These problems are designed to assess the performance of algorithms in handling various types of Pareto fronts and problem complexities.	Convex, concave, disconnected, and multimodal Pareto fronts - Varying degrees of difficulty and characteristics	Known Pareto-optimal frontiers	Widely used for benchmarking multi-objective optimization algorithms, especially in studies focusing on diverse problem characteristics and complexities.

(continued)

TABLE 13.2 (Continued)
Comparison of four widely used artificial benchmark functions for multi-objective optimization

Benchmark Function	Description	Characteristics	Optimal Solutions	Usage
Unconstrained Functions (UF)	The UF benchmark suite is composed of a set of scalable test problems designed to evaluate the performance of optimization algorithms in handling various optimization challenges such as multimodality, discontinuity, and irregular Pareto fronts. These problems are unconstrained and can have two or more objectives.	Multimodality - Discontinuity - Irregular Pareto fronts	Known Pareto-optimal frontiers	Widely used for benchmarking multi-objective optimization algorithms, especially in studies focusing on diverse problem characteristics and complexities.

Finally, we can say that benchmarking helps compare things by setting up rules that everyone follows. It makes it easier to see how well different things perform in the same conditions. This makes it fair and helps others check if the results are true. Benchmarking also gives insights and helps people work together to improve things in a specific area.

13.4 MATHEMATICAL CLASSIFICATION OF TEST FUNCTIONS

13.4.1 ZDT Test Functions

The ZDT test suite, named after Zitzler, Deb, and Thiele, includes a group of standard tests used to check how well multi-objective optimization algorithms perform. These tests have two goals and simulate various challenges like different shapes of optimal solutions. They help compare algorithms to see how well they find solutions and handle different types of problems, helping researchers develop better algorithms for complex optimization tasks [14].

All ZDT functions contain two objectives, which is the most common usage of Pareto optimization, especially in engineering applications. The specifics of the ZDT test suite are illustrated in Table 13.3. This table offers an extensive overview of the ZDT benchmark functions, which are frequently employed to evaluate the effectiveness of multi-objective optimization algorithms.

Each function within the ZDT suite possesses unique characteristics and challenges, making it well-suited for identifying the strengths and weaknesses of different optimization methods. Table 13.3 details the individual properties and mathematical

TABLE 13.3
ZDT test functions (bi-objective text functions)

		Objective functions				Characteristics			
ZDT	**d**	**Min F1(x)**	**Subject to**	**Where g(x)**	**Where h(f1, g)**	**Linear & NonLinear**	**Scalability**	**Convex & Non-Convex**	**Trade-off Between Objectives**
ZDT1[15]	30	$x1$	$0 <= x_i <= 1$ $i = 1, ..., n$ **Optimum** **0<=** $x_i^* <= 1$ **and** $x_i^* = 0$ ***for i*** $= 2,\ldots,$***n***	$1+\frac{9}{n-1}\sum_{i=2}^{n} x_i$	$1-\sqrt{\frac{f1}{g}}$	ZDT1 is known for its linear Pareto front, simplifying the optimization problem compared to others.	ZDT1 is designed to be scalable, accommodating varying numbers of decision variables (*n*) for comprehensive algorithm testing.	ZDT1's non-convexity in the objective space arises from the square root in *f2*(*x*), increasing the complexity for optimization algorithms.	ZDT1 exhibits a trade-off between *f1*(*x*) and f2(x), requiring solutions that balance conflicting objectives.
ZDT2[16]	30	$x1$	$0 <= x_i <= 1$ $i = 1, ..., n$ **Optimum** **0 < =** $x_i^* <= 1$ **and** $x_i^* = 0$ ***for i*** = $2,\ldots,$***n***	$1+\frac{9}{n-1}\sum_{i=2}^{n} x_i$	$1-\left(\frac{f1}{g}\right)^2$	ZDT2's nonlinear term (1-(*f1*/*g*)^2) in *f2*(*x*) introduces nonlinearity to the objective space, creating a more intricate optimization landscape.	ZDT2 is scalable, accommodating varying decision variables (*n*) for versatile algorithm testing across different problem dimensions.	ZDT2 is notable for its convex Pareto front, unlike ZDT1's linear front. This convex shape means improving one objective will affect others.	ZDT2, like other ZDT problems, requires balancing conflicting objectives between f1(x) and f2(x).

(continued)

TABLE 13.3 (Continued)
ZDT test functions (bi-objective text functions)

		Objective functions				Characteristics			
ZDT	**d**	**Min F1(x)**	**Subject to**	**Where g(x)**	**Where h(f1, g)**	**Linear & NonLinear**	**Scalability**	**Convex & Non-Convex**	**Trade-off Between Objectives**
ZDT3[17]	30	$x1$	$0 <= x_i <= 1$ $i = 1, ..., n$ **Optimum** $\mathbf{0 < = x_i^* < =0.85}$ **and** $x_i^* = 0$ ***for*** $\boldsymbol{i} = 2, \ldots, \boldsymbol{n}$	$1 + \frac{9}{n-1}\sum_{i=2}^{n} x_i$	$1 - \sqrt{\frac{f1}{g}} - \frac{f1}{g}\sin(10\pi f1)$	ZDT3 is a nonlinear problem due to the presence of trigonometric and square root functions in the second objective function.	ZDT3 is scalable, accommodating varying decision variables (n) for comprehensive algorithm testing across different problem dimensions.	ZDT3 features a non-convex Pareto front, making it difficult for optimization algorithms to navigate the complex trade-offs between objectives.	Like other ZDT problems, ZDT3 involves a trade-off between f1(x) and f2(x), requiring solutions that balance conflicting objectives
ZDT4[18]	10	$x1$	$0 <= x_i <= 1$ $-10 <= x_i <= 10$ $i = 2, ..., n$ **Optimum** $\mathbf{0 <= x_i^* <= 1}$ **and** $x_i^* = 0$ ***for*** $\boldsymbol{i} = 2, \ldots, \boldsymbol{n}$	$1 + 10(n-1) + \sum_{i=2}^{n}\left(x_i^2 - 10\cos(4\grave{A}x_i)\right)$	$1 - \sqrt{\frac{f1}{g}}$	The ZDT4 objectives involve nonlinear functions, including trigonometric functions like cosine.	The number of decision variables (n) can vary, allowing scalability testing of optimization algorithms.	Global minima in ZDT4 convex functions are also local minima, simplifying optimization.	ZDT4 has a known Pareto front that shows optimal trade-offs, and optimization algorithms work to approximate or find solutions on this front.

ZDT5	11	$x_1 + u(x1)$	**Optimum** **0<=u(x_1^*)<= 30** **and u(x_1^*) = 5** ***for i*** = 2,…,***n***	$\sum_{i=2}^{n} u(u(x_i))$ $u(u(x_i) = \begin{cases} 2+u(x_i) \\ if\, u(x_i)<5 \\ 1 \\ if\, u(x_i)=5 \end{cases}$	$\frac{1}{f1(x)}$	The ZDT5 combination of trigonometric functions, square roots, and the summation term results in a highly nonlinear problem.	ZDT5, scalability is often examined to understand how algorithms handle an increasing number of decision variables.	ZDT5 is non-convexity to the problem, making the optimization landscape challenging.	The ZDT5 problem is designed to exhibit a trade-off between its two objectives.
ZDT6 [19]	10	$1-exp(-4x_1)\, sin^6(6\pi x_1)$	$0 <= x_i <= 1$ $i = 1, \ldots, n$ **Optimum** **0 <= x_i^* <= 1** **and** $x_i^* = 0$ ***for i*** = 2,…,***n***	$1 + 9\left(\frac{\sum_{i=2}^{n} x_i}{n-1}\right)^{0.25}$	$1-(f1/g)2$	ZDT6 incorporates nonlinear relationships between the decision variables, adding complexity to the optimization problem.	The ZDT6 function, scalability can be assessed by varying the number of decision variables and observing how well the algorithm can handle larger-dimensional spaces.	ZDT6 is non-convex and exhibits a disconnected shape, making it challenging for optimization algorithms to find the optimal solutions.	Trade-offs in ZDT6 are typically visualized on a Pareto front, where each point represents a solution with a different trade-off between the two objectives.

formulations of these functions, serving as a valuable resource for researchers and practitioners in the domain.

13.4.2 DTLZ Test Functions

To address the limitation of the ZDT test suite (as well as many other bi-objective test problems), Deb et al. proposed another test suite: the DTLZ test suite. The DTLZ test functions are another set of benchmark problems used to evaluate the performance of multi-objective optimization algorithms. These functions are designed to assess the scalability of algorithms as the number of objectives and variables increases. The DTLZ test functions play a crucial role in assessing and advancing the capabilities of multi-objective optimization algorithms, particularly in scenarios involving high-dimensional and complex optimization problems [14]. Table 13.4 offers a detailed explanation of the DTLZ test suite. It provides an overview of the DTLZ benchmark functions, which are commonly used to evaluate multi-objective optimization algorithms. Each function is designed with specific characteristics and challenges, making them ideal for assessing different optimization techniques. Table 13.4 outlines the unique properties, mathematical formulations, and objectives of each function, serving as a valuable resource for researchers and practitioners in the field.

13.4.3 Objective Constrained Test Functions

Constrained Multi-Objective Test Functions are problems used to test how well optimization algorithms can balance multiple goals while meeting specific requirements. These tests simulate situations where solutions need to optimize several objectives while also respecting constraints, such as budget limits or design specifications. They're crucial for evaluating algorithms that can handle complex, real-world challenges where both objectives and constraints are important for decision-making.

Table 13.5 presents a variety of constrained test functions. This table offers detailed insights into several constrained benchmark functions, designed to assess the performance of optimization algorithms within restricted conditions.

Each function listed in Table 13.5 is distinguished by its unique constraints and goals, posing distinct challenges for optimization methods. The table details the mathematical formulations, constraint specifications, and objectives of these functions, serving as a valuable reference for researchers and practitioners dealing with constrained optimization issues.

13.4.4 Unconstrained Test Functions

As highlighted in numerous searches, initially, a collection of uniformly distributed solutions is created. These solutions are then categorized into foreground and background solutions. The primary focus of the search lies in foreground solutions, with some attention given to background solutions. Consequently, unconstrained minimization revolves around the task of locating a vector x that represents a local minimum for a scalar function $f(x)$: min x $f(x)$. The term "unconstrained" signifies that there are no limitations imposed on the range of x [21].

TABLE 13.4
DTLZ test functions (tri-objective text functions)

DTLZ	Objective function	Linear & nonlinear	Scalability	Convex & non-convex	Trade-off Between Objectives
DTLZ1[16]	$fi(x)=(1+g(x))*0.5\left(\prod_{i=1}^{M-1}Xi\right)$ $f_{m=2:M-1}=(1+g(x))*0.5\left(\prod_{i=1}^{M-m}X_i\right)*$ $(1-X_{M-m+1})$ $fm(x)=(1+g(x))*0.5(1-Xi)$ $g(x)=100$ $\left[k+\left(\sum_{i=1}^{k}(zi-0.5)^2-cos\,20\pi(z-0.5)\right)\right]$	DTLZ1 typically exhibits a linear Pareto front. This means that the set of non-dominated solutions forms a straight line in the objective space.	DTLZ1 is designed to be scalable, accommodating various numbers of decision variables (n) and objectives (m). This scalability allows researchers to evaluate optimization algorithms across a wide range of problem dimensions.	DTLZ1, the Pareto front is convex, meaning it forms a continuous and convex shape. This characteristic allows optimization algorithms to more easily navigate and identify optimal solutions across the Pareto front.	DTLZ1 involves a trade-off between conflicting objectives. Improving one objective typically leads to degradation in others, necessitating the discovery of solutions that represent a compromise.
DTLZ2 [19]	$fi(x)=(1+g)\left(\prod_{i=1}^{M-1}cos\left(X_i\frac{\pi}{2}\right)\right)$ $f_{m=2:M-1}=(1+g(x))*0.5\left(\prod_{i=1}^{M-m}cos\left(X_i\frac{\pi}{2}\right)\right)$ $\left(sin\left(x_{M-m+1}\frac{\pi}{2}\right)\right)$	Unlike DTLZ1, DTLZ2 features a spherical Pareto front, where non-dominated solutions form a curved surface in the objective space, representing a smooth and continuous trade-off between the objectives.	DTLZ2 is designed to be scalable, allowing it to accommodate varying numbers of decision variables (n) and objectives (m). This scalability enables comprehensive algorithm testing across different problem dimensions.	DTLZ2 is known for having a non-convex Pareto front, which means the set of non-dominated solutions forms a complex, curved shape in the objective space.	DTLZ2 requires a trade-off between conflicting objectives. Enhancing one objective often results in the degradation of others, highlighting the need to find solutions that strike a balance.

(continued)

TABLE 13.4 (Continued)
DTLZ test functions (tri-objective text functions)

DTLZ	Objective function	Linear & nonlinear	Scalability	Convex & non-convex	Trade-off Between Objectives
DTLZ3 [20]	As DTLZ2, except the equation for g is replaced by the one form DTLZ1	DTLZ3 is a nonlinear multi-objective optimization problem with complex interactions between decision variables and a Pareto front that doesn't follow a strict linear or nonlinear pattern.	DTLZ3 is useful for assessing an algorithm's scalability with many objectives. As the number of objectives exceeds three, the Pareto-optimal solutions lie within the first octant of the unit sphere in a three-objective plot.	DTLZ3 is non-convex, meaning it may contain irregularities, concave, convex, or flat regions.	The trade-off in DTLZ3 challenges algorithms to balance exploration and exploitation, requiring evaluation across multiple objectives to find the best compromise.
DTLZ4 [20]	As DTLZ2, except all $x_i \in x$ are replaced by x_i^a, where $\alpha > 0$	DTLZ4 is generally considered a nonlinear problem, but certain aspects can be interpreted as linear depending on the context.	DTLZ4 is scalable, accommodating various decision variables and objectives, ensuring a consistent problem structure across different dimensions, akin to linear scaling.	DTLZ4 is non-convex, containing irregularities and concave regions, adding complexity and nonlinearity to the optimization problem.	DTLZ4 involves multiple objective functions, typically denoted by m, representing the trade-off between conflicting objectives.

DTLZ5 [19]	As DTLZ2, except all $x_1, x_2, \ldots, x_{M-1} \in x$ are replaced by $\frac{1+2(gxi)}{2(1+g(x))}$	The DTLZ5 test function is inherently nonlinear due to the presence of multiple objective functions and the complex interaction between decision variables and objectives.	Scalability in DTLZ5 allows researchers to assess optimization algorithms' performance across different problem dimensions, ensuring consistent behavior and insights into algorithm robustness.	The Pareto front of DTLZ5 has both convex and concave regions, complicating the optimization process and making it harder for algorithms to navigate and identify the front accurately.	DTLZ5 features multiple objective functions, denoted by m, that illustrate the trade-offs between conflicting goals, influencing the shape and complexity of the Pareto front
DTLZ6 [20]	As DTLZ5, except the equation for g is replaced by $g = \sum_{i=1}^{k} z_i^{0.1}$	DTLZ 6 is typically considered a nonlinear test function. This is because the objective functions in DTLZ 6 are constructed using nonlinear transformations.	DTLZ6 is scalable, accommodating various problem dimensions. This feature enables researchers to evaluate algorithm performance across different sizes and complexities.	DTLZ6 is non-convex with irregularities and a mix of concave, convex, or flat regions, making optimization and algorithm performance challenging.	DTLZ6 has multiple objective functions, m, which influence the trade-offs between conflicting goals and shape the Pareto front.
DTLZ7 [20]	$f_{m=2:M-1} = x_m$ $f_M = (1+g(x)) * [M - \sum_{i=1}^{M-1}\left(\left(\frac{fi}{(1+g(x))}(1+\sin(3\pi f_i(x))\right)\right]$ $g = 1+9\ (\sum_{i=1}^{k}\left(\frac{z_i}{k}\right)$	The distinction between linear and nonlinear DTLZ7 lies in the nature of the inequality constraints imposed on the objective functions.	DTLZ7 is designed to be scalable, accommodating varying numbers of decision variables and objectives.	DTLZ7 is generally non-convex, exhibiting complex geometries that pose challenges for optimization algorithms.	DTLZ7 features multiple objective functions, denoted as m, which represent the trade-offs between conflicting objectives

TABLE 13.5
Constrained multi-objective test functions (tri-objective text functions)

Constrained Multi-Objective Test Functions

Function	Objective Functions	Decision Variables	Linear & Nonlinear	Scalability	Convex & non-convex	Trade-off Between Objectives
Rastrigin Function [24]	$f(x)=\sum_{i=1}^{D}\left(x_i^2-10\cos\left(2\pi x_i\right)+10\right)$		It is a typical example of nonlinear multimodal function	As dimensionality increases, the search space becomes larger and more complex, making it harder for optimization algorithms to find the global minimum.	The Rastrigin function is non-convex due to its multiple local minima and complex landscape.	Optimization often involves trade-offs between competing objectives.
Ackley Function [25]	$f(x)=-20\exp\left(-0.2\sqrt{\frac{1}{D}\sum_{i=1}^{D}x_i^2}\right)-exp\left(\frac{1}{D}\sum_{i=1}^{D}\cos(2\pi x_i\right)+20+e$		The function's multimodal nature and lack of linearity in its components further confirm its nonlinear behavior.	As the dimensionality increases, the search space becomes larger and more complex, posing challenges for optimization algorithms to find the global minimum.	The Ackley function is non-convex, challenging optimization algorithms due to its complex landscape and abundance of local minima.	The Ackley function serves as a benchmark for testing how algorithms balance exploration (finding global solutions) and exploitation (refining local solutions).

Griewank Function [26]	$f(x)=\sum_{i=1}^{D}\frac{x_i^2}{4000}-\prod_{i=1}^{D}\cos\left(\frac{x}{\sqrt{i}}\right)+1$	The Griewank function is a classic example of a nonlinear optimization problem often used for testing and benchmarking optimization algorithms.	As with other high-dimensional functions, scalability becomes a significant consideration as the dimensionality increases.	The Griewank function is non-convex because it contains both quadratic and trigonometric terms	it's possible to extend it to a multi-objective optimization problem by introducing additional objectives or constraints.
Rosenbrock Function [27]	$f(x)=\sum_{i=1}^{n-1}\left[100\left(x_{i+1}-x_i^2\right)^2+\left(1-x_i\right)^2\right]$	It is definitely nonlinear in nature.	This shape challenges optimization algorithms due to its long, narrow basin leading to the global minimum.	is a classic non-convex optimization problem	It's possible to extend it to a multi-objective optimization problem by adding more objectives or constraints.
Sphere Function [24]	$f(x)=\sum_{i=1}^{D}x_i^2,\; x\in[-100,100]^D$	The Sphere Function is a nonlinear function. It involves the squaring of each component of the input vector and then summing up these squared values.	The function's simplicity and lack of complex mathematical operations or intricate variable dependencies make it scalable and easy to evaluate in high-dimensional spaces.	is a convex function.	arise in multi-objective optimization problems, where the goal is to find solutions that simultaneously optimize multiple conflicting objectives

As per CEC 2009, these functions present unique optimization challenges, each with its own distinctive features. They form a test suite designated for competition within the special session and competition aimed at evaluating the performance of unconstrained multi-objective optimization algorithms. These functions vary in levels of multimodality, nonlinearity, and separability, serving as benchmarks to gauge the effectiveness of multi-objective optimization algorithms in terms of convergence, diversity, and robustness [22].

In the realm of algorithm generation, a curated set of functions is employed to put algorithms through their paces. While there are a total of 13 functions available [23], not all are utilized in every test comparison. Instead, some tests opt for a subset of these functions. Nonetheless, each function serves a unique purpose in honing the evaluation strategy for achieving Pareto optimality in multi-objective algorithms. Table 13.6 provides comprehensive details for each function. This table meticulously outlines each function's specific characteristics, mathematical formulations, and objectives. It serves as a valuable reference for understanding the unique properties and challenges associated with each function, offering detailed information crucial for evaluating optimization algorithms' performance. Researchers and practitioners can rely on Table 13.6 for an in-depth understanding of the functions, enabling more precise and effective optimization problem-solving.

13.5 DIVERSE COLLECTION OF GENERAL TEST FUNCTIONS

Benchmark functions play a crucial role in the evaluation of multi-objective optimization algorithms. These functions are designed to provide standardized testing tools that can assess the efficacy of optimization algorithms across various dimensions and characteristics. In this discussion, we delve into the compilation and categorization of these general test functions, their representation, and their practical utility in algorithmic evaluation.

13.5.1 Compilation and Categorization of General Test Functions

The compilation of benchmark functions for multi-objective optimization typically includes a diverse set of functions that encapsulate different optimization challenges. The ZDT, DTLZ, WFG, and UF test suites are some of the most commonly used in this context. Each suite has been crafted to test specific aspects of multi-objective optimization algorithms, such as handling nonlinearity, multimodality, scalability, and different Pareto front shapes.

The ZDT suite, comprising ZDT1–ZDT6, focuses on bi-objective problems [30]. These functions are specifically crafted to evaluate the performance of algorithms in managing conflicting objectives. For example, ZDT1 features a linear Pareto front, simplifying the optimization problem, whereas ZDT3 introduces multiple discontinuous Pareto fronts, adding complexity. ZDT5 and ZDT6 present non-convex and multi-modal optimization landscapes [31], respectively, challenging algorithms to find globally optimal solutions amid numerous local optima.

TABLE 13.6
Unconstrained multi-objective test functions (tri-objective text functions)

Unconstrained Multi-Objective Test Functions

Function	Objective Functions	Decision Variables	Linear & Nonlinear	Scalability	Convex & Non-Convex	Trade-off Between Objectives
SCH (Srinivas–Deb) SCH (Srinivas–Deb) Function[18]	$f_1(x)=x^2$ $f_2(x)=(x-2)^2$	x is typically constrained within the range [−10,10] or a similar interval.	The SCH functions (both linear and nonlinear) They provide a clear, well-understood set of objectives and decision variables that allow researchers to assess how well their algorithms handle the balance between convergence and diversity on both simple linear and more complex nonlinear Pareto fronts.	Does not inherently exhibit scalability in terms of dimensionality, its evaluation in the context of optimization algorithms often involves assessing how algorithms perform as problem size and complexity increase.	The SCH (Srinivas–Deb) function can be both convex and non-convex, depending on its formulation.	It provides a clear example of a trade-off between two objectives and is commonly used to test and demonstrate the performance of multi-objective evolutionary algorithms.

(continued)

TABLE 13.6 (Continued)
Unconstrained multi-objective test functions (tri-objective text functions)

Unconstrained Multi-Objective Test Functions

FON (Fonseca)[17]	$f_1(x)=1-exp\left(-\sum_{i=1}^{n}\left(x_i-\frac{1}{\sqrt{n}}\right)^2\right)$ $f_{21}(x)=1-exp\left(-\sum_{i=1}^{n}\left(x_i+\frac{1}{\sqrt{n}}\right)^2\right)$	$x_i \in [-4,4]$, *for* $i=1,2,\dots,n$ n represents the number of decision variables.	Linear or nonlinear, serves as a valuable test function in multi-objective optimization. It provides a controlled environment for evaluating and comparing the performance of optimization algorithms	Assessing the scalability of the FON function helps in understanding the performance of optimization algorithms on high-dimensional multi-objective optimization problems.	The FON function does not have a convex Pareto front, as the objectives involve exponential terms and the sum of squared terms of decision variables.	The objective space consists of two dimensions, allowing for easy visualization of trade-offs between objectives.
POL (Poloni)[18]	$f1(x)=1+A1-B1$ $f2(x)=1+A2-B2$ where: A1 = 0.5sin(1) – 2cos(1) +sin(2) – 1.5cos(2) A2 = 1.5sin(1) – cos(1) +2sin(2) – 0.5cos(2) B1 = 0.5sin(x1) –2 cos(1) +sin(x2) – 1.5cos(x2) B2 = 1.5sixn(x1) – cos(x1) + 2sin(x2) – 0.5cos(x2)	The variables $X1$ and $X2$ are typically constrained within a given range, often: $-\pi \le x_1,\ x_2 \le \pi$	The POL (Poloni) function, often referred to as the Poloniomial function, is typically a nonlinear function. Polynomial functions involve terms with variables raised to powers.	Scalability of the POL (Poloni) function refers to its ability to maintain certain properties and computational efficiency as the dimensionality of the search space increases.	The POL function is non-convex but meets the PL condition, making it amenable to optimization techniques that don't rely on convexity alone.	The trade-off between objectives in the POL function typically involves balancing competing interests or goals.

KUR (Kursawe)[17]	$f_1(x)=\sum_{i=1}^{n-1}\left(-10\ \mathbf{exp}\left(-0.2\sqrt{x_i^2+x_{i+1}^2}\right)\right)$ $f_2(x)=\sum_{i=1}^{n}\left(\lvert x_i\rvert^{0.8}+5\mathbf{sin}\left(x_i^3\right)\right)$ where $x=(x_1,x_2,\ldots,x_n)$ and typically n=3	The domain for each variable is: $-5\le x_i\le 5$, $i=1,2,3$	the KUR (Kursawe) function a nonlinear function.	The scalability of the KUR function can be examined by increasing the dimensionality n of the problem. As n increases, the search space becomes larger.	KUR function is **non-convex**, and its Pareto-optimal solutions require specialized techniques for efficient exploration.	This trade-off between objectives in the KUR function presents a challenge for optimization algorithms.
VNT (Viennet)[28]	$f_1(X)=0.5(x_1^2+x_2^2)+\sin(x_1^2+x_2^2)$ $f_2(X)=\frac{(3x_1+2x_2+4)^2}{8}+\frac{(x_1-x_2+1)^2}{27}+15$ $f_3(X)=\frac{1}{x_1^2+x_2^2+1}-1.1\exp\left(-(x_1^2+x_2^2)\right)$ Where: $-3\le x_1,\ x_2\le 3$	The decision variables in the VNT function are typically denoted as X= (X1,X2), X1,X2 are real numbers within certain bounds.	The VNT (Viennet) function is a nonlinear function. It is commonly used as a test problem in multi-objective optimization.	the VNT function can be analyzed by studying how its multimodal behavior and the complexity of its landscape change as the dimensionality increases.	the VNT function includes both convex and non-convex components, with the majority being non-convex.	The trade-off between these objectives in the VNT function is evident: improving one objective usually comes at the expense of the other.

(continued)

TABLE 13.6 (Continued)
Unconstrained multi-objective test functions (tri-objective text functions)

Unconstrained Multi-Objective Test Functions						
UF1 (Unconstrained Function 1) [29]	The objective function FUf1(x) is defined as the sum of squares of the decision variables. $f_{Uf1}(x) = \sum_{i=1}^{n} x_1^2$	The decision variables Xi are bounded within the range [−1,1].	The UF1 test function is nonlinear. It's actually a benchmark function used in optimization problems to evaluate the performance of optimization algorithms.	scalability in UF1 refers to the ability of optimization algorithms to maintain their performance as the dimensionality of the problem increases.	UF1 exhibits a convex Pareto-optimal front, suggesting that the set of Pareto-optimal solutions forms a convex shape	In the case of UF1, Pareto optimality means that improving one objective (e.g., reducing the number of function evaluations)
UF2 (Unconstrained Function 2) [29]	The mathematical expression for UF2 is not explicitly provided, but it follows a similar format as UF1 with different bounds on the decision variables. $f_{UF2}(\boldsymbol{x}) = \sum_{i=1}^{n} \boldsymbol{sin}^6\left(\frac{5\pi x_i}{3}\right)$	The decision variables Xi are bounded within the range [0,1].	The UF2 test function, typically used in benchmarking optimization algorithms, is nonlinear. It is often utilized to evaluate the performance of optimization algorithms in handling nonlinear optimization problems.	By evaluating an optimization algorithm's performance on UF2 across varying dimensions, researchers can assess its scalability, robustness, and generalization capabilities.	UF2 exhibits a non-convex Pareto-optimal front, suggesting that the set of Pareto-optimal solutions does not form a convex shape.	The trade-off between objectives in UF2 can be visualized on a Pareto front, where each point represents an optimally balanced solution

The DTLZ suite extends the complexity by including problems with more than two objectives, making it suitable for testing the scalability of algorithms [32]. DTLZ1, for instance, features a linear Pareto front, while DTLZ2 has a non-convex Pareto front, adding a layer of difficulty. DTLZ problems are scalable, allowing for a varying number of decision variables and objectives, which is crucial for testing algorithms in high-dimensional spaces.

The WFG suite includes functions with varying degrees of difficulty and characteristics such as convex, concave, disconnected, and multimodal Pareto fronts [32]. These functions are designed to assess an algorithm's ability to handle diverse types of Pareto fronts and problem complexities. The UF suite, on the other hand, focuses on unconstrained problems with multiple objectives, incorporating features like multimodality and irregular Pareto fronts.

13.5.2 Representation of Function Characteristics

Each benchmark function is characterized by specific properties that influence the optimization process. These characteristics include linearity, nonlinearity, convexity, non-convexity, scalability, and the trade-off between objectives, as shown in Table 13.7.

Linearity in benchmark functions, as seen in ZDT1 [33] and DTLZ1 [34], presents a straightforward optimization landscape where the Pareto front forms a straight line. Nonlinear functions, such as ZDT2 and ZDT3, introduce more complex relationships

TABLE 13.7
Properties of benchmark test functions in multi-objective optimization

Function Name	Linearity	Scalability	Convexity	Trade-off Between Objectives	Unique Properties
ZDT1	Linear	Yes	Non-convex	Simple	Linear Pareto front
ZDT2	Nonlinear	Yes	Convex	Moderate	Curved Pareto front
ZDT3	Nonlinear	Yes	Non-convex	Complex	Multiple discontinuous Pareto fronts
ZDT4	Nonlinear	Yes	Convex	Moderate	Multi-modal optimization landscape
DT5	Nonlinear	Yes	Non-convex	Complex	Binary, deceptive, and multi-modal
ZDT6	Nonlinear	Yes	Non-convex	Complex	Highly nonlinear, non-convex
DTLZ1	Linear	Yes	Convex	Simple	Linear Pareto front
DTLZ2	Nonlinear	Yes	Non-convex	Moderate	Spherical Pareto front
DTLZ3	Nonlinear	Yes	Non-convex	Complex	Many local optima
DTLZ4	Nonlinear	Yes	Non-convex	Moderate	Scalable difficulty
DTLZ5	Nonlinear	Yes	Non-convex	Complex	Degenerate Pareto front
DTLZ6	Nonlinear	Yes	Non-convex	Complex	Degenerate and scalable
DTLZ7	Nonlinear	Yes	Non-convex	Complex	Disconnected Pareto front

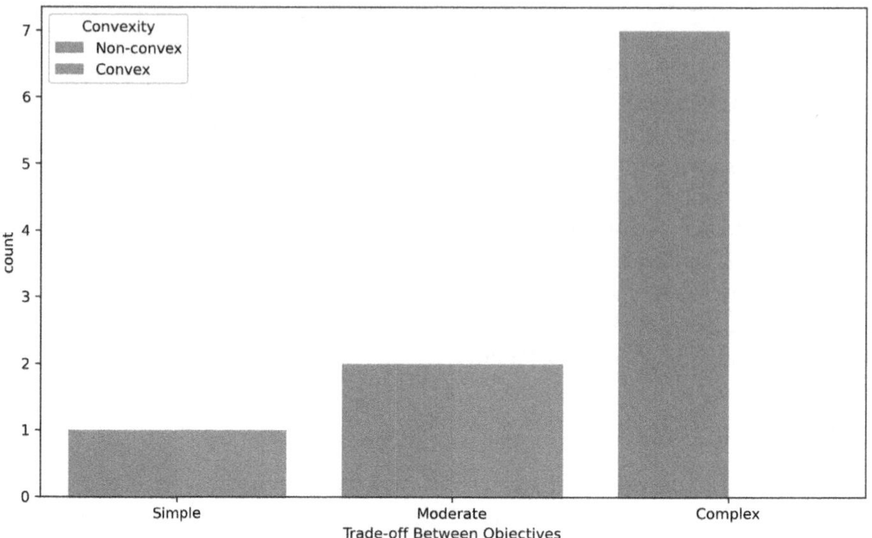

FIGURE 13.2 Trade-off between objectives based on convexity.

between decision variables and objectives, creating curved Pareto fronts. These nonlinearities challenge algorithms to navigate more intricate optimization landscapes.

Convex functions, like those in ZDT4 [35], have the property that any local minimum is also a global minimum, simplifying the optimization process. Non-convex functions, such as ZDT6 [36], present a more challenging scenario where multiple local optima exist, making it difficult for algorithms to find the global optimum. Convex Pareto fronts, as seen in some DTLZ functions, allow for easier navigation and identification of optimal solutions, whereas non-convex fronts require more sophisticated search strategies. Figure 13.2 illustrates non-convex functions, emphasizing the moderate to complex difficulty in balancing multiple conflicting objectives during function optimization.

Scalability is a critical aspect of benchmark functions, allowing them to be adapted for varying numbers of decision variables and objectives. This property is essential for testing the robustness of algorithms across different problem dimensions. For instance, the DTLZ and WFG suites are designed to be scalable, enabling comprehensive evaluation of an algorithm's performance in both low and high-dimensional spaces [13].

13.5.3 Practical Utility in Algorithmic Evaluation

Benchmark functions serve as a practical tool for evaluating the performance of multi-objective optimization algorithms. They provide a standardized framework for comparing different algorithms under consistent conditions, ensuring that the results are reliable and reproducible.

By using benchmark functions with known Pareto-optimal fronts, researchers can assess how closely an algorithm's solutions approximate the true optimal solutions.

Functions like ZDT1 and DTLZ1, with their well-defined optimal fronts, are particularly useful for this purpose. The performance metrics used in these evaluations include proximity to the true Pareto front and the diversity of solutions along the front.

Benchmark functions also help identify the strengths and weaknesses of optimization algorithms [37]. For instance, an algorithm that performs well on ZDT1 might struggle with ZDT6 due to the latter's non-convex and discontinuous nature. Similarly, an algorithm that excels in handling linear problems might need improvement when dealing with nonlinear or multimodal functions.

The insights gained from benchmarking studies can guide the development of new optimization algorithms. By understanding how existing algorithms perform on various benchmark functions, researchers can identify areas for improvement and develop new techniques that address specific challenges. For example, the development of algorithms like NSGA-II and MOEA/D was driven by the need to efficiently handle complex multi-objective problems [38].

13.6 DISCUSSION

Mathematics is frequently perceived as being entirely disconnected from everyday life. This perception creates a need for a bridge to connect the two. Many argue that mathematical representations serve this bridging function, enabling people to comprehend and convey mathematical concepts. Consequently, most multi-objective evaluations rely on mathematical functions to balance and solve real-world problems optimally. Representation can be categorized into internal and external types, but researchers typically focus on external representations. Real-world problems can be depicted using formulas, visuals, concrete objects, and other methods.

13.6.1 Overview of Significant Results and Insights

The benchmarks detailed in this chapter offer a comprehensive framework for evaluating multi-objective optimization algorithms. By providing standardized problems such as the ZDT Series and various constrained and unconstrained test functions, these benchmarks are essential for systematically comparing algorithm performance. Enhancing the accuracy of locating points in multi-objective algorithms involves several strategies and techniques, including selecting and hybridizing algorithms, improving initialization processes, enhancing fitness evaluations, applying refinement techniques, maintaining diversity, and utilizing parallelism and distributed computing. The results indicate that while some algorithms excel in handling specific problem types, others require enhancements to address diverse real-world scenarios. This insight is crucial for advancing the field of optimization by making algorithms more robust and versatile. Implementing these strategies can significantly improve the accuracy and reliability of multi-objective algorithms, leading to better performance and more dependable solutions.

13.6.2 Enhancing Real-World Applications

The practical utility of these benchmarks extends to real-world problems in design optimization, manufacturing processes, and structural health monitoring. In design optimization, benchmarks assist in evaluating structural integrity, material efficiency, and performance under different conditions. This collection of benchmarks encapsulates the most significant concepts in multi-objective optimization and provides a systematic study of the most frequently cited references in real-application engineering over recent years. It offers detailed insights into the primary multi-objective optimization algorithms and methods applied within this field, particularly in evaluation testing.

In manufacturing, benchmarks address challenges in welding, machining, and molding processes by improving precision, minimizing defects, and maximizing production efficiency. For welding, they assist in identifying optimal parameters for heat input, welding speed, and filler materials to achieve robust, high-quality welds. In machining, the benchmarks assess cutting speeds, tool paths, and cooling techniques to enhance surface finish, extend tool life, and improve material removal rates. Regarding molding, they evaluate aspects like mold design, cooling rates, and material flow to reduce defects and ensure uniform quality in molded parts. By systematically comparing optimization algorithms, these benchmarks reveal the most effective methods for refining manufacturing processes, resulting in superior product quality, lower costs, and higher productivity.

In structural health monitoring, modern algorithms using these benchmarks can process vast sensor data to predict and intervene in structural degradation more accurately. Problems in structural health monitoring focus on safeguarding structural safety and integrity through regular assessment and maintenance. While classic optimization methods historically played a key role in this field, advancements in computing have led to the adoption of new algorithms. These modern methods handle more variables, objectives, and nonlinearities more effectively. Capable of processing vast data from sensors, they offer more accurate predictions and timely interventions. This results in better detection of issues like cracks and physical degradation, ultimately enhancing the reliability and lifespan of structures. These applications demonstrate the benchmarks' role in refining algorithms for better performance and reliability in practical settings.

13.6.3 Comparison with Similar Studies

Similar studies in multi-objective optimization have advanced the use of benchmark functions, with notable suites such as ZDT, DTLZ, WFG, and UF providing foundational benchmarks [39, 40]. However, these studies often target specific aspects of multi-objective optimization. This chapter distinguishes itself by offering a more diverse and scalable set of functions, addressing a broader range of problem characteristics and complexities. Unlike many previous studies that evaluate specific algorithms like NSGA-II, MOEA/D, SPEA2, and MOPSO using a limited set of benchmarks [10, 39, 41], this chapter presents functions applicable to a wide range of algorithms, making it a versatile resource. Its comprehensive library and detailed classifications enable nuanced algorithm comparisons, helping researchers identify strengths and weaknesses across different scenarios. Additionally, the chapter's emphasis on real-world applicability, through characteristics such as modality and separability, ensures its benchmarks are relevant for practical optimization tasks.

The systematic overview and categorization provided serve as valuable guidance for researchers and developers in selecting the most appropriate benchmark functions for their specific needs.

While this study does compare itself to similar research, it offers several unique contributions that distinguish it. Unlike many existing studies that primarily use static, artificially constructed benchmark functions, this research develops benchmarks that more closely mimic real-world scenarios by incorporating dynamic changes, noise, and multi-modalities. This approach addresses the common gap of lacking real-world relevance. Additionally, whereas most studies focus on bi-objective problems, this research extends to high-dimensional objective spaces, allowing for a more comprehensive evaluation of algorithm scalability and performance in complex settings.

The study also proposes a comprehensive evaluation framework that is more advanced than those used in similar research, incorporating robust performance metrics and statistical methods to provide a thorough assessment of algorithms' adaptability and robustness. This addresses the limitations of current evaluation metrics, which may not fully capture real-world applicability.

Finally, by designing benchmark functions inspired by practical problems in diverse fields such as engineering, healthcare, and environmental management, this research ensures that evaluations are both theoretically sound and practically relevant. This unique approach enhances the real-world applicability and effectiveness of optimization algorithms, setting this study apart from other research in the field.

13.7 LIMITATIONS OF BENCHMARKING IN MULTI-OBJECTIVE OPTIMIZATION

Benchmarking in multi-objective optimization is fraught with limitations that undermine its effectiveness. The selection of benchmark problems often lacks diversity, failing to capture the wide range of real-world scenarios. This leads to problem selection bias and questions about generalizability.

Scalability issues arise as many benchmarks are small and do not reflect the computational complexity of real-world applications. Thus, they do not adequately test algorithm performance under increased problem sizes.

The selected performance metrics, such as hypervolume and Pareto front coverage, can be biased. They may favor specific algorithm characteristics and fail to adequately balance the trade-off between convergence and solution diversity.

Moreover, ensuring reproducibility is difficult due to variations in algorithm implementation, parameter settings, and computational environments. This leads to inconsistent and non-comparable results across studies. There is also a tendency for algorithms to be overfitted to benchmark problems rather than being designed for broader applicability, reducing their practical utility.

Static and deterministic benchmarks fail to test algorithm performance in dynamic and uncertain conditions typical of real-world problems. The high computational costs of extensive benchmarking and the lack of standardized practices in problem selection, performance metrics, and result reporting further exacerbate these issues. This makes it challenging to derive meaningful and generalizable conclusions from benchmarking studies.

13.8 CONCLUSION

Mathematical benchmark test functions are crucial for developing and evaluating optimization algorithms, providing standardized problems to systematically compare algorithm performance, identify strengths and weaknesses, guide improvements, and understand behavior across various problem types. These benchmarks ensure robustness, scalability, and efficiency by offering diverse challenges, although they may not always mirror real-world complexity. A curated collection of benchmarks for multi-objective optimization evaluates functions based on specific objectives and traits, including modality, separability, and scalability, encompassing the ZDT Series, and both constrained and unconstrained test functions tailored for real-world applications. The practical implications are significant for industries like logistics, finance, engineering, and artificial intelligence, allowing researchers and practitioners to better assess algorithm performance in scenarios mimicking real-world challenges, leading to more reliable and efficient solutions for optimizing supply chains, financial portfolios, engineering designs, and machine learning models. Our research offers valuable insights and tools for the optimization community, advancing the development of more effective solutions for complex multi-objective problems and enhancing the field's applications across various industries.

Future research should focus on developing comprehensive and diverse benchmarks that accurately represent real-world complexities, enhancing scalability tests for high-dimensional problems, improving performance metrics, ensuring reproducibility through standardization, creating shared repositories, integrating adaptive and self-tuning mechanisms, leveraging machine learning and data analytics, and promoting collaborative benchmarking among researchers and practitioners to drive progress in the field.

APPENDIX A

TABLE A.1
Brief explanations of key multi-objective optimization terms

Term	Explanation
Multi-Objective Optimization (MOO)	Optimizing two or more conflicting objectives simultaneously to find a balance, resulting in the Pareto front.s
Pareto Optimality	A state where no objective can be improved without worsening another. A solution is Pareto optimal if it isn't dominated by any other solution.
Benchmark Functions	Standardized test problems to evaluate and compare optimization algorithms, representing various problem characteristics.
ZDT Functions	Benchmark problems for two-objective MOO, focusing on challenges like nonlinearity and Pareto front shapes.
DTLZ Functions	Benchmark problems for assessing MOO algorithms in higher-dimensional spaces.

TABLE A.1 (Continued)
Brief explanations of key multi-objective optimization terms

Term	Explanation
NSGA-II	An evolutionary algorithm for MOO that sorts solutions by dominance and promotes diversity to explore the Pareto front.
MOEA/D	Decomposes MOO problems into simpler subproblems to find Pareto-optimal solutions efficiently.
SPEA2	An enhanced version of SPEA that balances solution quality and diversity to approach the Pareto front.
MOPSO	Particle Swarm Optimization algorithm for MOO, updating solutions based on performance and interaction over time.
MO-SHERPA	A method for optimizing hyperparameters in MOO by managing multiple objectives and generating optimal solutions.
UF Functions	A set of benchmark problems for MOO designed to test algorithms' ability to handle a variety of problem characteristics, including complexity and dimensionality.
WFG Functions	Benchmark problems for MOO focusing on scalable and flexible test problems with controllable complexity, useful for evaluating the performance of MOO algorithms across different problem scenarios.

REFERENCES

[1] I. R. Sapre Mandar and S. Kale, "A Brief Review of Bilevel Optimization Techniques and Their Applications," in *Handbook of Formal Optimization*, A. H. Kulkarni Anand J. and Gandomi, Ed., Singapore: Springer Nature Singapore, 2024, pp. 1179–1202. doi: 10.1007/978-981-97-3820-5_34.

[2] A. M. Aladdin and T. A. Rashid, "A new lagrangian problem crossover—A systematic review and meta-analysis of crossover standards," *Systems*, vol. 11, no. 3, p. 144, Mar. 2023, doi: 10.3390/systems11030144.

[3] N. Li, L. Ma, G. Yu, B. Xue, M. Zhang, and Y. Jin, "Survey on evolutionary deep learning: Principles, algorithms, applications, and open issues," *ACM Comput Surv*, vol. 56, no. 2, pp. 1–34, 2023.

[4] R. Kruse, S. Mostaghim, C. Borgelt, C. Braune, and M. Steinbrecher, "Introduction to Evolutionary Algorithms," in *Computational Intelligence: A Methodological Introduction*, Cham: Springer International Publishing, 2022, pp. 225–254. doi: 10.1007/978-3-030-42227-1_11.

[5] C. Picard and J. Schiffmann, "Realistic constrained multiobjective optimization benchmark problems from design," *IEEE Transac Evol Comput*, vol. 25, no. 2, pp. 234–246, 2021, doi: 10.1109/TEVC.2020.3020046.

[6] R. Cheng *et al.*, "A benchmark test suite for evolutionary many-objective optimization," *Complex Intell Syst*, vol. 3, no. 1, pp. 67–81, 2017, doi: 10.1007/s40747-017-0039-7.

[7] T. Bartz-Beielstein *et al.*, "Benchmarking in optimization: Best practice and open issues," Jul. 2020, [Online]. Available: http://arxiv.org/abs/2007.03488

[8] P. P. Ray, "Benchmarking, ethical alignment, and evaluation framework for conversational AI: Advancing responsible development of ChatGPT," *BenchCouncil Transac Benchmarks Stand Eval*, vol. 3, no. 3, p. 100136, Sep. 2023, doi: 10.1016/j.tbench.2023.100136.

[9] K. Farzane and D. Alireza, "A review and evaluation of multi and many-objective optimization: Methods and algorithms," *Glob J Ecol*, vol. 7, no. 2, pp. 104–119, Nov. 2022, doi: 10.17352/gje.000070.
[10] H. Ma, Y. Zhang, S. Sun, T. Liu, and Y. Shan, "A comprehensive survey on NSGA-II for multi-objective optimization and applications," *Artif Intell Rev*, vol. 56, no. 12, pp. 15217–15270, 2023, doi: 10.1007/s10462-023-10526-z.
[11] F. Karl *et al.*, "Multi-objective hyperparameter optimization -- An overview," Jun. 2022, [Online]. Available: http://arxiv.org/abs/2206.07438
[12] W. Huang and W. Zhang, "Adaptive multi-objective particle swarm optimization using three-stage strategy with decomposition," *Soft Comput*, vol. 25, no. 23, pp. 14645–14672, 2021, doi: 10.1007/s00500-021-06262-7.
[13] I. R. Meneghini, M. A. Alves, A. Gaspar-Cunha, and F. G. Guimarães, "Scalable and customizable benchmark problems for many-objective optimization," *Appl Soft Comput*, vol. 90, p. 106139, Jan. 2020, doi: 10.1016/j.asoc.2020.106139.
[14] R. Cheng, Y. Jin, M. Olhofer, and B. Sendhoff, "Test problems for large-scale multiobjective and many-objective optimization," *IEEE Trans Cybern*, vol. 47, no. 12, pp. 4108–4121, 2017, doi: 10.1109/TCYB.2016.2600577.
[15] J. Wan, P. Chu, Y. Jiao, and Y. Li, "Improvement of machine learning enhanced genetic algorithm for nonlinear beam dynamics optimization," *Nucl Instrum Methods Phys Res A,* vol. 946, pp. 162683, 2019. [Online]. Available: www.elsevier.com/open-access/userlicense/1.0/
[16] M. R. Sharifi, S. Akbarifard, K. Qaderi, and M. R. Madadi, "A new optimization algorithm to solve multi-objective problems," *Sci Rep*, vol. 11, no. 1, Dec. 2021, doi: 10.1038/s41598-021-99617-x.
[17] M. S. Noori, R. K. Z. Sahbudin, A. Sali, and F. Hashim, "Multi-objective multi-exemplar particle swarm optimization algorithm with local awareness," *IEEE Access*, p. 1, 2024, doi: 10.1109/ACCESS.2024.3426104.
[18] M. Alndiwee, M. M. Al-Mahairi, and R. Hamdan, "A novel bundle adjustment approach based on Guess-Aided and Angle Quantization Multiobjective Particle Swarm Optimization (GAMOPSO) for 3D reconstruction applications," *IEEE Access*, vol. 10, pp. 71508–71520, 2022, doi: 10.1109/ACCESS.2022.3187093.
[19] W.-H. Tan and J. Mohamad-Saleh, "MO-NFSA for solving unconstrained multi-objective optimization problems," *Eng Comput*, vol. 38, no. 3, pp. 2527–2548, 2022, doi: 10.1007/s00366-020-01223-4.
[20] A. Singh, Lekhraj, A. Kumar, and A. Kumar, "Multi-objective differential evolution algorithm with a new environmental parameter based mutation for solving optimization problems," *Trends Sci*, vol. 18, no. 20, Oct. 2021, doi: 10.48048/tis.2021.17.
[21] N. A. Rashed *et al.*, "Unraveling the versatility and impact of multi-objective optimization: Algorithms, applications, and trends for solving complex real-world problems," 2024. [Online]. Available: https://jscca.uotechnology.edu.iq/jsccaAvailableat:https://jscca.uotechnology.edu.iq/jscca/vol1/iss1/9
[22] Q. Zhang, A. Zhou, S. Zhao, P. N. Suganthan, W. Liu, and S. Tiwari, "Multiobjective optimization test instances for the CEC 2009 special session and competition," 2008.
[23] F. Hopfgartner *et al.*, "Evaluation-as-a-service for the computational sciences: Overview and outlook," *J Data Inf Qual*, vol. 10, no. 4, pp. 1–32, 2018.
[24] R. V. Rao, "Application of TLBO and ETLBO Algorithms on Complex Composite Test Functions," in *Teaching Learning Based Optimization Algorithm: And Its Engineering Applications*, Cham: Springer International Publishing, 2016, pp. 41–51. doi: 10.1007/978-3-319-22732-0_3.

[25] S. Goitom, T. Nagy, and T. Turányi, "Testing Various Numerical Optimization Methods on a Series of Artificial Test Functions," in *Annales Universitatis Scientiarum Budapestinensis de Rolando Eötvös Nominatae*. Sectio Computatorica, 2022.

[26] M. Abualhaj, A. Adel, S. Al-Khatib, and Q. Shambour, "MVF: A novel technique to reduce the Voip packet payload length," *Transp Telecommun J*, vol. 25, pp. 43–53, Jul. 2024, doi: 10.2478/ttj-2024-0005.

[27] F. Pagani, M. Wiegand, and S. Nadarajah, "An n-dimensional Rosenbrock distribution for Markov chain Monte Carlo testing," *Scand J Stat*, vol. 49, no. 2, pp. 657–680, 2022, https://doi.org/10.1111/sjos.12532.

[28] L. A. Márquez-Vega, J. G. Falcón-Cardona, and E. Covantes Osuna, "On the adaptation of reference sets using niching and pair-potential energy functions for multi-objective optimization," *Swarm Evol Comput*, vol. 83, p. 101408, 2023, https://doi.org/10.1016/j.swevo.2023.101408.

[29] R. Venkata Rao and G. G. Waghmare, "A comparative study of a teaching–learning-based optimization algorithm on multi-objective unconstrained and constrained functions," *J King Saud Univ Comp Inf Sci*, vol. 26, no. 3, pp. 332–346, 2014, https://doi.org/10.1016/j.jksuci.2013.12.004.

[30] Y. Liu, J. Liu, S. Tan, Y. Yang, and F. Li, "A bagging-based surrogate-assisted evolutionary algorithm for expensive multi-objective optimization," *Neural Comput Appl*, vol. 34, no. 14, pp. 12097–12118, 2022, doi: 10.1007/s00521-022-07097-5.

[31] Y. Cui, Z. Geng, Q. Zhu, and Y. Han, "Review: Multi-objective optimization methods and application in energy saving," *Energy*, vol. 125, pp. 681–704, 2017, https://doi.org/10.1016/j.energy.2017.02.174.

[32] D. Brockhoff, A. Auger, N. Hansen, and T. Tušar, "Using well-understood single-objective functions in multiobjective Black-Box optimization test suites," *Evol Comput*, vol. 30, no. 2, pp. 165–193, Jun. 2022, doi: 10.1162/evco_a_00298.

[33] A. M. Khalid, H. M. Hamza, S. Mirjalili, and K. M. Hosny, "MOCOVIDOA: A novel multi-objective coronavirus disease optimization algorithm for solving multi-objective optimization problems," *Neural Comput Appl*, vol. 35, no. 23, pp. 17319–17347, 2023, doi: 10.1007/s00521-023-08587-w.

[34] M. Wang, F. Ge, D. Chen, and H. Liu, "A Many-objective Evolutionary Algorithm using Determinantal Point Process in Potential Region," in *Proceedings of the 6th International Conference on Control Engineering and Artificial Intelligence*, 2022, pp. 83–91.

[35] M. Liang, L. Wang, L. Wang, and Y. Zhong, "A hypervolume-based cuckoo search algorithm with enhanced diversity and adaptive scaling factor," *Appl Soft Comput*, vol. 151, p. 111073, 2024, https://doi.org/10.1016/j.asoc.2023.111073.

[36] H. M. Sheikh and P. S. Marcus, "Bayesian optimization for mixed-variable, multi-objective problems," *Struct Multidiscip Optim*, vol. 65, no. 11, p. 331, 2022, doi: 10.1007/s00158-022-03382-y.

[37] P. Sharma and S. Raju, "Metaheuristic optimization algorithms: A comprehensive overview and classification of benchmark test functions," *Soft Comput*, vol. 28, no. 4, pp. 3123–3186, 2024, doi: 10.1007/s00500-023-09276-5.

[38] H. Li, K. Deb, Q. Zhang, P. N. Suganthan, and L. Chen, "Comparison between MOEA/D and NSGA-III on a set of novel many and multi-objective benchmark problems with challenging difficulties," *Swarm Evol Comput*, vol. 46, pp. 104–117, 2019, https://doi.org/10.1016/j.swevo.2019.02.003.

[39] J. G. Falcón-Cardona, R. Hernández Gómez, C. A. Coello Coello, and M. G. C. Tapia, "Parallel multi-objective evolutionary algorithms: A comprehensive survey," *Swarm Evol Comput*, vol. 67, p. 100960, 2021, https://doi.org/10.1016/j.swevo.2021.100960.

[40] J. L. J. Pereira, G. A. Oliver, M. B. Francisco, S. S. Cunha, and G. F. Gomes, "A review of multi-objective optimization: Methods and algorithms in mechanical engineering problems," *Arch Comput Methods Eng*, vol. 29, no. 4, pp. 2285–2308, 2022, doi: 10.1007/s11831-021-09663-x.

[41] J. Singh, "A Review and Comparison of Two Archive Based Algorithms: SPEA2 and PAES," in *AIP Conference Proceedings*, AIP Publishing, 2023.

14 CSANA

Cosine Similarity Ant Nesting Algorithm Based on an Improved Ant Nesting Algorithm

Esra Muhannad Shawkat, Tarik A. Rashid and Aso M. Aladdin

14.1 INTRODUCTION

The Ant Nesting Algorithm (ANA) was developed in 2021 by Hama Rashid and Rashid. ANA was inspired by the behavior of Leptothorax ants while building a new nest. The model entities mimic the ant's actions. The algorithm's main component is derived from ant behavior in the search for a suitable place to deposit grain. Worker ants are represented as artificial search agents. Worker ants drop grains around the queen ant and the grains' positions correspond to potential solutions explored by the search agents. ANA is utilized to address optimization problems by iteratively exploring to determine the best solution. ANA is a Fitness Dependent Optimizer (FDO)-based model, but it uses the Pythagorean theorem to generate weights [1]. CSANA is an enhanced version of the original ANA. ANA has some drawbacks, such as slow convergence rates due to its heavy reliance on the Pythagorean theorem during generating weights for worker ants while searching for the deposition position. Moreover, ANA lacks an exploration phase (global best variable); it only has an exploitation phase (local best variable). There exists no balance between the exploration and exploitation phases. The fundamental work of this study is providing another mathematical equation for generating weights as they are utilized for driving search agents during the search phases, that is, cosine similarity. Additionally, nested cosine similarity hybridized with the Euclidean distance is used, which improves the algorithm's convergence speed, efficiency, performance, and execution time. CSANA represents an enhanced version of ANA, as it follows a similar procedure for updating agent positions but in a distinct method. Additionally, a balance is introduced between exploration and exploitation phases in ANA. This study presents CSANA to minimize computing time, enhance algorithm speed, make the algorithm clearer to understand, and reduce iterations in the algorithm, which are the major reasons for algorithm speed reduction. Section 14.2 discusses ANA in detail, with its mathematical modeling, Section 14.3 provides CSANA with its mathematical model,

DOI: 10.1201/9781003601555-14

Section 14.4 discusses the results of CSANA and compares them to other results, and Section 14.5 concludes the chapter.

14.2 ANT NESTING ALGORITHM

The Leptothorax is a social insect that lives in colonies that can grow to a size of several hundred individuals. It prefers horizontal crevices and builds nests in flat areas of rocks. These places provide the ants with a haven from physical threats. To keep themselves protected from the elements, the ants build a barrier composed of small pieces of stone or grains around their chosen crevice. This type of wall is ideal for the size of a colony of Leptothorax ants, which usually includes a queen and several workers. It has a narrow opening that allows ants to defend themselves against physical and biological threats. Worker ants, which are about 2 to 3 millimeters long, are responsible for building the nest. Unlike other insects, the Leptothorax doesn't build a floor or roof. Instead, it focuses on creating a protective barrier surrounding its queen and broods, which includes its eggs, larvae, pupae, and the queen's brood. Their arrangement encircling the brood cluster acts as a guide that helps the ants identify the ideal locations for their nest walls [2, 3]. Figure 14.1 shows a Leptothorax ant nest built by ants in the laboratory.

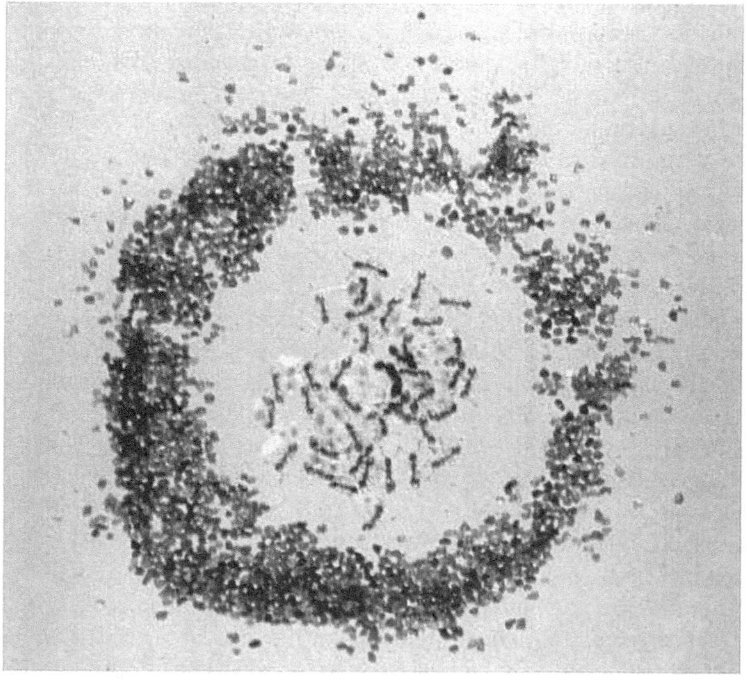

FIGURE 14.1 Leptothorax ant nest in the laboratory [2].

ANA involves worker ants depositing grain in positions randomly within specified limits during each iteration. In the first generation, all ants start with the same initial positions. The algorithm compares each ant's position to the local best, calculating changes based on alignment with the best solution. New positions are derived and fitness is evaluated. If a superior solution is found, it is accepted; otherwise, the algorithm continues with the current solution. Each worker ant that explores a particular deposition position exhibits a potential solution and the entire process of choosing the deposition position from the many favorable options is regarded as approaching optimality [1]. In ANA, each worker ant that explores a particular deposition position exhibits a potential solution, and the entire process of choosing the deposition position from the many favorable options is regarded as approaching optimality. The artificial workers utilize the deposition mechanism to explore the environment randomly. The artificial worker ant's current position $X_{t,i}$ is automatically updated whenever it finds a new ideal position, it is represented as $X_{t,i+1}$ ($t = 1,2,3, \ldots, m$ and $i = 1,2,3, \ldots, n$), where "i" denotes the current worker ant, "t" denotes the current iteration, and "X" represents the position where the artificial worker ant deposits grains (deposition position). It does so by adding the rate of change ($\Delta X_{t,\, i+1}$) of the new position to its existing position (current deposition position $X_{t,i}$) using the following expression:

$$X_{t,i+1} = X_{t,i} + \Delta X_{t,i+1} \tag{1}$$

The rate at which the deposition position changes $\Delta X_{t,i+1}$ relies on two factors: the deposition weight dw and the difference between the position of the local best-known worker ant $X_{t,ibest}$ and the deposition position of the current worker ant $X_{t,i}$. Hence, the calculation of $\Delta X_{t,i+1}$ is determined by the following equation:

$$\Delta X_{t,\, i+1} = dw \times \left(X_{t,i\ best} - X_{t,i} \right) \tag{2}$$

In ANA, while the worker ant performs a search in the search space, several conditions must be satisfied for the algorithm to calculate the rate of change of the deposition position $\Delta X_{t,i+1}$ as follows:

If the current worker is regarded as the best-known ant:

$$\Delta X_{t,\, i+1} = r \times X_{t,i} \tag{3}$$

If the current and previous deposition position are the same, the following equation is employed:

$$\Delta X_{t,\, i+1} = r \times \left(X_{t,i\ best} - X_{t,i} \right) \tag{4}$$

Worker ants perform random walks while searching for deposition grains. These random walks are expressed as a deposition weight dw, which is obtained from the tendency rates (current and previous) for depositing grains at a particular position, where $T_{previous}$ refers to the previous tendency rate and T refers to the current tendency rate. To regulate the deposition weight, ANA uses a deposition factor denoted by r, which is a random value ranging from -1 to 1. Among various methods of generating random numbers, ANA specifically uses the Levy flight mechanism. Due to its favorable distribution curve, it provides more consistent and stable movements [4]. The dw is calculated as follows:

$$dw = r \times \left(\frac{T}{T_{previous}} \right) \tag{5}$$

The values of T and $T_{previous}$ are determined using the Pythagorean theorem:

$$T = \sqrt{\left(X_{t,i\ best} - X_{t,i}\right)^2 + \left(X_{t,i\ best\ fitness} - X_{t,i\ fitness}\right)^2} \tag{6}$$

$$T_{previous} = \sqrt{\left(X_{t,i\ best} - X_{t,i\ previous}\right)^2 + \left(X_{t,i\ best\ fitness} - X_{t,i\ previous\ fitness}\right)^2} \tag{7}$$

The working mechanism for ANA is as follows: For each worker ant, the algorithm begins by randomly setting deposition positions $X_{t,i}$ (where t = 1, 2, 3, 4, ..., m and i = 1, 2, 3, 4, ..., n) within the specified lower and upper limits. At the beginning, since it is the first generation, the previous deposition position $X_{t,iprevious}$ is set the same as $X_{t,i}$ for every worker ant. During every iteration, the local best deposition position $X_{t,ibest}$ is determined then a random value r is generated within the interval -1 to 1. Next, the

TABLE 14.1
ANA's mathematical notation [1]

Notation	Description
t	Current iteration
i	Current worker ant
m	Iteration number
n	Worker ant number
$X_{t,i}$	Worker ant's current deposition position
$X_{t,ibest}$	Worker ant's local best deposition position
$X_{t,iprevious}$	Worker ant's previous deposition position
$X_{t,i}\,fitness$	Worker ant's current deposition position's fitness
$X_{t,ibest}\,fitness$	Worker ant's local best deposition position fitness
$X_{t,iprevious}\,fitness$	Worker ant's previous deposition position fitness
$X_{t,i+1}$	Worker ant's new deposition position
$\Delta X_{t,i+1}$	Worker ant's deposition position's rate of change
T	Worker ant's current deposition tendency rate
$T_{previous}$	Worker ant's previous deposition tendency rate
dw	Deposition weight
r	Random number in the [-1, 1] range

algorithm compares $X_{t,i}$ to $X_{t,ibest}$ for every worker ant. $\Delta X_{t,i+1}$ is calculated using Eq. (3) if the current worker ant's deposition position aligns with the best solution found thus far. If the previous deposition position is equal to the current deposition position, $\Delta X_{t,i+1}$ is calculated through the utilization of Eq. (4); otherwise, *T*, *Tprevious*, *dw*, and $\Delta X_{t,i+1}$ are calculated employing Eqs. (6), (7), (5), and (2), respectively. Following this, a new solution $X_{t,i+1}$ is derived using Eq. (1). Whenever an artificial worker ant discovers a new solution, it evaluates the fitness of the newly found solution using the fitness function. If the recently discovered solution is superior to the current solution, it is approved, and the former solution is preserved as $X_{t,iprevious}$. By contrast, if the newly discovered solution is not superior, the algorithm continues with the current solution until the upcoming iteration. Table 14.1 presents a summary of all the notation utilized in ANA.

14.3 COSINE SIMILARITY ANT NESTING ALGORITHM

The Cosine Similarity ANA (CSANA) is an enhanced version of the original ANA. ANA was inspired by the behavior of ants. It has some drawbacks, such as slow convergence rates due to its heavy reliance on the Pythagorean theorem during generating weights for worker ants while searching for the deposition position. Moreover, ANA lacks an exploration phase (global best variable); it only has an exploitation phase (local best variable). There exists no balance between the exploration and exploitation phases. However, in CSANA, these limitations are improved. CSANA addresses the slow convergence rates by changing the reliance of the Pythagorean theorem to another mathematical theorem. It also introduces the balance between the exploration and exploitation phases. Additionally, CSANA integrates the storage of past rate changes that are found by worker ants for potential reuse in future steps, which enhances ANA's performance. In the upcoming subsections, each of the enhancements is explained more in detail.

14.3.1 COMPREHENSIVE EXPLORATION AND EXPLOITATION BALANCE

Attaining an optimal balance between the exploration phase and exploitation phase in an optimizing procedure is very important for ensuring that the process operates properly since it impacts the performance of the process. The precision of the obtained solution, which measures the ability to discover optimal solutions, and the convergence speed, which determines how quickly satisfactory solutions are reached, are factors that are affected by this balance. The inability to achieve a balanced approach can result in various issues, for example, if the diversification phase (exploration) is too high, this can result in slowing down the speed of the convergence rate as it investigates the search space extensively in a random manner. However, if the exploration search strength is too low, the risk of local optima increases and it may cause the algorithm to trap local optima. Similarly, if the intensification phase (exploitation) is too high, it may cause a premature convergence rate, which leads to the development of biased solutions that do not provide much value as the search area is not extensively explored. Conversely, if the intensification is too low, it decreases the speed of the convergence rate. In the optimization world, metaheuristic algorithms tend to encounter issues when trying to reach the global optimal. They are often trapped in local optima; multiple local optima are typically encountered in optimization problems. These are solutions that seem to be optimal in a limited area, but when it comes to global areas, they are not optimal. The introduction of a global variable can help to solve this issue. It can provide an efficient reference point for the exploration process and help the algorithm to find a promising solution that goes beyond local optima, which allows the algorithm to do more effective exploration and avoids getting stuck in suboptimal solutions. By introducing global search and having a global optimum solution or near optimum solution, the metaheuristic algorithm can recognize and discard local optima solutions, enabling searching for potentially better solutions within the search space, resulting in improved efficiency and performance in metaheuristic algorithms, which can result in reduced computation time and a faster convergence rate. Therefore, one of the most important aspects of a successful metaheuristic algorithm is balancing between exploitation and exploration searches [5].

In ANA, there is an absence of a balance between the search phases (exploration and exploitation). The algorithm focuses only on identifying local bests (exploitation phase), which are the worker ant's local best deposition position and the fitness of the worker ant's local best deposition position, as well as comparing them to the current parameters, which are the worker ant's current deposition position and its fitness value, and the previous parameters represented, which are the worker ant's previous deposition position and its fitness value. However, there is no consideration for a global best variable, which means that ANA lacks an exploration phase.

In CSANA, during the search process, a balance between exploration and exploitation phases is established. "$X*_{i,tbest\ fitness}$" indicates the fitness function value of the best global solution obtained by artificial worker ants so far. The global variable is used to measure the fitness value of the best solution during the entire search process within the entire search space. By contrast, the "$X_{t,ibest}$" parameter refers to the worker ant's local best deposition position. It depicts the best deposition position found in each iteration of the search process and is considered as a local variable. By using both the global best fitness value and the local best deposition position, CSANA ensures the balance between exploration and exploitation is achieved. The global best fitness value, which corresponds to the exploration phase, allows the algorithm to search for new and diverse solutions beyond local optima. Meanwhile, the local best deposition position, which is referred to as an exploitation phase, has a great role in refining the current solutions.

This balance between the exploration and exploitation phases that CSANA achieves makes the algorithm progress more effectively and efficiently, and increases the convergence rate, which leads to optimal or near-optimal solutions, as the enhanced results of CSANA show in Section 4.

14.3.2 A Novel Formula to Generate the Deposition Weight

ANA heavily depends on the Pythagorean theorem, which is the algorithm's fundamental mathematical representation. The deposition weight (*dw*) is generated by using the Pythagorean theorem. The *dw* is a random weight assigned to a worker ant for depositing grain at a particular location during its random walk and has a great role in the worker ant's decision-making factor while the worker ant selects a location to deposit grains. The Pythagorean theorem is mainly focused on the distance between points, but it has some limitations that can affect the algorithm. One of these limitations is that it does not take into consideration the relationships and directional aspects between vectors and this drawback can prevent the algorithm from effectively searching for solutions in the search space. Additionally, the Pythagorean theorem applies to right-angled triangles, as one of the angles measures 90 degrees. This causes a problem when trying to find solutions in complex situations, such as when an ant is searching in crevices and constructing a nest within crevices found in rocks. Since the right-angled triangles may not be formed all the time, this leads to the Pythagorean theorem not being able to provide a precise measurement of distances. CSANA can address these issues by introducing nested cosine similarity, which is hybridized with the Euclidean distance. The angle-based measure (cosine similarity) is mainly used to capture the directional elements of data points through their angles. The distance metric (Euclidean distance) provides information about the distance between the data points, regardless of their orientation. By hybridizing distance and similarity measures, CSANA develops a more comprehensive understanding of the relationships between data points. This integration makes the algorithm an overview of both the differences and similarities between the data points, and performs a more detailed analysis of relationships and patterns among the data points [6]. The concept shown in Figure 14.2, and Leptothorax ant nest built by ants in the laboratory, as shown in Figure 14.3.

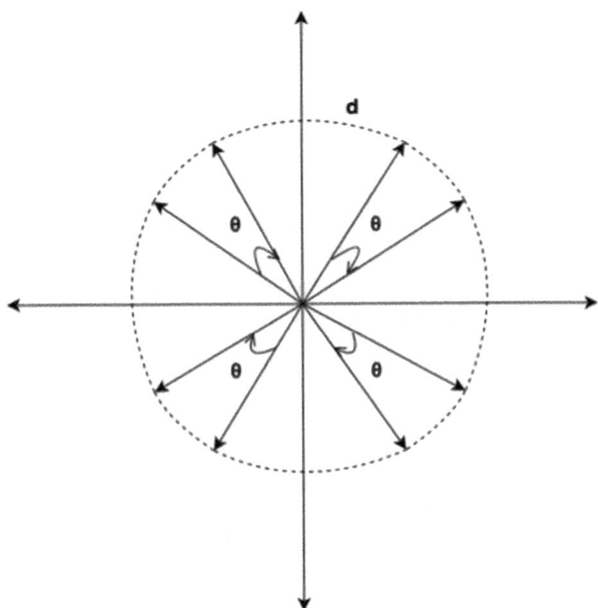

FIGURE 14.2 Cosine similarity (•) and Euclidean distance (*d*).

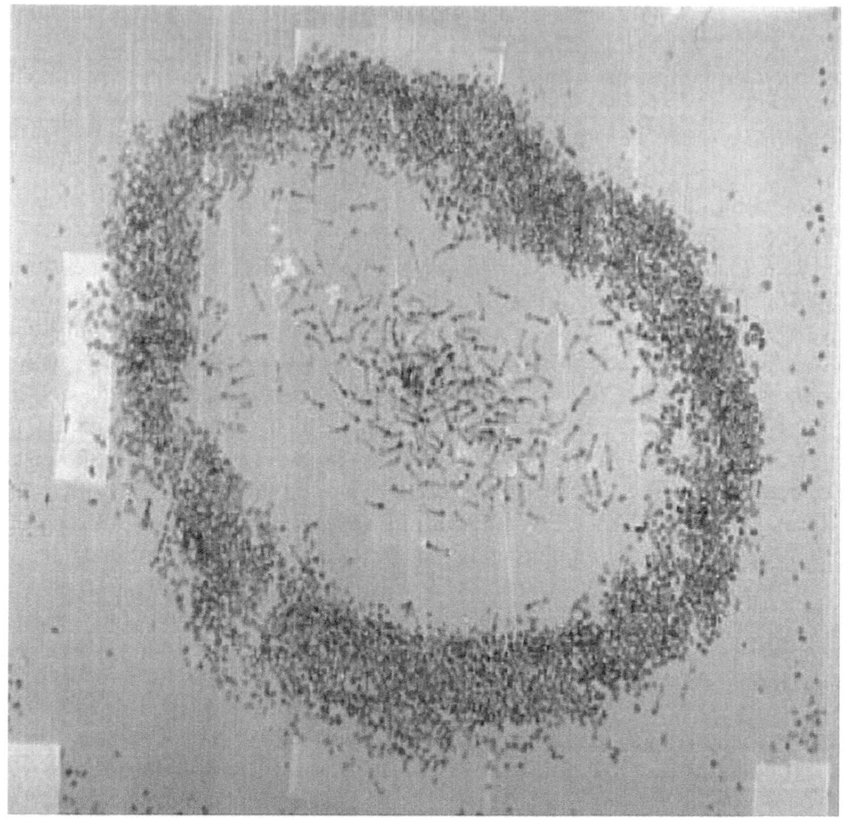

FIGURE 14.3 Leptothorax ant nest in the laboratory [2].

As CSANA changes, the mathematical theorem first finds the nested cosine similarity for the current tendency rate, which is denoted by *T*, and the previous tendency rate, which is denoted by $T_{previous}$. To find the current tendency rate (*T*), first, cosine similarity between ($X_{t,ibest}$, $X_{t,i}$) is presented in Figure 14.4 (a). The equation below shows its mathematical representation:

$$cosine\ similarity\left(X_{\oint t,ibest}, X_{t,i}\right) = \frac{\left(X_{t,ibest}\right)\cdot\left(X_{\mathrm{t,i}}\right)}{\|X_{t,ibest}\|\ \|X_{\mathrm{t,i}}\|} \tag{8}$$

The cosine similarity between ($X^*_{t,ibest\ fitness}$, $X_{t,i\ fitness}$) is shown in Figure 14.4 (b). Its calculation is presented in the following equation:

$$\text{cosine similarity}\left(\mathrm{X}^*_{\text{t,ibest fitness}}, \mathrm{X}_{\text{t,i fitness}}\right) = \frac{(\mathrm{X}^*_{\text{t,ibest fitness}})\cdot(\mathrm{X}_{\text{t,i fitness}})}{\|\mathrm{X}^*_{\text{t,ibest fitness}}\|\ \|\mathrm{X}_{\text{t,i fitness}}\|} \tag{9}$$

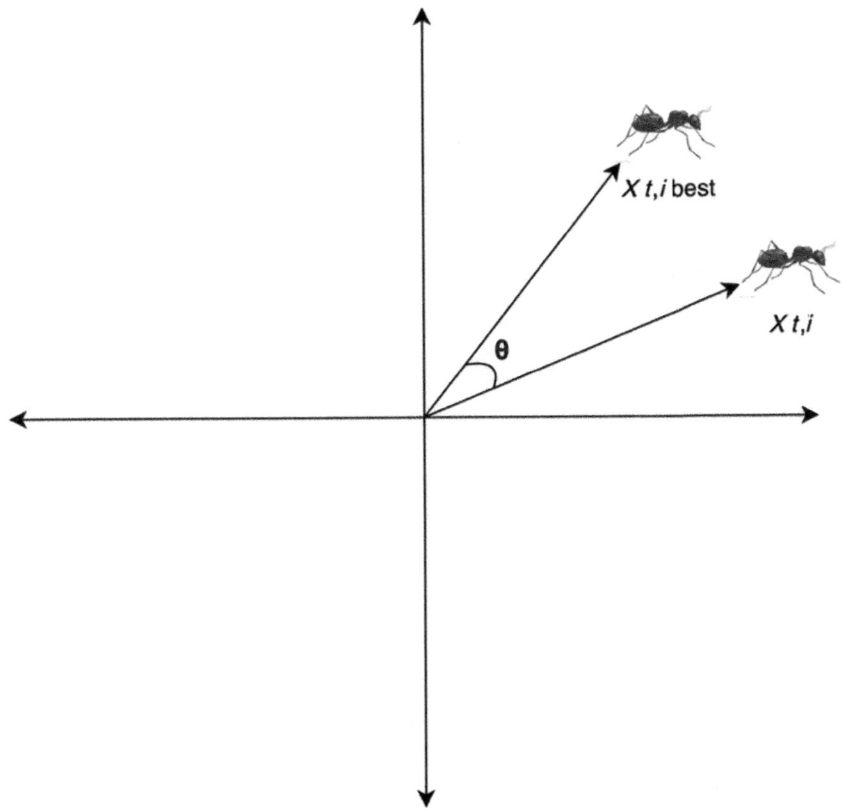

FIGURE 14.4(a) Cosine similarity angle between the local best deposition position and current deposition position of worker ants.

The current tendency rate is solved by calculating the cosine similarity between ($X_{t,ibest}$, $X_{t,i}$) as point (A) and ($X^*_{\text{t,ibest fitness}}$, $X_{t,i\,fitness}$) as point of (B), as shown in Figure 14.4(c). The equation below represents its mathematical expression:

$$\text{Current tendency rate} = \frac{(\text{A}) \cdot (\text{B})}{\|\text{A}\| \ \|\text{B}\|} \tag{10}$$

- $\text{A} = (X_{t,ibest}, X_{t,i})$
- $\text{B} = (X^*_{t,ibest\,fitness}, X_{t,i\,fitness})$

After CSANA calculates the result of the current tendency rate, it uses the nested cosine similarity to calculate the previous tendency rate as follows:

First, the cosine similarity between ($X_{t,\ ibest}$, $X_{t,\ iprevious}$) is shown in Figure 14.5(a). The equation below is its mathematical representation:

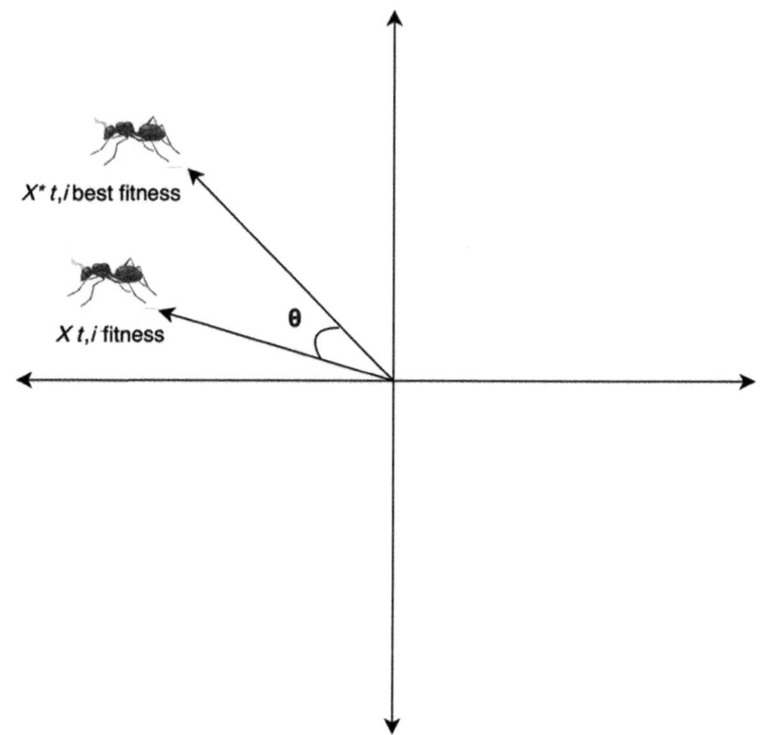

FIGURE 14.4(b) Cosine similarity angle between the global best deposition position fitness and current deposition position's fitness of worker ants.

$$\text{cosine similarity}\left(X_{t,ibest}, X_{t,iprevious}\right) = \frac{\left(X_{t,ibest}\right)\cdot\left(X_{t,iprevious}\right)}{\| X_{t,ibest} \| \; \| X_{t,iprevious} \|} \tag{11}$$

Additionally, the cosine similarity between ($X^*_{t,\ ibest\ fitness}$, $X_{t,iprevious\ fitness}$) is calculated as illustrated in Figure 14.5(b). The calculation is performed as follows:

$$\text{cosine similarity}\left(X^*_{t,i\ best\ fitness}, X_{t,i\ previous\ fitness}\right) = \frac{(X^*_{t,ibest\ fitness})\cdot(X_{t,iprevious\ fitness})}{\| X^*_{t,ibest\ fitness} \| \; \| X_{t,iprevious\ fitness} \|} \tag{12}$$

Then, the cosine similarities of ($X_{t,\ ibest}$, $X_{t,\ iprevious}$) as point (A) and ($X^*_{t,\ ibest\ fitness}$, $X_{t,iprevious\ fitness}$) as point (B) are obtained, as shown in Figure 14.5(c). This result is referred to as the previous tendency rate and its calculation is as follows:

$$\text{Previous tendency rate} = \frac{(\mathrm{A})\cdot(\mathrm{B})}{\| \mathrm{A} \| \; \| \mathrm{B} \|} \tag{13}$$

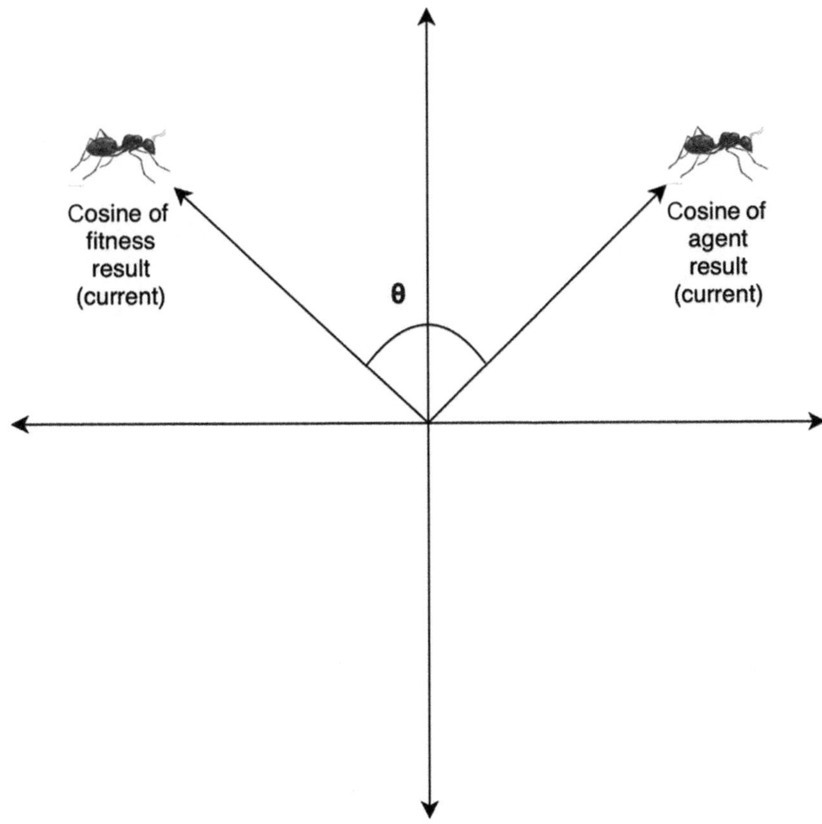

FIGURE 14.4(c) Cosine similarity angle between the cosine similarity result of worker ant's deposition position and worker ant's deposition position's fitness.

- $A = (X_{t,\ ibest}, X_{t,\ iprevious})$
- $B = (X^*_{t,\ ibest\ fitness}, X_{t,iprevious\ fitness})$

Once CSANA calculates the current tendency rate and previous tendency rate using nested cosine similarity, the Euclidean distance is utilized, which is used to find the distance between the two points in the search area, as presented in Figure 14.6. The equation below is its mathematical representation:

$$d = \sqrt{(\text{Current tendency rate})^2 + (\text{Previous tendency rate})^2} \tag{14}$$

So, the deposition weight "*dw*" equation is calculated as follows and the flowchart is shown in Figure 14.7, while Figure 14.8 presents the pseudocode:

$$dw = r^*d \tag{15}$$

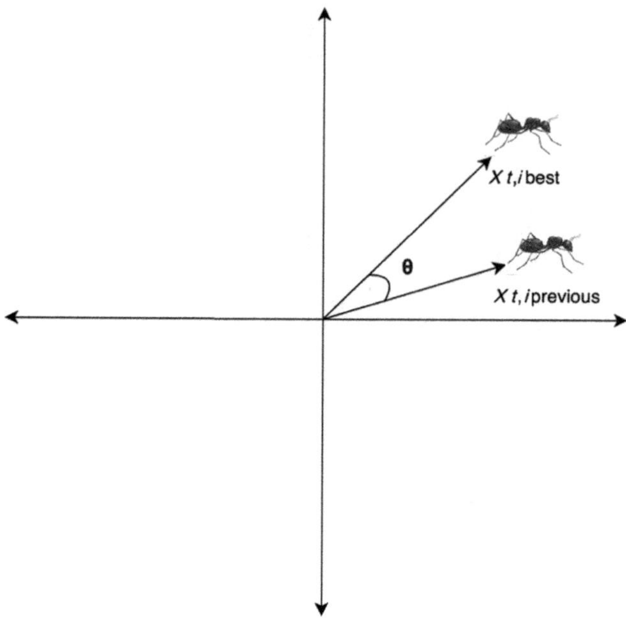

FIGURE 14.5(a) Cosine similarity angle between the local best deposition position and previous deposition position of worker ants.

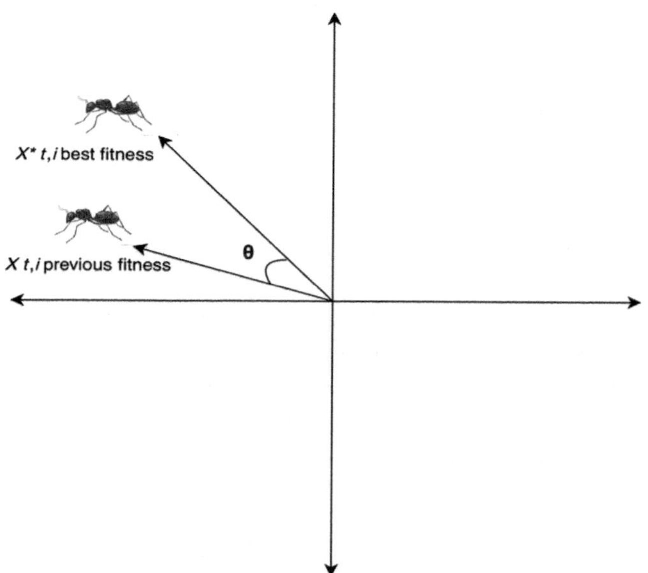

FIGURE 14.5(b) Cosine similarity angle between the global best deposition position fitness and previous deposition position's fitness of worker ants.

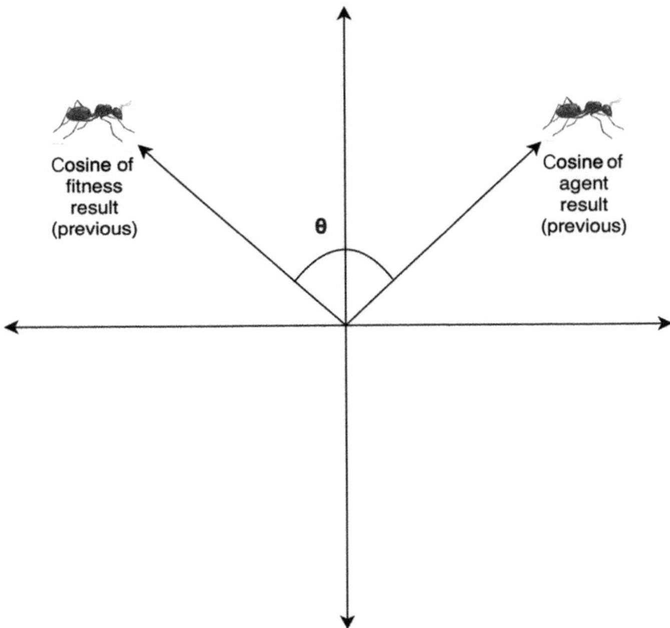

FIGURE 14.5(c) Cosine similarity angle between the cosine similarity result of worker ant's deposition position and worker ant's deposition position's fitness.

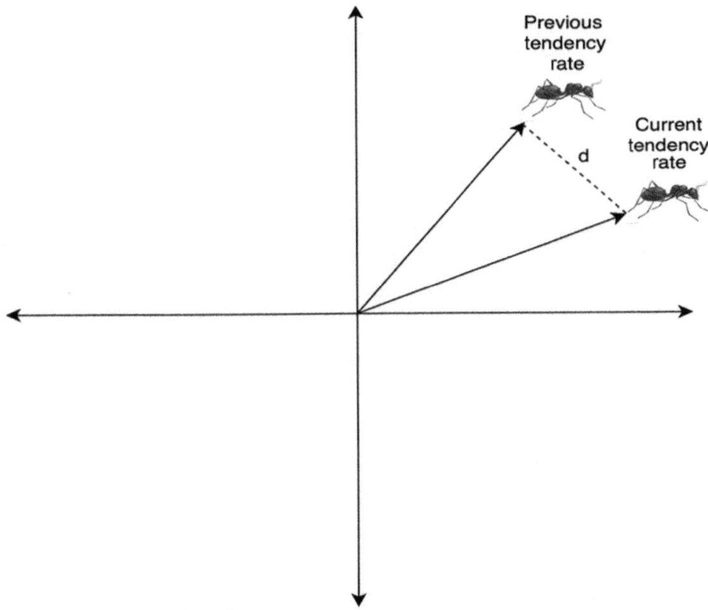

FIGURE 14.6 Euclidean distance between points of the current tendency rate and previous tendency rate.

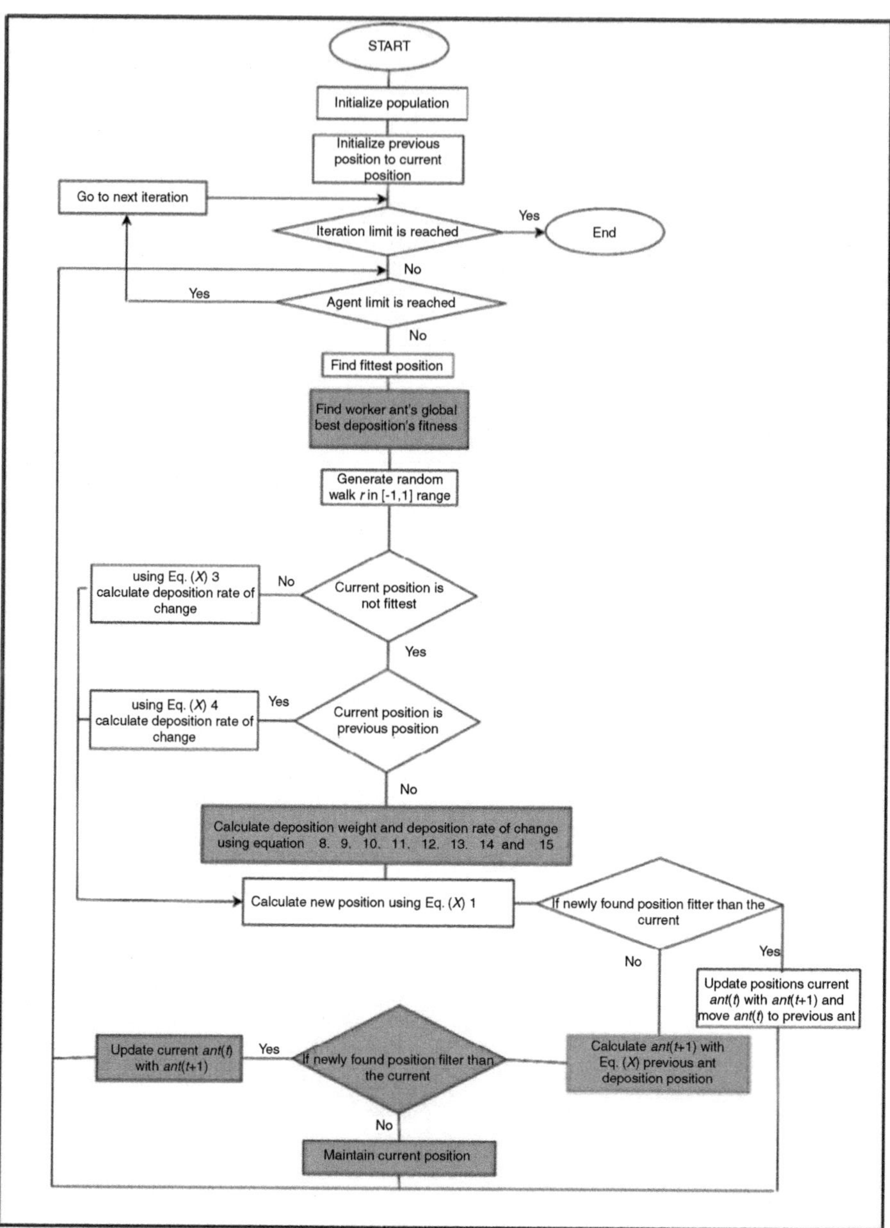

FIGURE 14.7 Flowchart of CSANA.

14.3.3 Utilization of Previous Solutions for Non-Fitter Solutions

The last enhancement of CSANA is utilizing the previous deposition position of a worker ant when the newly found solution is not superior. In ANA, whenever the worker ant discovers a new solution $X_{t,i+1}$ by using the fitness function, it checks

```
Initialize worker ant population Xi (i=1,2,3...n)
Initialize worker ant previous position X_previous
Find worker ant's global best deposition's fitness
  While iteration (t) limit not reached
    For each artificial worker ant X_t,j
      Find best artificial worker ant Xt,ibest
        Generate random walk run [-1,1] range
          If (X_t,j = =X_t,ibest)
            Calculate ΔX_t,i+1 using equation 3
          else if (X_t,j==X_t,iprevious)
            Calculate ΔX_t,i+1 using equation 4
          else
            Calculate T using equation 8, 9, 10.
            Calculate T_previous using equation 11, 12, 13.
            Calculate distance between T and T_previous using equation 14.
            Calculate dw using equation 15.
          end if
            Calculate Xt,i+1 using equation 1.
          if (X_t,ifitness < X_t,ifitness)
            Move accepted and assign X_t,i to X_t,iprevious
          else
            Calculate X_t,i+1 using equation: 1. With X_t,iprevious
          if (X_t,ifitness < X_t,ifitness)
          Move accepted and assign X_t,i to X_t,iprevious
      else
      maintain current position
    end if
end for
```

FIGURE 14.8 Pseudocode of CSANA.

whether the newly found solution surpasses the current solution. If the obtained new solution surpasses the current solution, it is approved and the prior solution is preserved as $X_{t,iprevious}$; otherwise, the current solution is utilized until the next iteration. One of the limitations of ANA is that it does not consider any previous solutions. However, CSANA checks for previous solutions too in case the previous solution guides the worker ants to a superior result. In CSANA, after a better new solution is obtained and it is not fitter than the current solution, contrary to ANA, CSANA continues with using previous solutions, but only if this leads to superior results; otherwise, CSANA retains the current solution until the upcoming iteration. CSANA keeps track of the progress of its previous solution, which allows a more detailed search in the search area and more accurate results will be achieved.

14.4. RESULTS AND DISCUSSION

This section introduces a range of methodologies for gauging the feasibility and performance of algorithms, encompassing both real-world applications and test functions. The section delineates distinct testing mechanisms essential for the comprehensive analysis and evaluation of the obtained results. It is noteworthy that CSANA was developed utilizing the Java programming language, and its evaluation was conducted employing identical parameter settings, test functions, and iteration numbers as those employed in the testing of ANA. Moreover, the performance of CSANA was evaluated against six state-of-the-art algorithms: ANA, Dragonfly Algorithm (DA), Genetic Algorithm (GA), Particle Swarm Optimization (PSO), Salp Swarm Algorithm (SSA), and Whale Optimization Algorithm (WOA). The results of the tests of the 16 classical standard test functions and CEC-C06 tests for the different algorithms were taken from the original ANA work [7]. In addition, one real-world application was optimized using CSANA. Therefore, this section consists of seven parts.

14.4.1 Classical Benchmark Test Functions

CSANA undergoes performance evaluation using three categories of test functions. These functions exhibit diverse characteristics, including unimodal, multimodal, and composite features. These distinct test groups are employed to assess specific outcomes of the algorithm. The first type of test function is the unimodal test function which refers to having only a single optimum. Unimodal functions are used in the assessment of an algorithm's exploitation capability. Instead of focusing on the final results, the unimodal functions concentrate more on the algorithm's rate of convergence. The second type of test function is the multimodal test function. It indicates that there is more than one optimum, and as the problem dimension increases, the quantity of local optima rapidly rises. Multimodal functions are used in the assessment of exploration capability, which is utilized for avoiding becoming trapped in local optima. The third type of test function is the composite test function. As indicated by its name, composite test functions are derived by combining, shifting, rotating, and introducing biased versions of the remaining test functions [1, 8]. Various patterns with numerous local optima are offered by composite test functions, which allow for an assessment of the balance between an algorithm's exploration and exploitation phases (see Tables Appendix 14.A, Appendix 14.B, and Appendix 14.C) [1, 9, 10]. To perform verification and analysis, CSANA was compared with other algorithms, such as ANA, DA, PSO, and GA. Common parameters were set for all the cases as follows: 30 agents were selected as the population size, 500 iterations were selected as the maximum iterations, which is defined as the termination condition, and 10 was set as a dimension of the benchmark functions. After that, 30 runs were tested for each algorithm, and then the mean and standard deviation were computed. The outcomes from testing CSANA, ANA, DA, PSO, and GA using the standard benchmark functions are presented in Table 14.2 and from the benchmark functions, each of the algorithm's test functions is minimized toward 0.0 [1, 9]. The research highlights CSANA as an enhanced version of ANA, demonstrating the effectiveness of employing cosine similarity and hybridizing it with the Euclidean distance over the Pythagorean theorem. CSANA outperformed ANA across various test functions (F1, F2, F3, F4, F5, F7, F9, F12, F13, F14, F15, F16, and F18), achieving optimal results in F4 and F14. However, CSANA showed comparable performance to ANA in

F10 and F11, despite exhibiting a better standard deviation in F11. Overall, CSANA surpasses other algorithms (ANA, DA, PSO, and GA) in F2, F4, F9, F13, F14, F15, F16, and F18, demonstrating an improved exploitation ability and convergence rate in unimodal test functions, enhanced exploration ability in multimodal functions, and balanced exploration-exploitation phases in composite functions, as evidenced by the results presented in Table 14.2.

14.4.2 CEC-C06 2019 Benchmark Functions

To better evaluate CSANA, a set of ten modern CEC benchmark functions was utilized. Then the obtained results were compared to those of ANA, DA, WOA, and SSA. For benchmarking single-objective optimization problems, a set of test functions was developed and referred to as "The 100-Digit Challenge" [11]. The CEC-06 2019 test functions utilized in benchmarking CSANA are shown in Table (Appendix 14.D). Every test function within CEC-06 2019 exhibited scalability. However, the functions CEC01–CEC03 remained unshifted and unrotated, whereas the functions CEC04–CEC10 went through shifting and rotation. In terms of the parameters, the default test function parameters were set by the CEC benchmark developers, and are defined as follows: nine dimensions were set for the CEC01 function as a minimization problem within the boundary range [-8192, 8192], 16 dimensions within the boundary range [-16384, 16384] were set for CEC02, and 18 dimensions within the boundary range [-4,4] were set for CEC03. The rest of the test functions from CEC04–CEC10 were configured as ten-dimensional minimization problems within the range [-100, 100], as shown in Table (Appendix 14.D). To provide more convenience, 1.0 was utilized as the global optimum for each of the CEC functions [1]. The obtained test outcomes of CSANA were contrasted with those of the original ANA, DA, WOA, and SSA as they are modern optimization algorithms and taken from the papers by Hama Rashid and Rashid [1] and Abdullah and Rashid [7]. It is worth mentioning that the same parameter settings were used. For each algorithm, 500 iterations and 30 agents were used. Additionally, the algorithm was executed 30 times, and the mean and standard deviation were calculated for every test function. The obtained test outcomes of CSANA, ANA, DA, WOA, and SSA on the CEC-C06 2019 are shown in Table 14.3. For CEC02, CEC03, and CEC10, CSANA obtained the same result as ANA. Additionally, CSANA couldn't yield desired results for CEC01 and CEC06 due to the excessive run time, which made it inapplicable. However, CSANA surpassed ANA and other competitor algorithms in the case of CEC04, CEC05, CEC07, CEC08, and CEC09. This demonstrates that CSANA had outstanding performance and efficiency, as the obtained results show in Table 14.3.

14.4.3 Execution Time

This analysis examines the duration of execution for standard benchmark tests and CEC-C06 2019 benchmark test functions. For every test function, nanoTime() method was used, which is a method used to measure the execution time. By finding the difference between the start and end timestamps, the elapsed time with high precision can be calculated. After the execution time was calculated, it was converted to

TABLE 14.2
Standard benchmark results [1]

Test	CSANA		ANA		DA		PSO		GA	
function	Mean	SD	Mean	SD	Mean	SD	Mean	SD	Mean	SD
F1	2.98E-05	1.34E-04	0.016299162	0.062457134	**2.85E-18**	**7.16E-18**	4.20E-18	1.31E-17	748.5972	324.9262
F2	**3.21E-11**	**2.62E-11**	2.36E-05	2.15E-05	1.49E-05	3.76E-05	0.003154	0.009811	5.971358	1.533102
F3	0.694197	1.004615	1.239172335	1.035406371	**1.29E-06**	**2.10E-06**	0.001891	0.003311	1949.003	994.2733
F4	**0.0**	**0.0**	2.37E-16	8.86E-16	0.000988	0.002776	0.001748	0.002515	21.16304	2.605406
F5	8.037669	7.23598	14.95306041	30.63521072	**7.600558**	**6.786473**	63.45331	80.12726	133307.1	85,007.62
F7	0.675914	0.304193	0.770696384	0.366042632	0.010293	0.004691	**0.005973**	**0.003583**	0.166872	0.072571
F9	**5.954444**	**2.932746**	25.46221873	4.313900987	16.01883	9.479113	10.44724	7.879807	25.51886	6.66936
F10	1.25E-11	2.75E-11	**5.54E-15**	**2.37E-15**	0.23103	0.487053	0.280137	0.601817	9.498785	1.271393
F11	0.469905	0.070694	0.411189712	0.073239096	0.193354	0.073495	**0.083463**	**0.035067**	7.719959	3.62607
F12	0.009662	0.022822	3.219956841	2.52657309	0.031101	0.098349	**8.57E-11**	**2.71E-10**	1858.502	5820.215
F13	**5.48E-37**	**2.95E-36**	1.76E-23	9.47E-23	103.742	91.24364	150	135.4006	130.0991	21.32037
F14	**0.0**	**0.0**	4.26E-14	1.54E-13	193.0171	80.6332	188.1951	157.2834	116.0554	19.19351
F15	**1.19E-11**	**1.08E-11**	4.89E-06	3.31E-06	458.2962	165.3724	263.0948	187.1352	383.9184	36.60532
F16	**23.72224**	**0.042355**	23.76092355	0.048390796	596.6629	171.0631	466.5429	180.9493	503.0485	35.79406
F17	223.5567	**0.003135**	223.5622125	0.008813889	229.9515	184.6095	136.1759	160.0187	1**18.438**	51.00183
F18	**31.49604**	**0.007719**	31.51015225	0.020777872	679.588	199.4014	741.6341	206.7296	544.1018	13.30161

TABLE 14.3
CEC-C06 2019 benchmark results [1]

Test function	CSANA		ANA		DA		WOA		SSA	
	Mean	SD	Mean	SD	Mean	SD	Mean	SD	Mean	SD
CEC01	-	-	-	-	5.43E+10	6.69E+10	4.11E+10	5.42E+10	**6.05E+09**	**4.75E+09**
CEC02	**4**	**0.0**	4	2.87E-14	78.0368	87.7888	17.3495	0.0045	18.3434	0.0005
CEC03	13.7024	**2.81E-10**	13.70240422	2.01E-11	13.7026	0.0007	**13.7024**	**0**	13.7025	0.0003
CEC04	**25.4523**	**7.572358**	38.50887822	10.07245727	344.356	414.098	394.675	248.563	41.6936	22.2191
CEC05	**1.185772**	**0.102313**	1.224598709	0.114632394	2.5572	0.3245	2.7342	0.2917	2.2084	0.1064
CEC06	-	-	-	-	9.8955	1.6404	10.7085	1.0325	6.0798	1.4873
CEC07	**109.6135**	**5.20754**	116.5962143	8.825046006	578.953	329.398	490.684	194.832	410.396	290.556
CEC08	**5.032718**	0.625383	5.472814997	0.429461877	6.8734	0.5015	6.909	0.4269	6.3723	0.5862
CEC09	**2.000105**	**0.000396**	2.000963996	0.00341781	6.0467	2.871	5.9371	1.6566	3.6704	0.2362
CEC10	**2.718281828**	**4.44E-16**	2.718281828	4.44E-16	21.2604	0.1715	21.2761	0.1111	21.04	0.078

TABLE 14.4
ANA vs. CSANA execution time

	Execution Time			Execution Time	
Test Function	**ANA**	**CSANA**	**Test Function**	**ANA**	**CSANA**
F1	40	10	F11	190	34
F2	64	14	F12	152	35
F3	60	15	F13	292	53
F4	30	9	F14	4	4
F5	46	13	F15	38	15
F6	47	11	F16	174	35
F7	33	41	F17	100	21
F8	166	33	F18	111	23
F9	129	28	F19	121	24
F10	21	27			

TABLE 14.5
ANA vs. CSANA execution time CEC-C06

	Execution time	
Test Function	**ANA**	**CSANA**
CEC02	58	17
CEC03	581	75
CEC04	242	46
CEC05	326	58
CEC07	381	67
CEC08	359	52
CEC09	174	32
CEC10	28	37

seconds using the "TimeUnit.NANOSECONDS.toSeconds" method, which is used for converting nanoseconds into seconds. For both ANA and CSANA, tests were conducted on a computer with the following properties: processor,11th Gen Intel®, Core™ i7-1165G7 @ 2.80 GHz, RAM 8.00 GB. Table 14.4 presents the execution time for CSANA versus ANA for standard test functions. As shown, CSANA surpassed ANA in all standard test functions, except F7 and F10. Only in F14 did both CSANA and ANA have the same result. Table 14.5 presents the execution time for CSANA versus ANA for CEC-C06 2019 benchmark functions. As presented, CSANA required less execution time than ANA in all modern CEC benchmark functions, except CEC10. From Tables 14.4 and 14.5, it can be concluded that CSANA was significantly faster than ANA when it came to executing tasks and CSANA had a superior minimum execution duration as it minimized the execution duration by 81.85% in F13 and 87.10% in CEC03, which indicates that CSANA had a superior minimum duration.

14.4.4 Quantitative Measurement of the Convergence Rate

To do more investigation and detailed observation to examine the algorithm's performance in reaching optimality, the rate at which the global best agent converged over iterations was measured for three of the standard benchmark functions. The outcomes are presented in Figure 14.9. F4, which is the unimodal benchmark function, F10 which is the multimodal benchmark function, and F13 which is the composite benchmark function, were chosen for the experiment, each with 500 iterations and 30 search agents. The experimental results indicated that, with an increase in the number of iterations, the accuracy of the global best improved. Additionally, the global best was achieved during the very initial stages of iteration, as shown in Figs. 14.9 (a), (b), and (c).

14.4.5 Statistical Test

To establish the statistical significance of the findings that were presented in the previous section, and to interpret and verify the research outcomes, p-values serve

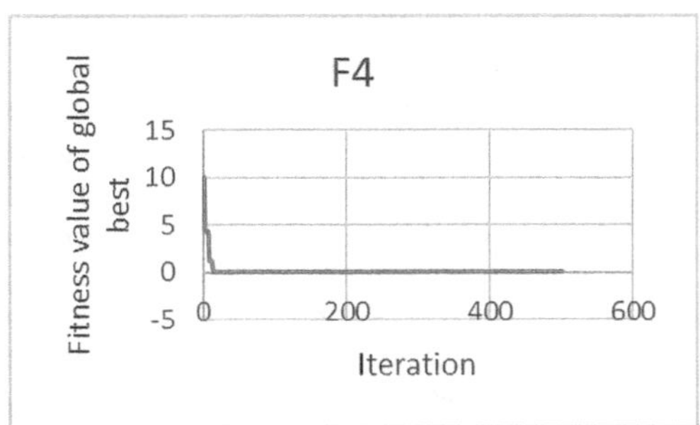

FIGURE 14.9(a) Convergence curve of CSANA on the unimodal test function.

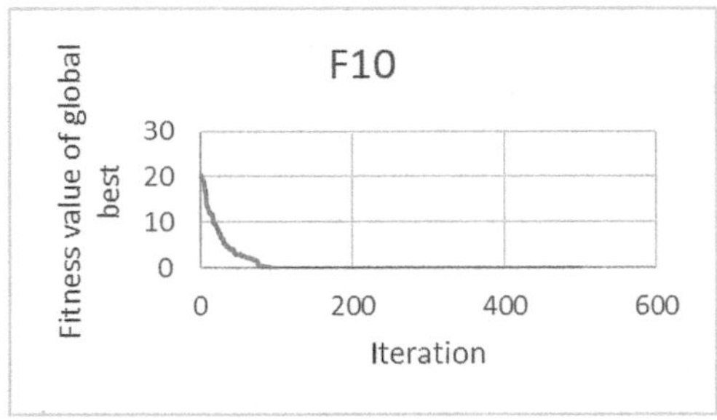

FIGURE 14.9(b) Convergence curve of CSANA on the multimodal test function.

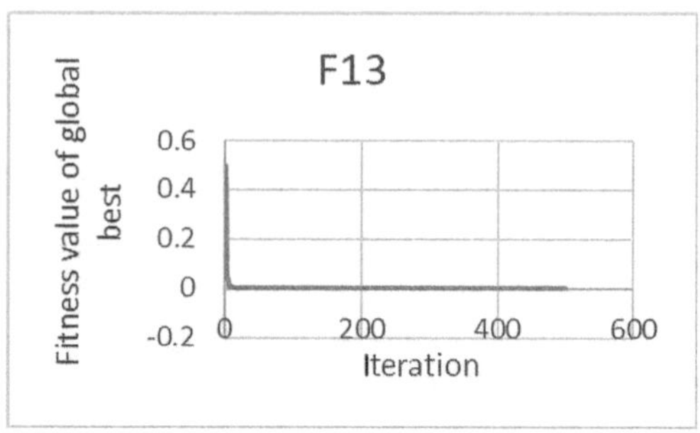

FIGURE 14.9(c) Convergence curve of CSANA on the composite test function.

TABLE 14.6
Student's *t*-Test and Wilcoxon Signed-Rank Test of CSANA and ANA on standard benchmark functions

Test Function	Student's *t*-Test	Wilcoxon Signed-Rank Test
F1	0.066424723	**0.000505488**
F2	**8.07305E-06**	**1.86E-09**
F3	**0.001084266**	0.0120476
F4	0.156903053	0.0222541
F5	0.095630973	0.515848
F7	**0.004991579**	0.014538
F9	**1.31928E-29**	**1.86E-09**
F10	**7.54979E-15**	**1.86E-09**
F11	**0.003859386**	0.0105983
F12	**8.00547E-07**	**1.86E-09**
F13	0.160730498	**4.27632E-05**
F14	0.057345772	**3.73E-09**
F15	**5.04468E-06**	**1.86E-09**
F16	**5.79842E-08**	**9.98378E-07**
F17	**0.005451299**	0.0164312
F18	**0.000285867**	**0.0008718**

as a useful tool [12]. The result of the *p*-value obtained from both the Student's *t*-Test and Wilcoxon Signed-Rank Test for the entire set of standard benchmark functions is presented in Table 14.6. A comparison was performed only between CSANA and ANA because ANA had already been compared to FDO in [1], while DA, PSO, and GA were compared to FDO in [7]. As shown in Table 14.6, the results obtained from CSANA were regarded as significant in 11 statistical tests of the Student's *t*-Test, that is, F2, F3, F7, F9, F10, F11, F12, F15, F16, F17, and F18, as well as in 10 statistical tests of the Wilcoxon Signed-Rank Test, that is, F1, F2, F9, F10, F12, F13, F14, F15, F16, and F18. All these results are significant as their *p*-values are below 0.05.

14.4.6 Comparative Study

Various techniques and measures are utilized to compare the effectiveness of algorithms. In this section, a comparative evaluation of the average of the global best solutions of CSANA, ANA, DA, PSO, and GA on standard benchmark functions is utilized as they were used for testing CSANA. The average global best solution of the algorithms on the standard benchmark function was obtained from Table 14.2 and assigned from 1 to 5, where 1 indicated the algorithm yield minimum result and 5 indicated the algorithm yield maximum value. Table 14.7 presents the ranking of algorithms on the 16 standard benchmark functions. The total number of 1st, 2nd, 3rd, 4th, and 5th rankings of the algorithms is presented in Table 14.8. From Tables 14.7 and 14.8, it can be concluded that CSANA took first place, with a total of eight first

TABLE 14.7
CSANA, ANA, DA, PSO, and GA ranking on standard benchmark functions

Test Function	CSANA	ANA	DA	PSO	GA
F1	3	4	1	2	5
F2	1	3	2	4	5
F3	3	4	1	2	5
F4	1	2	3	4	5
F5	2	3	1	4	5
F7	4	5	2	1	3
F9	1	4	3	2	5
F10	2	1	3	4	5
F11	4	3	2	1	5
F12	2	4	3	1	5
F13	1	2	3	5	4
F14	1	2	5	4	3
F15	1	2	5	3	4
F16	1	2	5	3	4
F17	4	3	5	2	1
F18	1	2	4	5	3

TABLE 14.8
CSANA, ANA, DA, PSO, GA total number of ranking on standard benchmark functions

Rank	CSANA	ANA	DA	PSO	GA
First	8	1	3	3	1
Second	3	6	3	4	0
Third	2	4	5	2	3
Fourth	3	4	1	5	3
Fifth	0	1	4	2	9

TABLE 14.9
CSANA, ANA, DA, PSO, and GA ranking by type and total

Test Function Type	CSANA	ANA	DA	PSO	GA
Unimodal	14	21	10	17	28
Multimodal	9	12	11	8	20
Composite	9	13	27	22	19
Total	32	46	48	47	67
Total Ranking/ No. of Function	32/16 = 2	46/16 = 2.875	48/16 = 3	47/16 = 2.9375	67/16 = 4.1875
Ranking (1–5)	1	2	4	3	5

and no fifth rankings in comparison to the most popular algorithms, like ANA, DA, PSO, and GA.

To facilitate a more detailed assessment of the algorithm, the ranking of CSANA was determined according to the specific benchmark function type and the total of each type in comparison to ANA, DA, PSO, and GA, and then ranked from 1 to 5, as shown in Table 14.9. As demonstrated, CSANA had a minimum total compared to other algorithms, which indicates that CSANA achieved great performance in comparison to these well-known algorithms. Nevertheless, it is important to note that some algorithms perform excellently on some problems, whereas others do not. There is no single algorithm that can attain optimal performance for every optimization problem [13].

14.4.7 Real-World Application of CSANA

Aperiodic antenna array design is a real-world engineering problem subjected to CSANA to measure the algorithm's feasibility. Several types of antennas have been developed over time to adapt to the needs of the industry. One of the most important requirements in the design of an antenna is the capability to arrange the non-uniform elements of the array to achieve the peak sidelobe level (SLL). Various methods are used for the placement of elements of an array, such as numerical optimization and thinning. However, despite the use of the latest tools, the design process remains difficult. An optimization algorithm can be utilized to optimize the positions of elements and minimize SLL to reduce the complexity of the array's design [14]. A non-uniform isotropic array is composed of ten elements and optimization requires only four of them. This means that the optimization process involves four dimensions. The arrangement of a ten-element array is presented in Figure 14.10 [15]. For a comprehensive understanding of this issue, interested readers can see [15].

The following equation presents the problem's fitness function:

$$f = max\left\{20\log\left|AF(\theta)\right|\right\} \quad (16)$$

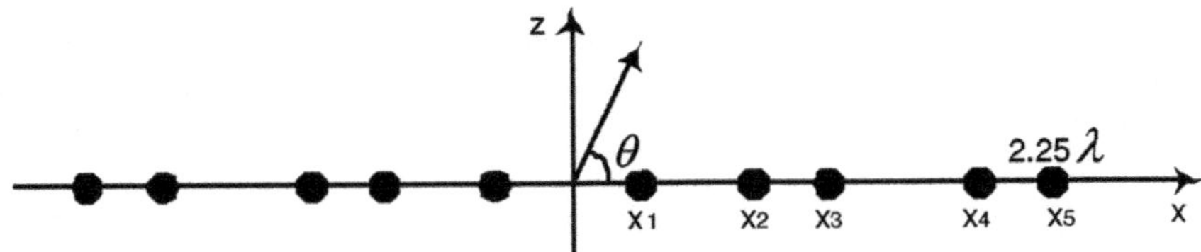

FIGURE 14.10 Arrangement of a ten-element array [15].

where

$$AF(\theta) = \sum_{i=1}^{4} \cos\left[2\pi x_i\left(\cos\theta - \cos\theta_s\right)\right] + \cos\left[2.25 \times 2\pi\left(\cos\theta - \cos\theta_s\right)\right] \quad (17)$$

Consider $\theta_s = 90^o$ [15]

To enhance the peak SLL within non-uniform arrays, it is essential to optimize the positions of elements using a real number vector. Furthermore, to reduce the grating lobe level, specific limitations on element spacing must be implemented. Nevertheless, the limitation is that every element must be between 0.125 λ_0 and 2.0 λ_0, which indicates that no element is less than 0.125 λ_0 or greater than 2.0 λ_0. As a result of these restrictions, each element is bounded within the range of 0 and 2.25 λ_0 due to the element at 2.25 λ_0 remaining fixed and adjacent elements cannot be closer than 0.25 λ_0.

The constraints of the problem are represented by

1. $x_i \in (0, 2.25)\left|x_i - x_j\right| > 0.25\lambda_0$
2. $min\left\{x_i\right\} > 0.125\lambda_0.\ i = 1,2,3,4.\ i \neq j$ (18)

The aperiodic antenna array design problem was subjected to several metaheuristic algorithms to solve this problem. Additionally, it was subjected to CSANA and successfully attained the optimal solution by using 20 search agents and conducting 200 iterations. The algorithm achieved its optimum state in the 23rd iteration, displaying element positions of {0.8069, 0.2221, 0.4994, 1.5799}. This result also indicates that CSANA can perform better than ANA, even in real-world engineering problems, as ANA could obtain the optimal state in the 57th iteration with element positions {1.5959, 0.3081, 0.8747, 0.6072}.

14.5 CONCLUSION

This work introduced an improved version of ANA known as CSANA, which is a novel swarm intelligent algorithm that was developed to optimize single-objective problems. ANA was inspired by the behavior of Leptothorax ants during constructing a new nest. It mimics the collective actions of worker ants as they search for a suitable location for depositing grains around their queen ant. Furthermore, the search process is conducted through the performance of a random walk. ANA utilizes the Pythagorean theorem to measure the slope sides, which are used to generate weights. These weights direct each of the search agents toward optimal solutions. Additionally, in all search phases, ANA relies on the randomization mechanism. The improved version of ANA, known as CSANA, utilizes many modifications, which significantly enhance the algorithm's performance. First, CSANA achieved an effective balance between the exploitation and exploration phases, as this is a fundamental aspect of all metaheuristic algorithms, through integrating global variables that are utilized for the exploration phase. Then, the Pythagorean theorem was replaced by another formula, that is, the nested cosine similarity hybridized with the Euclidean distance, to generate weights. The combination of an angle-based measure (cosine similarity) and distance metric (Euclidean distance) facilitated a more comprehensive understanding of data point relationships, which enhanced the algorithm's performance. The last enhancement employed was utilizing previous depositions in case of the occurrence of non-fitter solutions. CSANA tracks a record of previous solutions, which empowers a more thorough exploration within the search space that generates more accurate and precise results. To assess the algorithm's effectiveness and performance, 26 single-objective test functions were chosen. The initial set of 16 test functions was from the standard benchmark test functions, with various characteristics such as unimodal, multimodal, and composite. The remaining ten test functions were from modern CEC-C06 2019 test functions. CSANA demonstrated remarkable performance and effectively minimized toward zero in all 16 test cases of the standard benchmark test functions. After that, the obtained results were compared to ANA and three other well-known metaheuristic algorithms: DA, PSO, and GA. CSANA outperformed the other four algorithms in eight cases on the standard benchmark test functions, while achieving highly competitive results in the others. For further evaluation of the algorithm's performance, the execution time of CSANA was compared to ANA. The comparison revealed that CSANA outperformed ANA significantly with a much shorter execution time. To perform further analysis, a statistical experiment was performed on the global best solution over many iterations. An investigation of the test results was from unimodal, multimodal, and composite benchmark test functions for assessing the algorithm's convergence toward optimality. CSANA reached optimality in the very early stages of iteration, which is another indicator that CSANA achieved outstanding performance. To provide a comprehensive analysis of CSANA's performance, CSANA was compared to ANA. CSANA showed highly comparable performance to ANA on 16 standard benchmark test functions. Lastly, to assess feasibility and evaluate its performance, CSANA was applied to a well-known real-world application to aperiodic antenna array design. The algorithm achieved successful optimization and delivered quite good results. Based on CSANA's results and performance, it is noticeable that CSANA obtained better results than ANA. The weaknesses observed in ANA were successfully handled and improved in CSANA. The conclusion is that CSANA now has a better balance between exploration and exploitation phases, as well as a better convergence rate, which is one of the most important aspects of a powerful algorithm. Additionally, it was explicitly shown that CSANA was superfast and required less execution time, which is a critical aspect for time-sensitive applications.

APPENDIX

TABLE APPENDIX 14.A
Unimodal test [1, 10]

Function	Dimension	Range	Shift Position	f_{min}
$F1(x)=\sum_{i=1}^{n} x_i^2$	10	[-100, 100]	[-30, -30, ... -30]	0
$F2(x)=\sum_{i=1}^{n}\lvert x_i\rvert+\prod_{i=1}^{n}\lvert x_i\rvert$	10	[-10,10]	[-3, -3, ... -3]	0
$F3(x)=\sum_{i=1}^{n}\left(\sum_{j-1}^{i} x_j\right)^2$	10	[-100, 100]	[-30, -30, ... -30]	0
$F4(x)=\max_i\{\lvert x\rvert, 1\le i\le n\}$	10	[-100, 100]	[-30, -30, ... -30]	0
$F5(x)=\sum_{i=1}^{n-1}\left[100\left(x_{i+1}-x_1^2\right)^2+\left(x_i-1\right)^2\right]$	10	[-30,30]	[-15, -15, ... -15]	0
$T6(x)=\sum_{i=1}^{n}\left(\left[x_i+0.5\right]\right)^2$	10	[-100, 100]	[-750, ... -750]	0
$F7(x)=\sum_{i=1}^{n} ix_i^4+\text{random}[0, 1]$	10	[-1.28,1.28]	[-0.25, ...-0.25]	0
$F8(x)=\sum_{i=1}^{n} -x_i^2 \sin\left(\sqrt{\lvert x_i\rvert}\right)$	10	[-500, 500]	[-300, ... -300]	-418.9829

TABLE APPENDIX 14.B
Multimodal test functions [1, 10]

Function	Dimension	Range	Shift Position	f_{min}
$F9(x)=\sum_{i=1}^{n}\left[x_i^2-10\cos(2\pi x_i)+10\right]$	10	[-5.12,5.12]	[-2, -2, ...-2]	0
$F10(x)=-20exp\left(-0.2\sqrt{\sum_{i=1}^{n} x_i^2}\right)-exp\left(\frac{1}{n}\sum_{i=1}^{n}\cos(2\pi x_i)\right)+20+e$	10	[-32, 32]	[0, 0, ... 0]	0
$F11(x)=\frac{1}{4000}\sum_{i=1}^{n} x_i^2-\prod_{i=1}^{n}\cos\left(\frac{x_i}{\sqrt{i}}\right)+1$	10	[-600, 600]	[-400, ... -400]	0

(continued)

TABLE APPENDIX 14.B (Continued)
Multimodal test functions [1, 10]

Function	Dimension	Range	Shift Position	f_{min}
$F12(x)=\frac{\pi}{n}\left\{10\sin(\pi y_1)+\sum_{i=1}^{n-1}(y_i-1)^2\left[1+10\sin^2(\pi y_{i+1})\right]+(y_n-1)^2\right\}+\sum_{i=1}^{n}u(x_i,10,100,4)$	10	[-50,50]	[-30, 30, … 30]	0
$y_i=1+\frac{x+1}{4}$.				
$u(x_i,a,k,m)=\begin{cases}k(x_i-a)^m & x_i>a\\ 0 & -a<x_i<a\\ k(-x_i-a)^m & x_i<-a\end{cases}$				

TABLE APPENDIX 14.C
Composite test functions [1, 10]

Function	Dimension	Range	f_{min}
F13 (CF1) $f1,f2,f3\ldots f10=$ Sphere function $\delta1,\delta2,\delta3\ldots\delta10=[1,1,1,\ldots.1]$ $\lambda1,\lambda2,\lambda3\ldots\lambda10=\left[\frac{5}{100},\frac{5}{100},\frac{5}{100},\ldots\frac{5}{100}\right]$	10	[-5, 5]	0
F14 (CF2) $f1,f2,f3\ldots f10=$ Griewank's function $\delta1,\delta2,\delta3\ldots\delta10=[1,1,1,\ldots.1]$ $\lambda1,\lambda2,\lambda3\ldots\lambda10=\left[\frac{5}{100},\frac{5}{100},\frac{5}{100},\ldots\frac{5}{100}\right]$	10	[-5, 5]	0
F15 (CF3) $f1,f2,f3\ldots f10=$ Griewank's function $\delta1,\delta2,\delta3\ldots\delta10=[1,1,1,\ldots.1]$ $\lambda1,\lambda2,\lambda3\ldots\lambda10=[1,1,1,\ldots.1]$	10	[-5, 5]	0
F16 (CF4) $f1,f2=$ Ackley's function $f3,f4=$ Rastrigin's function $f5,f6=$ Weierstrass function $f7,f8=$ Griewank's function $f9,f10=$ Sphere function $\delta1,\delta2,\delta3\ldots\delta10=[1,1,1,\ldots.1]$ $\lambda1,\lambda2,\lambda3\ldots\lambda10=\left[\frac{5}{32},\frac{5}{32},1,1,\frac{5}{0.5},\frac{5}{0.5},\frac{5}{100},\frac{5}{100},\frac{5}{100},\frac{5}{100}\right]$	10	[-5, 5]	0

TABLE APPENDIX 14.C (Continued)
Composite test functions [1, 10]

Function	Dimension	Range	f_{min}
F17 (CF5) $f1, f2$ = Rastrigin's function $f3, f4$ = Weierstrass function $f5, f6$ = Griewank's function $f7, f8$ = Ackley's function $f9, f10$ = Sphere function $\delta1, \delta2, \delta3 \ldots \delta10 = [1,1,1,\ldots.1]$ $\lambda1, \lambda2, \lambda3 \ldots \lambda10 =$ $\left[\frac{1}{5}, \frac{1}{5}, \frac{5}{0.5}, \frac{5}{0.5}, \frac{5}{100}, \frac{5}{100}, \frac{5}{32}, \frac{5}{32}, \frac{5}{100}, \frac{5}{100}\right]$	10	[-5, 5]	0
F18 (CF6) $f1, f2$ = Rastrigin's function $f3, f4$ = Weierstrass function $f5, f6$ = Griewank's function $f7, f8$ = Ackley's function $f9, f10$ = Sphere function $\delta1, \delta2, \delta3 \ldots \delta10 = [0.1, 0.2, 0.3, 0.4, 0.5, 0.6, 0.7, 0.8, 0.9, 1]$ $\lambda1, \lambda2, \lambda3 \ldots \lambda10 = \left[0.1 * \frac{1}{5}, 0.2 * \frac{1}{5}, 0.3 * \frac{5}{0.5}, 0.4 * \frac{5}{0.5}, 0.5 * \frac{5}{100}, 0.6 * \frac{5}{100}, 0.7 * \frac{5}{32}, 0.8 * \frac{5}{32}, 0.9 * \frac{5}{100}, 1 * 5/100\right]$	10	[-5, 5]	0

TABLE APPENDIX 14.D
CEC-C062019 test functions [1, 11]

Function No.	Function Name	Dimension	Range	f_{min}
CEC01	Storn's Chebyshev Polynomial Fitting Problem	9	[-8192, 8192]	1
CEC02	Inverse Hilbert Matrix Problem	16	[-16384, 16384]	1
CEC03	Lennard-Jones Minimum Energy Cluster	18	[-4,4]	1
CEC04	Rastrigin's Function	10	[-100, 100]	1
CEC05	Griewangk's Function	10	[-100, 100]	1
CEC06	Weierstrass Function	10	[-100, 100]	1
CEC07	Modified Schwefel's Function	10	[-100, 100]	1
CEC08	Expanded Schaffer's F6 Function	10	[-100, 100]	1
CEC09	Happy Cat Function	10	[-100, 100]	1
CEC10	Ackley Function	10	[-100, 100]	1

REFERENCES

[1] Hama Rashid, D.N., Rashid, T.A. and Mirjalili, S., 2021. ANA: ant nesting algorithm for optimizing real-world problems. *Mathematics*, 9(23), p.3111.

[2] Franks, N.R. and Deneubourg, J.-L., 1997. Self-organizing nest construction in ants: individual worker behaviour and the nest's dynamics. *Animal Behaviour,* 54(4), pp. 779–796.

[3] Sumpter, D.J.T., 2010. Structures. In: *Collective Animal Behavior.* Princeton, NJ: Princeton University Press, pp. 151–172.

[4] Yang, X.S., 2010. *Nature-inspired Metaheuristic Algorithms.* Luniver press.

[5] Geem, Z.W. ed., 2009. *Music-inspired Harmony Search Algorithm: Theory and Applications* (Vol. 191). Springer.

[6] Mohd, W.R.W. and Abdullah, L., 2018, June. Similarity measures of Pythagorean fuzzy sets based on combination of cosine similarity measure and Euclidean distance measure. In *AIP Conference Proceedings* (Vol. 1974, No. 1, p. 030017). AIP Publishing LLC.

[7] Abdullah, J.M. and Ahmed, T., 2019. Fitness dependent optimizer: inspired by the bee swarming reproductive process. *IEEE Access*, 7, pp. 43473–43486.

[8] Muhammed, D.A., Saeed, S.A. and Rashid, T.A., 2020. Improved fitness-dependent optimizer algorithm. *IEEE Access*, 8, pp. 19074–19088.

[9] Mirjalili, S., 2015. Dragonfly algorithm: a new meta-heuristic optimization technique for solving single-objective, discrete, and multi-objective problems. *Neural Computing and Applications*, 27, p. 1053–1073.

[10] Li, J. and Tan Y., 2015 May. Orienting mutation based fireworks algorithm. in *Proceedings of IEEE Congress Evolutionary Computation (CEC)*, pp. 1265–1271.

[11] Price, K.V., Awad, N.H., Ali, M.Z. and Suganthan, P.N., 2018. Problem definitions and evaluation criteria for the 100-digit challenge special session and competition on single objective numerical optimization. In *Technical Report.* Singapore: Nanyang Technological University.

[12] Thiese, M.S., Ronna, B. and Ott, U., 2016. P value interpretations and considerations. *Journal of Thoracic Disease*, 8(9), p. E928.

[13] Cortés-Toro, E.M., Crawford, B., Gómez-Pulido, J.A., Soto, R. and Lanza-Gutiérrez, J.M., 2018. A new metaheuristic inspired by the vapour-liquid equilibrium for continuous optimization. *Applied Sciences*, 8(11), p. 2080.

[14] Diao, J., Kunzler, J.W. and Warnick, K.F., 2017. *Sidelobe Level and Aperture Efficiency Optimization for Tiled Aperiodic Array Antennas.* IEEE.

[15] Jin, N. and Rahmat-Samii, Y., 2007. Advances in particle swarm optimization for antenna designs: real-number binary single-objective and multiobjective implementations. *IEEE Transaction on Antennas and Propagation*, 55(3), pp. 560–562.

15 A Systematic Evaluation of the Shrike Optimization Algorithm for the Multi-objective Multi-modal Function

Hanan K. AbdulKarim and Tarik A. Rashid

15.1 INTRODUCTION

Nowadays, rather than focusing on a single objective, most challenges handle many objectives. In such cases, Multi-Objective Optimization (MOO) is a powerful tool for finding solutions that achieve a good balance between conflicting goals and meet several constraints. MOO is applied in many fields of industrial, science, engineering, and product design. MOO cannot find an exact solution, therefore MOO deals with three techniques to make decisions over solutions. First, consolidate all the objectives into a unified scalar value, usually a weighted sum, and optimize the scalar value [1,2]. Second, solve the goals in a hierarchical manner, prioritizing the optimization of the first objective. If there are several solutions, optimize them for the second objective [1]. Third, gather a variety of non-dominant alternative solutions [1,2].

The MOO has the potential to implement the decision-making mechanism in industrial process synthesis, design, operation, and control [3]. The multi-objective problem of chemical product design is solved using the Pareto set technique of a posteriori method for decision-making [4]. Selecting high-quality data from many sensor systems is difficult, and is an NP-hard problem, because it has many objectives and many criteria on which to make a decision; to achieve that, a multi-objective evolutionary algorithm is used for decision-making and numerical studies are used to evaluate the process [5].

Recently, constrained MOO evolutionary algorithms have become popular. They have many constraints with many objectives for solving optimization problems [6]. Many MOO problems, which have many constraints with many conflicting objectives, are called constraint MOO problems. The Pareto front is applied to find the diversity distance to measure infeasible solutions to get an optimal result [7].

Furthermore, Evolution-Guided Bayesian Optimization is applied for self-driving lab material, which is a constrained multi-objective to find high throughput [8].

DOI: 10.1201/9781003601555-15

Minimizing the cost and deflection of reinforced concrete is the goal of a multi-objective model that makes heavy use of evolutionary algorithms for its solution. To prevent under- or over-design issues, the Pareto front of the model helps decision-makers to tune the design parameters [9].

Pareto optimal prediction is applied to a learning model in a machine learning strategy to improve multi-objective evolutionary algorithm performance with less execution time [10]. MOO is also applied in laser-plasma interaction using Bayesian optimization [11]. The Fitness-dependent Optimization Algorithm using the Pareto set model applied to a multi-objective design problem finds the minimum distance between Pareto sets as the algorithm convergence result [12]. The Multi-objective Particle Swarm Optimizer (MPSO) hybridized with Firefly is used to solve load balancing problems in cloud computing [13].

Modified MPSO and MPSO were applied to bi or tri objective functions, and Pareto set distances were compared as results [14]. Constraint multi-objective PSO was used to solve many constraint MOO problems [7]. The Multi-objective Artificial Bee Colony (MOABC) was applied to passive power filters and compared with other algorithms in some cases [15]. The Grey Wolf Optimizer was applied to multi-objective problems and compared with multi-objective evolutionary and PSO algorithms to show the algorithms' performance [16]. The multi-objective Bacterial Foraging algorithm was applied to multi-criteria grid scheduling problems [17], and evolutionary algorithms were analyzed and designed to solve multi-objective problems [18].

A scalarizing function is used to solve MOO problems in a single objective, where many framework models with different functions are used to compare the MOO algorithm's performance [19]. The weighted sum and modified version were applied to multi-objective queries for cloud databases, and the proof showed that the weighted sum always selects the best from the Pareto optimal set without additional overhead [20]. A Teacher-Learning-Based Optimization Algorithm was applied to solve a multi-objective flexible job shop scheduling problem [21]. The non-dominated sorting genetic algorithm and strength Pareto evolutionary algorithm were applied to solve binary image segmentation in image processing. The result showed that the type of image input affected the Pareto set and algorithm performance [22].

The main contribution of this study is to implement the multi-objective Shrike Optimization Algorithm (SHOA) using multiple case studies to show the performance of the enhancement over different parameters and the conversion of SHOA to adapt to multi-objective problems.

In addition to this section, a simple explanation of the SHOA methodology is presented in Section 15.2; a multi-objective case study with an implementation is shown in Section 15.3; a discussion of the results is presented in Section 15.4; and finally, the conclusion and future study are explained in the last section.

15.2 SHRIKE OPTIMIZATION ALGORITHM

Shrike birds were studied and the behavior inspired the creation of an optimization algorithm called SHOA [23]. SHOA begins by initializing parameters, where N represents the quantity of nests in the population, B represents the number of eggs considered nestlings in each nest, and α represents a constant that is regarded as a natural factor impacting the bird. SHOA's search space starts with a population of N nests. Two randomly initialized parent birds populate each nest initially. Once the population is formed and the nests are prepared, B nestlings are produced. The population is described by

$$Population(N) = \begin{pmatrix} \begin{bmatrix} p_{im} & p_{if} \\ n_{ij} & n_{ij} \end{bmatrix} & \cdots & \begin{bmatrix} p_{im} & p_{if} \\ n_{ij} & n_{ij} \end{bmatrix} \\ \vdots & \ddots & \vdots \\ \begin{bmatrix} p_{im} & p_{if} \\ n_{ij} & n_{ij} \end{bmatrix} & \cdots & \begin{bmatrix} p_{im} & p_{if} \\ n_{ij} & n_{ij} \end{bmatrix} \end{pmatrix} \quad (1)$$

where a population is a group of nests and each nest is a representative subgroup of solutions, as parents and offspring are considered the solution of SHOA, where i ranges from 1 to N, and the following equation is used to generate parents:

$$p_i = LB + rand[0,1] * (UB - LB) \quad (2)$$

After the nests are initialized, the most fit bird is chosen as the dominant one. M_{parent} refers to the male parent, whereas F_{parent} refers to the female parent. During the breeding period, each nest produces B nestlings using

$$\Delta egg_j = (F_{parent} - M_{parent}) + r \quad (3)$$

$$nestling_j = F_{parent} + \Delta egg_j \quad (4)$$

If Δegg_j is produced by both parents and r is a random number within the range of -1 to 1, then Δegg_j is utilized to create $nestling_j$, where i ranges from 1 to B. Following initialization, each nest and parameter parents should be given based on their goal function. Subsequently, the value of r is created for each dimension using

$$r = e^{-2xt/T_{max}} \quad (5)$$

The "r" parameter is a fundamental aspect of feeding; x is the dimension variable for $bird_j$; t represents the current iteration; and T_{max} represents the maximum iteration permitted for performing SHOA. Subsequently, by employing the following equation, every male parent bird independently feeds itself:

$$\Delta food_j = bird_j \times r \quad (6)$$

When other birds get fed in the nest, $\Delta food$ is produced using the following equation, where $bird_j$ represents the current condition of the bird and M parent refers to the male parent who is responsible for providing food:

$$\Delta food_j = r \times \left(bird_j - M_{parent}\right) + M_{parent} \tag{7}$$

Nestlings, unable to survive on the male parent's food, attempt to rely on the female parent using

$$\Delta food_j = r \times \left(bird_j - F_{parent}\right) + \sin(\alpha) \tag{8}$$

which is similar to formula (7), but with the variable r ranging from -1 to 1 and the inclusion of the sine function sin(α), where α is a constant component.

The equation

$$bird_j^{t+1} = bird_j^t + \Delta food_j \tag{9}$$

represents the current condition of birds obtaining food and determines the birds' subsequent status on food generation.

To determine the fitness of each bird, it is necessary to compare their present condition with an improved state. Only the fittest bird survives to the next generation, as not all birds receive food simultaneously. If a bird fails to get food from its parent, it sustains itself by utilizing

$$bird_j^{t+1} = bird_j^t + \left(r \times bird_j + \sin(\alpha)\right) \tag{10}$$

to produce $\Delta food_j$. This equation randomly creates variable r in the range [-1, 1]. Another variable parameter α, is randomly selected in the range [0, dimension] and used to introduce additional randomness. This step aims to explore the solution space by identifying new solutions that are significantly different from the current state. In this phase, the current solution diverges from the local best solution to produce a new solution that is significantly different from its parent.

The SHOA selects the most optimal solution from each nest as the local best, and then the population chooses the best solution from all the local best solutions as the global best. After each repetition, the nest updates the previous nestlings. After birds have completed their nesting stage, the algorithm selects only the top two birds as parents and excludes all other birds. Each parent produces a new offspring, resulting in a new generation that improves on the existing solution [23]. The pseudocode SHOA that specifies the process is presented in Figure 15.1. The algorithmic process persists until it reaches a specified stopping condition [23], as shown in Figure 15.2.

15.3 CASE STUDY

Optimization problems can be classified as either maximization or minimization. In this study, we focus on a minimization problem represented in Table 15.1 [5], where D denotes the dimension of the problem in this specific example. The variable x is constrained to the specified range for each function as the [minimum, maximum]

1. Parameter Initialization N,,Max_Iteration, B, α, k objective _function
2. Initialize $nest_i$, (i=1,2,…,N) using equation (2), and find fitness
3. While (! Max_Iteration)
    a. if iteration%k = =0
        i. Generate nestling using equation (3 & 4)
    b. Find M_{parent}, F_{parent} from ($nest_i$)
    c. for each $bird_j$ in $nest_i$
        i. Generate random r using equation(5)
        ii. if ($bird_j$ == MParent)
        iii. Calculate $\Delta food_j$ using equation(6)
        iv. else
        v. Calculate $\Delta food_j$ using equation(7)//with male bird
        vi. Calculate $\Delta food_j$ using equation(8))//with female bird
        vii. Calculate $bird_j^{(t+1)}$ using equation(9))//with step (v) or with (vi)
        viii. if ($Fit_j^{(t+1)}$ better than $Fit_j^{(t)}$) keep $bird_j^{(t+1)}$
        ix. else
        x. calculate $bird_j^{(t+1)}$ using eauation (10) &apply elitism// for nestling
    d. Keep local best of $nest_i$, and global best from all local bests
4. end While

FIGURE 15.1 SHOA pseudocode.

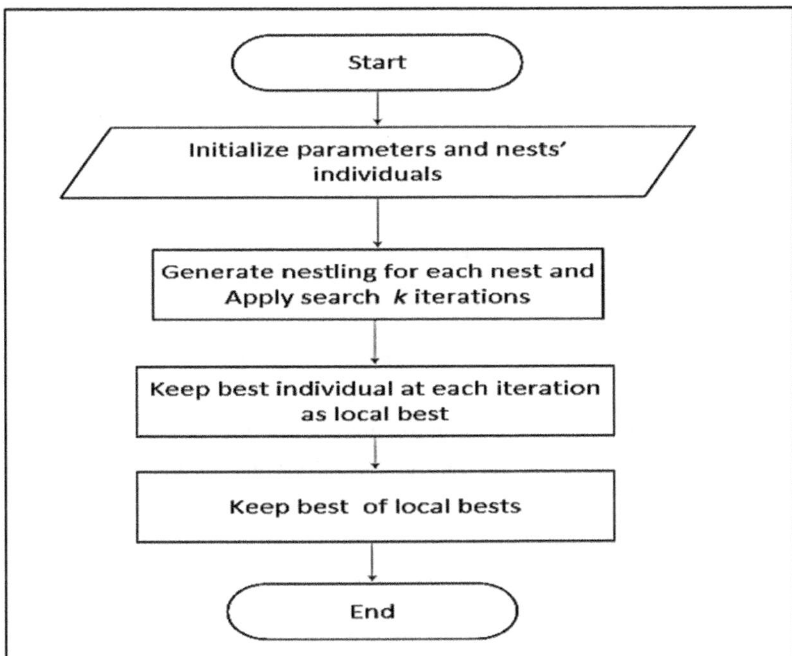

FIGURE 15.2 SHOA processing step diagram.

TABLE 15.1
Multi-objective functions

Name	Functions and constraint	D	Range
ZDT1	$g(x)=1+9\left(\sum_{i=2}^{D}x_i\right)/(n-1)$ $F_1(x)=x_1$ $F_2(x)=g(x)\left[1-\sqrt{x_1/g(x)}\right]$	2	[0,1]
ZDT2	$g(x)=1+9\left(\sum_{i=2}^{D}x_i\right)/(n-1)$ $F_1(x)=x_1$ $F_2(x)=g(x)\left[1-\sqrt{\left(x_1/g(x)\right)^2}\right]$	2	[0,1]
ZDT3	$g(x)=1+9\left(\sum_{i=2}^{D}x_i\right)/(n-1)$ $F_1(x)=x_1$ $F_2(x)=g(x)\left[1-\sqrt{x_1/g(x)}-x_1/g(x)\sin(10\pi x_1)\right]$	5	[0,1]
ZDT4	$g(x)=91+\left(\sum_{i=2}^{D}x_i^2-10\cos(4\pi x_i)\right)$ $F_1(x)=x_1$ $F_2(x)=g(x)\left[1-\sqrt{x_1/g(x)}\right]$	10	$x_1\in[0,1]$ $x_i\in[-5,5]$
ZDT5	$g(x)=1+9\left(\left(\sum_{i=2}^{D}x_i\right)/(n-1)\right)^{0.25}$ $F_1(x)=1-exp(-4x_1)\sin(6\pi x_1)^6$ $F_2(x)=g(x)\left[1-\left(f_{1(x)}/g(x)\right)^2\right]$	5	[0,1]

acceptable range. We consider 100 maximum iterations and the population size $N = 10$, and weights [0.743, 0.257] are generated randomly. The objective function using the weighted sum is generated using

$$f(x)=min\sum_{i=1}^{y}f(x_i)w_i \tag{11}$$

where y is the number of multi-objective functions, $i = 1, 2, \ldots, y$, and w_i is the weight for the corresponding function i. The result of SHOA is presented in the following section.

TABLE 15.2
Parameter initialization

Number of nests (N)	10
Number of birds (B)	3
Number of iterations (Max_iteration)	100
Number of iterations for the new nest (k)	3, 10, 30
Dimension	Depends on the function
α	Upper boundary
Upper boundary limit	Maximum acceptable value
Lower boundary limit	Minimum acceptable value

TABLE 15.3
Multi-objective function with different k values for SHOA

	k = 3		k = 10		k = 30	
Fun	**mean**	**std**	**mean**	**std**	**mean**	**std**
ZDT1	**2.46E-06**	**2.49E-06**	1.80E-05	1.77E-05	7.72E-05	9.77E-05
ZDT2	**1.07E-05**	**8.36E-06**	7.30E-05	6.41E-05	3.14E-04	2.70E-04
ZDT3	**-0.19876**	**7.09E-08**	-0.19875	6.18E-06	-0.19875	1.64E-05
ZDT4	**0.008547**	**0.037207**	0.01189	5.17E-02	0.012416	0.053805
ZDT5	0	0	0	0	0	0

All the needed parameters and functions are initialized at the beginning, as shown in Tables 15.1 and 15.2. Some of the parameters, like α, change during execution time depending on the conditions.

15.4 RESULTS AND DISCUSSION

The results of SHOA were applied for simple multi-objective test functions using the weighted sum, as shown in Table 15.3, where fun represents functions of (ZDT1–ZDT5) [5], and mean and std are the mean and standard deviation of 20 iterations, respectively.

Table 15.3 represents the result of SHOA applied to five of the multi-objective test functions, where the value of k varies between (3, 10, 30). The smallest k value gets the best solution, which is bold in Table 15.3, compared with other k values, but in ZDT5, all get the same result.

In Figure 15.3, the mean values are presented for different k values for each ZDT_n, where (n = 1, 2,…, 5). For each k = 3, the performance of SHOA is either better or equal compared with the performance when k = 10 or 30.

Table 15.4 shows SHOA's performance against the FOX [24] Optimization Algorithm with the same specification using the weighted sum mechanism. The

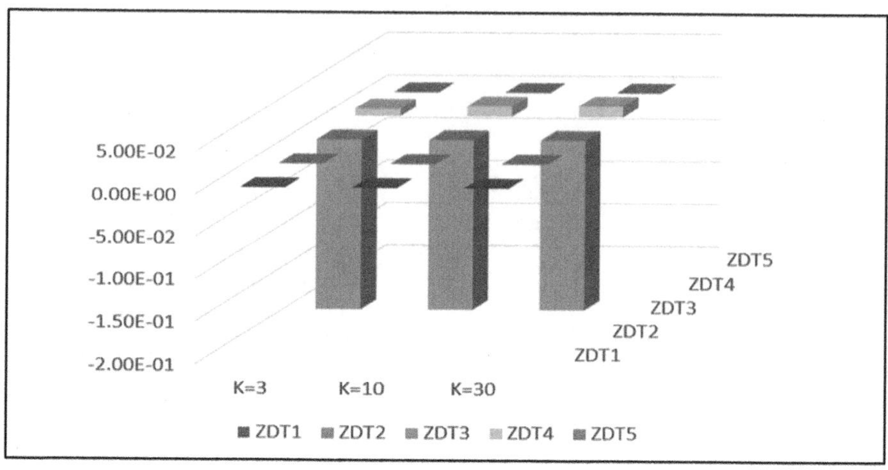

FIGURE 15.3 Comparison results across different k values.

TABLE 15.4
Comparison results for SHOA and FOX

	SHOA		FOX	
Fun	**mean**	**std**	**mean**	**std**
ZDT1	**2.46E-06**	2.49E-06	5.99E-03	7.01E-03
ZDT2	**1.07E-05**	8.36E-06	1.04E-03	8.69E-04
ZDT3	**-0.19876**	7.09E-08	-0.17667	1.53E-02
ZDT4	**0.008547**	0.037207	1.503488	1.04E+00
ZDT5	**0**	0	0.023106	0.014426

mean and std are presented for 20 rounds. The performance of SHOA was better than that of FOX in all cases; all best performances are in bold, whereas in the ZDT4 problem, there was a high difference between both algorithm's outcomes, where SHOA outperformed FOX.

In Figure 15.4, the performances of both SHOA and FOX are shown clearly by column for each algorithm across all study cases from (ZDT1 to ZDT5).

15.5 CONCLUSIONS

This study outlined the application of SHOA for multi-objective problems using a weighted sum, with different k values. We described a case study with fixed parameters, and weighted sum values were generated randomly, but we can adjust these parameters for other problems or new case studies to enhance the results for current or new study cases. The results demonstrate that SHOA improved its results with small k values and progressed toward the best solutions. At $k = 3$, the algorithm

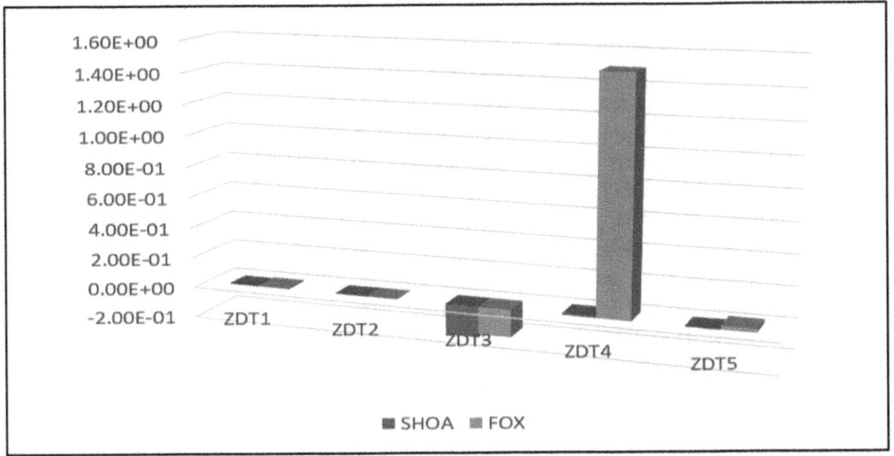

FIGURE 15.4 SHOA and FOX performance.

abandoned the old solution and generated new solutions that were far from optimal in order to discover the search space. Conversely, at larger k values, the algorithm iteratively searched around the solution based on the k value, resulting in low exploration and high algorithmic exploitation. SHOA outperformed the FOX Optimization Algorithm under the same criteria for the multi-objective model functions, with the same parameters. In the future, the Pareto optimal strategy will be applied. Then the convergence distance between Pareto set elements will be compared with further algorithms. Additionally, SHOA's working strategy will be applied to clustering problems. This will demonstrate the effectiveness of subpopulation solutions and grouping them into both local and global optimal solutions to show SHOA's performance.

REFERENCES

[1] Mumford-Valenzuela CL. *A Simple Approach to Evolutionary Multiobjective Optimization.* London: Springer London; 2005. https://doi.org/10.1007/1-84628-137-7_4.

[2] Abdel-Basset M, Mohamed R, Abouhawwash M. Balanced multi-objective optimization algorithm using improvement based reference points approach. *Swarm Evol Comput* 2021;60:100791. https://doi.org/10.1016/j.swevo.2020.100791.

[3] Cerda-Flores SC, Rojas-Punzo AA, Nápoles-Rivera F. Applications of multi-objective optimization to industrial processes: A literature review. *Processes* 2022;10:1–15. https://doi.org/10.3390/pr10010133.

[4] Ng LY, Chemmangattuvalappil NG, Dev VA, Eden MR. Mathematical principles of chemical product design and strategies. 2016;39. https://doi.org/10.1016/B978-0-444-63683-6.00001-0.

[5] Gong W, Duan Q, Li J, Wang C, Di Z, Ye A, et al. Multiobjective adaptive surrogate modeling-based optimization for parameter estimation of large, complex geophysical models. *Water Resour Res* 2016;52:1984–2008. https://doi.org/10.1002/2015WR018230.

[6] Long W, Dong H, Wang P, Huang Y, Li J, Yang X, et al. A constrained multi-objective optimization algorithm using an efficient global diversity strategy. *Complex Intell Syst* 2023;9:1455–78. https://doi.org/10.1007/s40747-022-00851-1.

[7] Wang H, Cai T, Li K, Pedrycz W. Constraint handling technique based on Lebesgue measure for constrained multiobjective particle swarm optimization algorithm [Formula presented]. *Knowledge-Based Syst* 2021;227:107131. https://doi.org/10.1016/j.knosys.2021.107131.

[8] Low AKY, Mekki-Berrada F, Gupta A, Ostudin A, Xie J, Vissol-Gaudin E, et al. Evolution-guided Bayesian optimization for constrained multi-objective optimization in self-driving labs. *NPJ Comput Mater* 2024;10. https://doi.org/10.1038/s41524-024-01274-x.

[9] Afshari H, Hare W, Tesfamariam S. Constrained multi-objective optimization algorithms: Review and comparison with application in reinforced concrete structures. *Appl Soft Comput J* 2019;83:105631. https://doi.org/10.1016/j.asoc.2019.105631.

[10] Wenbin L, Zi'An C, Liu G. Expensive multiobjective optimization algorithm based on equivariate component analysis. *IEEE Access* 2022;10:73835–46. https://doi.org/10.1109/ACCESS.2022.3188248.

[11] Irshad F, Karsch S, Döpp A. Multi-objective and multi-fidelity Bayesian optimization of laser-plasma acceleration. *Phys Rev Res* 2023;5:13063. https://doi.org/10.1103/PhysRevResearch.5.013063.

[12] Abdullah JM, Rashid TA, Maaroof BB, Mirjalili S. Multi-objective fitness-dependent optimizer algorithm. *Neural Comput Appl* 2023;35:11969–87. https://doi.org/10.1007/s00521-023-08332-3.

[13] Devaraj AFS, Elhoseny M, Dhanasekaran S, Lydia EL, Shankar K. Hybridization of firefly and improved multi-objective particle swarm optimization algorithm for energy efficient load balancing in cloud computing environments. *J Parallel Distrib Comput* 2020;142:36–45. https://doi.org/10.1016/j.jpdc.2020.03.022.

[14] bin Mohd Zain MZ, Kanesan J, Chuah JH, Dhanapal S, Kendall G. A multi-objective particle swarm optimization algorithm based on dynamic boundary search for constrained optimization. *Appl Soft Comput J* 2018;70:680–700. https://doi.org/10.1016/j.asoc.2018.06.022.

[15] Yang NC, Mehmood D. Multi-objective bee swarm optimization algorithm with minimum Manhattan distance for passive power filter optimization problems. *Mathematics* 2022;10. https://doi.org/10.3390/math10010133.

[16] Mirjalili S, Saremi S, Mirjalili SM, Coelho LDS. Multi-objective grey wolf optimizer: A novel algorithm for multi-criterion optimization. *Expert Syst Appl* 2016;47:106–19. https://doi.org/10.1016/j.eswa.2015.10.039.

[17] Kaur M. Elitist multi-objective bacterial foraging evolutionary algorithm for multi-criteria based grid scheduling problem. *2016 International Conference on Internet of Things and Application IOTA* 2016:431–6. https://doi.org/10.1109/IOTA.2016.7562767.

[18] Jose S, Vijayalakshmi C. Design and analysis of multi-objective optimization problem using evolutionary algorithms. *Procedia Comput Sci* 2020;172:896–9. https://doi.org/10.1016/j.procs.2020.05.129.

[19] Chugh T. Scalarizing functions in Bayesian multiobjective optimization. *2020 IEEE Congr Evol Comput CEC 2020 – Conf Proc* 2020. https://doi.org/10.1109/CEC48606.2020.9185706.

[20] Helff F, Gruenwald L, D'Orazio L. Weighted sum model for multi-objective query optimization for mobile-cloud database environments. *CEUR Workshop Proc* 2016;1558:1–6.

[21] Sreekara Reddy MBS, Ratnam C, Rajyalakshmi G, Manupati VK. An effective hybrid multi-objective evolutionary algorithm for solving real time event in flexible job shop scheduling problem. *Meas J Int Meas Confed* 2018;114:78–90. https://doi.org/10.1016/j.measurement.2017.09.022.
[22] Lezcano C, Vázquez Noguera JL, Pinto-Roa DP, García-Torres M, Gaona C, Gardel-Sotomayor PE. A multi-objective approach for designing optimized operation sequence on binary image processing. *Heliyon* 2020;6:e03670. https://doi.org/10.1016/j.heliyon.2020.e03670.
[23] Abdulkarim HK, Rashid TA. In search of excellence: SHOA as a competitive Shrike Optimization Algorithm for multimodal problems. *IEEE Access* 2024;12:1–19. https://doi.org/10.1109/ACCESS.2024.3427632.
[24] Mohammed H, Rashid T. FOX: A FOX-inspired optimization algorithm. *Appl Intell* 2023;53:1030–50. https://doi.org/10.1007/s10489-022-03533-0.

16 Artificial Cardiac Conduction System

Simulating Heart Function for Advanced Computational Problem Solving

Rebaz Mohammed, Nawzad K. Al-Salihi, Tarik A. Rashid, Aso M. Aladdin, Mokhtar Mohammadi and Jafar Majidpour

16.1 INTRODUCTION

Over the past three decades, metaheuristic optimization techniques have been drawing many researchers' attention. A metaheuristic is a high-level framework for problem-independent algorithms that combines intelligently different approaches to developing heuristic optimization algorithms [1]. Some of these algorithms, such as the Artificial Immune System (AIS) [2], Ant Colony Optimization (ACO) [3], Particle Swarm Optimization (PSO) [4], and Dragonfly Algorithm (DA) [5], are surprisingly well-known among scientists from a wide variety of disciplines, not only computer scientists. In addition to a vast number of theoretical works, these optimization techniques have been utilized in numerous academic disciplines. There is a question here regarding why metaheuristics have become so prevalent. This question has a four-part answer: simplicity, flexibility, derivation-free mechanism, and avoidance of local optima [6].

First, metaheuristics are quite simple due to the originating metaheuristic frameworks from natural phenomena. They have been influenced primarily by straightforward concepts. Typically, the sources of inspiration are physical facts, animal behavior, or evolutionary concepts. Second, metaheuristics do not have any restrictions when formulating an optimization problem, such as predefining constraints or objective functions of the decision variables [7]. Third, the majority of metaheuristics contain processes that do not require derivation. Metaheuristics optimize problems stochastically, as opposed to gradient-based techniques. The optimization procedure begins with random solutions, and it is not necessary to calculate the derivative of search spaces to locate the optimal solution. This makes metaheuristics particularly applicable to real-world issues involving costly or uncertain derivative data. Finally, metaheuristics are superior to conventional optimization techniques in

 DOI: 10.1201/9781003601555-16

avoiding local optima. Due to the stochastic character of metaheuristics, they can avoid stagnation in local solutions and exhaustively scan the whole search space. The search space of real-world problems is typically unknown and highly complicated, containing many local optima; hence, metaheuristics are excellent possibilities for optimizing these problematic real-world problems.

It is worth mentioning that no universal algorithm can guarantee to get the optimal solution for all problems, and this statement has been mathematically proven by the No Free Lunch theorem (NFL) [8]. NFL also proves that there is still room for improvement and enhancement since most algorithms may have good results on some problems while performing poorly on other problems. This motivates our efforts to formulate a novel metaheuristic based on cardiac conduction system behavior. Generally, metaheuristics can be divided into three fundamental classes based on the method by which solutions are manipulated [9]. Local search metaheuristics search for suitable solutions iteratively by conducting small changes to a single solution. Constructive metaheuristics construct the best possible solutions at each iteration rather than improving complete solutions. Population-based metaheuristics find optimum solutions by combining selecting and existing solutions into a new one called the population. Population-based metaheuristics offer several benefits over single-solution-based algorithms; multiple candidate solutions share information about the search space, resulting in abrupt leaps into the most promising part of the search space. Multiple candidate solutions collaborate to eliminate locally optimum solutions. In general, population-based metaheuristics are more exploratory than single-solution-based algorithms.

One of the main branches of population-based metaheuristics is bio-inspired algorithms. Bio-inspired algorithms are inspired by scientific methods that conduct biological metaphors to find and develop new engineering solutions to many complex optimization and real-world problems. Some of the most popular bio-inspired algorithms are the Whale Optimization Algorithm (WOA) [10], PSO [4], AIS [2], Genetic Algorithm (GA) [11], Artificial Bee Colony (ABC) [12], Lagrange Elementary Optimization (Leo) [13], and Ant Nesting Algorithm (ANA) [14]. A thorough literature review of bio-inspired algorithms is presented in the next section. Regardless of the differences between the metaheuristics, exploration and exploitation are the most important two key factors common in all bio-inspired metaheuristic algorithms. Exploration randomly searches for a new solution space to find global optima and exploitation leads to finding local optima in the explored solution space [15]. Hence, the balance between these two vital processes in all bio-inspired algorithms is vital because intense exploration and deep exploitation may have the effect of not giving optimal solutions and being trapped in the local optima of algorithms, respectively. This work proposes a new bio-inspired algorithm inspired by the human cardiac conduction system.

The rest of the paper is organized as follows: Section 16.2 presents a literature review of bio-inspired algorithms. Section 16.3 states the biological motivation for the Artificial Cardiac Conduction System (ACCS). Section 16.4 outlines the proposed ACCS algorithm. The experimental performance evaluation and discussion of benchmark functions are presented in Section 16.5. Finally, Section 16.6 concludes the paper and makes recommendations for future studies.

16.2 LITERATURE REVIEW

Metaheuristic nature-inspired optimization algorithms are classified into three categories: physics-based algorithms, Swarm Intelligence (SI) algorithms, and bio-inspired (non-SI-based) algorithms. Physical-based optimization algorithms typically imitate physical rules. Some of the most popular algorithms are Simulated Annealing [16], Harmony Search [17], Gravitational Search Algorithm (GSA) [18], Big-Bang Big-Crunch [19], Black Hole [20] algorithm, and Curved Space Optimization [19]. These algorithms were inspired by physical laws such as using ray casting, inertia forces, weights, and electromagnetic force.

For example, the Simulated Annealing algorithm uses the concept of annealing within metallurgy. It is referred to as the process used to modify certain metal alloys by heating beyond their melting points, holding their temperature, and then regulating the cooling process to enhance ductility and hardness. The procedural idea behind the Simulated Annealing algorithm is that, in the same way the metal atom attempts to reach its ideal state during cooling down after the annealing process, Simulated Annealing provides an ideal (best) solution. Similarly, the annealing process discards all solutions that are not ideal for the metal; instead, the Simulated Annealing algorithm uses the probability function to decide whether to discard the inferior solution or accept it. With the metal cooling down the process, the atoms are more likely to reach their ideal state than the state where the metal is in the melting point state. Thus, the Simulated Annealing algorithm can be described in the terminology of metallurgy, that is, it starts in the hot state and cools down over time (convergence state) [21]. GSA is another physics-based algorithm. Rashedi, Nezamabadi-pour, and Saryazdinin proposed it in 2009. It was inspired by the Newtonian law of gravity and mass interactions. GSA conducts a search using a collection of agents whose masses are proportional to the value of a fitness function. Throughout iteration, the masses are drawn to one another by gravitational forces. The attracting force increases proportionally with mass. Therefore, the heaviest mass, which may be near to the global optimum, attracts the other masses proportionally to their distances.

The second main branch of metaheuristics is SI algorithms. These algorithms mostly imitate the social behavior of insects like bees, ants, and termites, as well as the social behavior of other animals, such as a group of fish or birds. The technique is nearly identical to a physics-based algorithm, but the search agents navigate by simulating organisms' collective and social intelligence. Some of the most popular SI algorithms are PSO [4], ACO [3], DA [13], WOA [9], ABC [11], ANA [14] and Firefly Algorithm [22]. The PSO algorithm was proposed by Kennedy and Eberhart in 1995, and was motivated by the social behavior of the birds flying as a flock searching for food. The population of birds moves on some specific path to arrive at its food destination, and the shortest path defined by a bird is considered the best local known solution position or particle best solution. In other words, the movement of a particle takes into account both its optimal solution and the optimal solution obtained by the swarm.

Another popular SI algorithm is ACO, introduced by Dorigo et al. [3]. The basic concept of this algorithm came from ants, small creatures that can search intelligently for a common source of food through a simple indirect low-level communication channel called pheromone trails. The possible solutions evolve a pheromone matrix by repeated iteration. This characteristic enables ants to be seen as multiple agent procedures that solve complex optimization problems. The WOA is another popular algorithm rooted in the hunting behavior of humpback whales. Mirjalili and Lewis developed it in 2016 [10]. The observed and considered behaviors of humpback whales are bubble-net and spiral bubble-net feeding, which have been mathematically modeled to obtain the WOA algorithm. ABC is another popular SI algorithm proposed by Karaboga in 2005 [12]. The algorithm was inspired by foraging patterns of honeybees during the collection of nectar from a flower as a feeding source for the hive. The ABC algorithm uses three groups of bees to analyze the algorithm's working procedure: onlookers, scouts, and employed bees. Employed and onlooker bees perform the exploitation process in the search space, while scouts control only the exploration process in the search space. Thus, scouts search for a source of food randomly and transfer the nectar amount of each specific food position to onlookers.

The third subclass of metaheuristic nature-inspired algorithms is bio-inspired (not SI-based) algorithms. These algorithms mostly mimic biological metaphors to find and develop new engineering solutions to many complex optimization and real-world problems since they demonstrate robust heterogeneous and fascinating phenomena. The most popular algorithm in this branch is GA. This algorithm was introduced by Holland in 1992 [11], and acquired its motivation from the process of evolution and procreation in biology and the theory of evolution presented by Charles Darwin. The idea behind this algorithm is the possibility of combining two solutions called parents into new solutions called children. The children compare with each other how they get some features from their parents and the repetition of this comparison occurs until the best children (i.e., optimal solution) can be indicated as the better solution. Each child inherited from the best parts of its parents to improve the one started with. This process is called optimization.

Another popular bio-inspired algorithm is the Bacterial Foraging Optimization Algorithm (BFOA), initially presented by Kevin M. Passino in 2002 [23]. During foraging, biologists discovered that the movement of bacteria is not random. It moves toward certain stimuli and far away from others, which is known as bacterial chemotaxis. This behavior became the base for developing BFOA. Differential Evolution (DE) is another algorithm initially developed by Storn and Price in 1995 [24]. The DE concept is similar to GA because both use individual populations to look for the optimal solution for an optimization problem. On the other hand, the leading principal difference between GA, Leo, and the DE Algorithm is that GA depends on crossover, a probability method used to exchange information via solutions to locate a better solution, whereas DE strategies use mutation as the primary search method to find a better solution. Additionally, the main operators of Leo and DE are mutation and selection. At the start of the evolution process, exploration occurs through a mutation operator and then the mutation operator enables exploitation. Thus, mutation is incremented until the best optimal value is satisfied based on the stage of the evolutionary process.

To sum up, despite very intense work accomplished on developing new algorithms, metaheuristic algorithms are still considered to be recent compared to well-established conventional algorithms. Besides the advantages and peak performance of the current mentioned nature-inspired algorithms, it can be said that each algorithm has its drawback that motivates researchers to investigate and look forward to developing new nature-inspired metaheuristic algorithms. To the best of our knowledge, however, there is no bio-inspired algorithm in the literature that mimics the functional behavior of the human cardiac conduction system. This motivated our attempt to mathematically model the involuntary behavior of human heart electrical nodes, propose a new bio-inspired algorithm inspired by the cardiac conduction system, and investigate its ability to solve benchmarks.

16.3 ALGORITHM INSPIRATION

In this section, the inspiration for the proposed method is discussed.

16.3.1 Cardiac Conduction System

The Cardiac Conduction System (CCS) is what makes the heart go [25]. CCS is part of the Autonomic Nervous System (ANS). The ANS supplies, influences, and controls the system for regulating the function of internal organs [26]. However, CCS is a group of heart muscle cells in the myocardium layer of the heart that causes blood to move throughout the body and produces and controls the heart rate through muscle cells that contract in waves [27]. This means that whenever the first group of cells is stimulated. They have the effect of stimulating adjacent cells and this chain reaction continues until all cells are stimulated to enable the heart to contract. Specifically, an electrical impulse is generated due to the distribution of ions (primarily calcium, sodium, potassium, and chloride) in cardiac cells [28]. The ions move in and out of the cell via dedicated ion channels in the cell membrane, which produces a wave of electricity flow across the cardiac tissue. The electrical wave propagates so that the ventricles and atria contract in a harmonious rhythm.

Furthermore, CCS contains pacemaker cells with three unique characteristics: automaticity, conductivity, and contractility [27]. Automaticity is the ability of the cardiac tissue to produce an electrical impulse without any initiation factors, whereas conductivity is the capability to navigate the electrical impulse to the next neighbor cell. Contractility is the property of the cells to shorten the fibers in the heart while receiving the electrical impulse [29]. In addition, pacemaker cells with these mentioned unique characteristics coordinate the rhythmic relaxation and contraction of the heart's chambers, called the cardiac cycle. The currents that exert a powerful effect on the heart cells are less than a millionth of an ampere [30]. The cardiac cycle has two main phases, diastole and systole, also called repolarization and depolarization. Diastole is the state of the heart when the heart's ventricles are relaxed and takes longer than systole, which is about two-thirds of the cycle. Systole is the state of the heart during which blood is pumped out of the ventricles and takes up the remaining one-third of

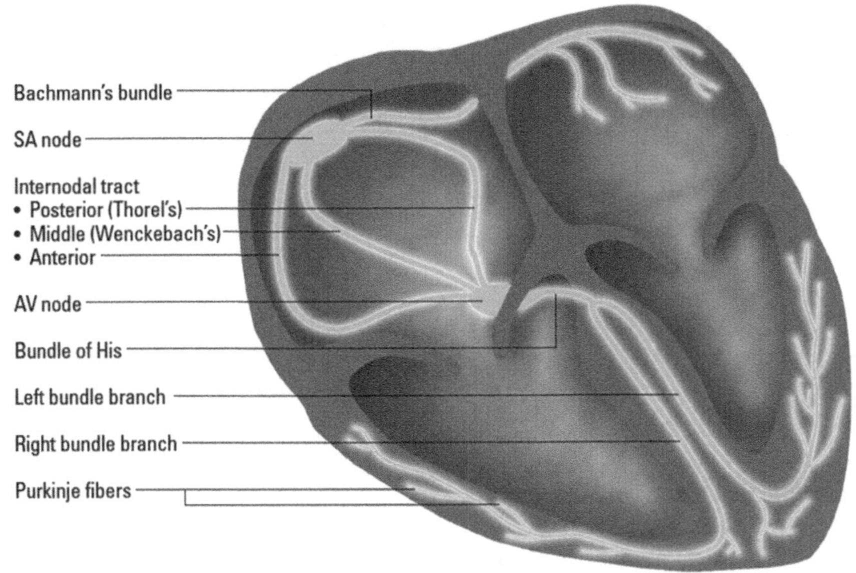

FIGURE 16.1 Cardiac Conduction System elements' anatomy [28].

the cycle [29]. Hence, the process of systole and diastole enables the heart to pump blood. The volume of blood distributed by the heart in one minute is known as the cardiac output, which CCS produces. The significant elements of CCS are the sinoatrial or sinus node (SA), atrioventricular node (AV), bundle of His (BoH), and Purkinje fibers (PF), as shown in Figure 16.1.

CSS generates and controls the heart rate among its elements through four main steps:

1. First step (SA node): It is the dominant heart pacemaker that is located on the endocardial surface of the right atrium [27]. The SA node generates an impulse through the heart wall that enables atrial contraction [28].
2. Second step (AV node): It can delay conduction between the ventricles and atria by a tenth of one second [25]. Hence, this node acts as a resistor to allow enough time for the contracting atria to transfer all blood to the ventricles before the ventricles contract (i.e., lower chambers of the heart) [28].
3. Third step (BoH): From the AV node, the impulse is passed to BoH and branches off the impulse into the left and right bundles with fewer charges [25].
4. Fourth step (PFs): The impulse arrives at the PFs located at the distal portions of the right and left bundle branches. The fibers across the surface of the ventricles cover the endocardium to the myocardium, which causes ventricular depolarization [31].

Moreover, the SA node is the heart's normal pacemaker that determines the heart's rate. If any abnormality occurs at this node, the heart pace will face defeat in keeping

TABLE 16.1
Conduction rate in cardiac tissue

Conduction tissue	Conduction rate (M/S)
SA node	0.05
Atrial pathway	1
AV node	0.05
BoH	1
PF	4
Ventricular muscle	1

Source: [29].

the heart rate balanced [31]. As mentioned, the AV node acts as a resistor and slows conduction due to three main factors:

1. diminished number of gap junctions between the successive cells in the conducting pathways;
2. cells are made up of smaller diameter fibers; and
3. presence of multiple sub-branches.

As a result, excellent resistance to the conduction of excitatory ions is produced from one conducting fiber cell to the next. Moreover, the electrical impulse is navigated in cardiac tissue with different conduction rates in a meter per second, as shown in Table 16.1. Hence, the time taken for the cardiac impulse to reach each node varies based on the speed of the conduction rate at each node. The time required for the impulse to travel from the SA node to AV node is 0.03 seconds, between the AV node and BoH is 0.13 seconds, then from BoH to PF is 0.03 seconds, and finally from PF to the endocardial and epicardial surface of ventricles is 0.03 seconds. However, any disturbance in the simulation sequence and timing of any node in the conduction system is covered by the next node; otherwise, the heart rate will face a problem that is called conduction abnormality, which is discussed in the next section.

16.3.2 Clinical Significance of CCS

A disturbance in the arrangement of the simulation of the specialized cardiac tissue leads to a rhythmic disturbance called arrhythmia or conduction abnormality known as heart block [28]. In the case of rhythmic disturbance, CCS has two built-in protection and safety mechanisms that cover the failure of any node to navigate an impulse [27]. Thus, if the natural heart pacemaker (SA node) fails to fire, the AV node, called the secondary heart pacemaker, generates an impulse between 45 and 60 beats per minute. The AV node takes the lead in generating an impulse under two conditions: if the SA node is out of order and there is a blockage of conduction from the SA node. The second safety mechanism is ventricles that can generate impulses between 30 and 45 beats per minute if the SA node and AV node fail to fire [30]. To clarify this more, BoH generates an impulse between 40 and 45 beats per minute at the ventricles and PF can generate its impulse between 30 and 40 beats per minute. As a result, these safety mechanisms enable the heart to be one of the most precise scheduling organs that fights to the last minute to maintain balance for pumping blood to and from the body.

As mentioned, the heart is innervated and affected by the ANS. This innervation sourced from the autonomic center in the brainstem cannot simulate a heartbeat, but it can affect either decreasing or increasing the heartbeat rate [31]. Furthermore, the heart's innervation consists of two parasympathetic and sympathetic components, collectively pointed to the coronary plexus in the brain. On the one hand, sympathetic innervation increases the permeability of the heart muscle fibers to calcium ions, increasing the force and heart contraction rate. On the other hand, parasympathetic innervation increases the permeability of heart muscle fibers to potassium ions, leading to a decrease in impulse generation and heart rate. Still, generally, it does not affect the force of the contractions. However, increasing or decreasing blood supply is required based on the situation during physical exertion or heart block. In this case, there are natural methods for increasing heart power, such as physical exercise and deep breathing. At the same time, chemical methods, like taking drugs to activate cardiac nerves (i.e., sympathetic innervation) increase the heart rate and vice versa [31]. Therefore, the heart rate range varies depending on the chemical and physical factors to regulate and provide the best performance suitable for the body. To sum up, the above behaviors of CCS and factors that affect CCS can be used to create a novel metaheuristic optimization algorithm to find an optimal solution for real work and NP problems.

16.4 ARTIFICIAL CCS

From the biological facts mentioned in Section 16.3, the behavior of CCS can be concluded in the following points:

1. rhythmical electrical impulse generation
2. built-in safety mechanism
3. conduction rate
4. AV node delay
5. ANS effect (i.e., sympathetic and parasympathetic)

The above behaviors have been translated into a novel metaheuristic bio-inspired optimization algorithm that we named ACCS. The details of the algorithm construction can be seen in Subsections 16.4.6 and 16.4.7, where we used CCS behaviors to construct the ACCS algorithm, which searches for the best solution in a loop using metaheuristic techniques to reach the optimal solution. ACCS is proposed at an abstract high level in this chapter and can be thought of as a natural little brain occurring in the learning metaheuristic system. However, one of the main characteristics of CCS is the deep scheduling procedure that enables the heart to have two built-in protection mechanisms to backup any failure node to generate or pass impulses through it.

This biological feature can be described as a self-organization feature, a spontaneous algorithmic approach adopted by the system that means that without external intervention, it adjusts its behavior using a local agent inside the system. Hence, CCS is an intelligent collective of nodes that strive to keep the heart pacing in the normal range. The behavior of this system is used to develop ACCS and is detailed in the following subsections.

16.4.1 Rhythmical Electrical Impulse Generation

CCS muscles can generate electrical impulses as they have pace characteristics to enable the heart to pump blood from the heart to/from the body. The optimal heart rate value in the search space is not known a priori since CCS is controlled via not one but four major nodes inside the myocardium layer of the heart. This behavior is applied in the ACCS algorithm assuming that the current best candidate solution is the target heartbeat value or close to the optimal heartbeat for an effective heart rating. CCS is the exploration process used to optimize the heartbeat. After the best heart rate is defined, the other elements in the search area move and update toward the best search agent found, which is near the optimal solution. The ACCS algorithm starts to choose four random candidate solutions among all candidates over the search spaces (exploration) based on the problems to represent the heart rate at each SA node, AV node, BoH node, and PF node. ACCS performs a random search for each agent in the search space and moves toward the best solution in a cooperative search with other agents across the search space. Therefore, R is used as a random factor with a value greater than 1 or less than I_{HR} to force the conditions and search agent to move far away from a reference heart rate and explore the search space more efficiently. This behavior is formulated as

$$R_{SA,AV,BoH,PF} = randi\left[1\ I_{HR}\right] \quad (1)$$

where I_{HR} is an index number for the maximum number of decision variables and the same equation is used for all nodes' conditions. The heart works in a very strategic manner to find the optimal factors that generate the most precise heartbeats.

For ACCS to seek an optimal solution, the best candidate heart rate solution is selected, and each heartbeat moves toward its previous best (HR_{PB}) solution concerning the current global best ($X^*(t)$) candidate heart rate solution using

$$HR_{PB}(i,t) = arg_{k=1,\ldots,t}\left[f\left(HR_i(k)\right)\right],\ i \in \{1,2,\ldots,N_{HR}\} \quad (2)$$

$$X^*(t) = arg_{i=1,\ldots,N_{HR}\ k=1,\ldots,t}\left[f\left(HR_i(k)\right)\right], i \in \{1,2,\ldots,N_{HR}\} \quad (3)$$

where i denotes the heart rate index, N_{HR} is the total number of heart rates, t is the current iteration number, f is the fitness function, and HR is the heart rate value.

16.4.2 Built-in Safety Mechanism

CCS has a deep scheduling procedure among four major nodes of the conduction system that enable the heart to fight to the last point to keep pace to pump blood from the heart to the body. Each node can cover the other node in the case of a conduction blockage from one node to another that navigates the impulses through the heart. However, the four randomly chosen candidate solutions in the rhythmical electrical impulse generation stage are compared against the random number to determine whether the heart rate is normal or not at each node. This behavior is represented by the following equation, which is inspired by the equation used for calculating the heart rate from the EKG waveform:

$$Imp_{SA,AV,BoH,PF} = \frac{LB}{UP} * r \tag{4}$$

where *LB* is the lower boundary of the problem, *UP* is the upper boundary of the problem, and r is a random number in [0, 1]. It is worth mentioning that (*Imp*) is updated by a random number that changes in each iteration. In addition, at each heart node, a heart rate is generated randomly using the same equation. Later, the heart rate indicates if the heart node is at a normal pace or if the upcoming node needs to generate a heartbeat independently.

16.4.3 Conduction Rate

Another intercepting point is that sometimes dysfunction within the nodes themselves causes these nodes to be either out of order or unable to generate an impulse properly. Hence, this condition is obeyed for checking along with conduction blockage using either the option to check if the heart is in a normal condition or not using a flag-like value via

$$flag_{T,\, SA,\, AV,\, BoH,\, PF} = \begin{cases} r_1, & [0,0.5] \\ r_2, & [0.6,1] \end{cases} \tag{5}$$

where r_1 is flagged if there isn't any dysfunction at any of the four nodes and r_2 is flagged in the case of dysfunction at any of the four nodes. Hence, after passing through these two filters, the electrical impulses start to navigate through the heart muscles, causing the heart to start polarization and depolarization to force the pumping operation. The current heart rate is updated considering the optimal heart rate obtained so far by this solution, which represents the exploitation mechanism of the algorithm to achieve better results in optimizing the solutions. The heart rate at the SA node, BoH node, and PF node, if the heart rate is in its normal state at these nodes, is updated using

$$\vec{X}(t+1) = \vec{R}.\left(\vec{X}(t) - r * X^*(t)\right) - \vec{X}(t) \tag{6}$$

where t indicates the current iteration, $\vec{R}$ is a random vector in [0,1], r is a random number in [0,1], X^* is the best heart rate solution obtained so far, $\vec{X}$ is the heart rate vector, and represents element-by-element multiplication. Moreover, it is worth mentioning that X^* should be updated if a better candidate solution is obtained in each iteration; otherwise, if the heart rate at the SA node, BoH node, and PF node is not normal, then an additional factor is added to the heart to increase the efficiency of the blood pumping process, which is heart power. This behavior is formulated as

$$\vec{X}(t+1) = \vec{R}.\left(\vec{X}(t) - r * X^*(t)\right) - HP.\vec{X}(t) \tag{7}$$

where t indicates the current iteration, $\vec{R}$ is a random vector in [0,1], r is a random number in [0,1], X^* is the best heart rate solution obtained so far, $\vec{X}$ is the heart rate vector, represents element-by-element multiplication, and HP is the heart power factor. HP is calculated as

$$HP = 0.985 * C \tag{8}$$

where HP is linearly decreased from 0.985 to 0 throughout the iterations (exploitation) and C is a constant number (i.e., 0.999) chosen via trial and test to decrease the convergence time of the ACCS algorithm. In addition to Eq. (6) and (7), the impulse at the AV node is updated with the delay factor. The details and mathematical model for this update are in the following section.

16.4.4 AV Node Delay

In CCS, the electrical impulse at the AV node faces a delay of a tenth of one second to allocate enough time to navigate blood efficiently from the heart's upper chambers to the heart's lower chambers (ventricles). Therefore, a delay factor should be added to each of Eqs. (6) and (7) for updating the heart rate at the AV node if the heart is at a normal state. However, if the heart rate at the AV node is normal, then this updating behavior is formulated as

$$\vec{X}(t+1) = \vec{R}.\left(\vec{X}(t) - r * X^*(t)\right) - D_1.\vec{X}(t) \tag{9}$$

where t indicates the current iteration, $\vec{R}$ is a random vector in [0,1], r is a random number in [0,1], X^* is the best heart rate solution obtained so far, $\vec{X}$ is the heart rate vector, represents element-by-element multiplication, and $D_1$is a normal delay factor, which is a random number in [0.1,0.2]. If the heart rate is not normal at the AV node, then the random delay value is increased to widen the search space and the considered heart power factor. This behavior is illustrated as follows to formulate nodes:

$$\vec{X}(t+1) = \vec{R}.\left(\vec{X}(t) - r * X^*(t)\right) - D_2.HP.\vec{X}(t) \tag{10}$$

where t indicates the current iteration, $\vec{R}$ is a random vector in [0,1], r is a random number in [0,1], X^* is the best heart rate solution obtained so far, $\vec{X}$ is the heart rate vector, represents element-by-element multiplication, HP is the heart power factor generated using Eq. (8), and D_2 is a delay factor that is a random number in [0.2,0.3].

16.4.5 ANS Effect

The ANS affects the decrease or increase in heart rate per minute. The sympathetic nervous system (SNS) and parasympathetic nervous system (PNS) are parts of the ANS that activate the fight or flight response. Hence, the ANS affects the heart rate level of humans either by using physical or chemical factors that activate each SNS and PNS. This behavior is achieved by adding the heart power (HP) factor in Eqs. (7) and (10). Note that the heart power acts as a random number to increase the search power over the search space for both equations to reach an optimal value in less time and prevent the case of local optima. On the one hand, in the exploration phase, when the heart rate is not normal, the moving electrical impulses from one node to another are blocked. In such a case, the heart power action causes the candidate to search the problem space efficiently. Some candidate solutions keep repeating the movement at the close circle without any movement and activity in the search space without this action. However, this factor enforces the impulses to go away from this blockage and start a new search, increasing the chance to find global optima.

On the other hand, ACCS is searching and updating the candidate solutions through four levels of checking for the same candidate solution in each iteration, which makes ACCS effective in exploiting the explored solutions. Hence, this ensures the global convergence or exploration of the algorithm to prevent ACCS from being trapped in a local solution. In addition, at later iterations of the ACCS algorithm, when passing through the four levels of updating and decreasing the HP value, this increases the exploitation capability of the algorithm. However, in the proposed ACCS algorithm, the following steps in Subsection 16.4.6 should be compiled and repeated based on the description above for the algorithm to find the optimal solution for the problem, and the pseudocode details are reported as shown in Subsection 16.4.7. Therefore, in the next section, benchmarking for ACCS is accomplished using standard test functions to demonstrate its performance.

16.4.6 Execution of the ACCS Algorithm

The flowchart for the execution of the ACCS algorithm is presented in Figure 16.2.

16.4.7 Pseudocode for the ACCS Algorithm

The pseudocode of the ACCS algorithm is presented in Figure 16.3, which illustrates the states of the solution to get the optimal values.

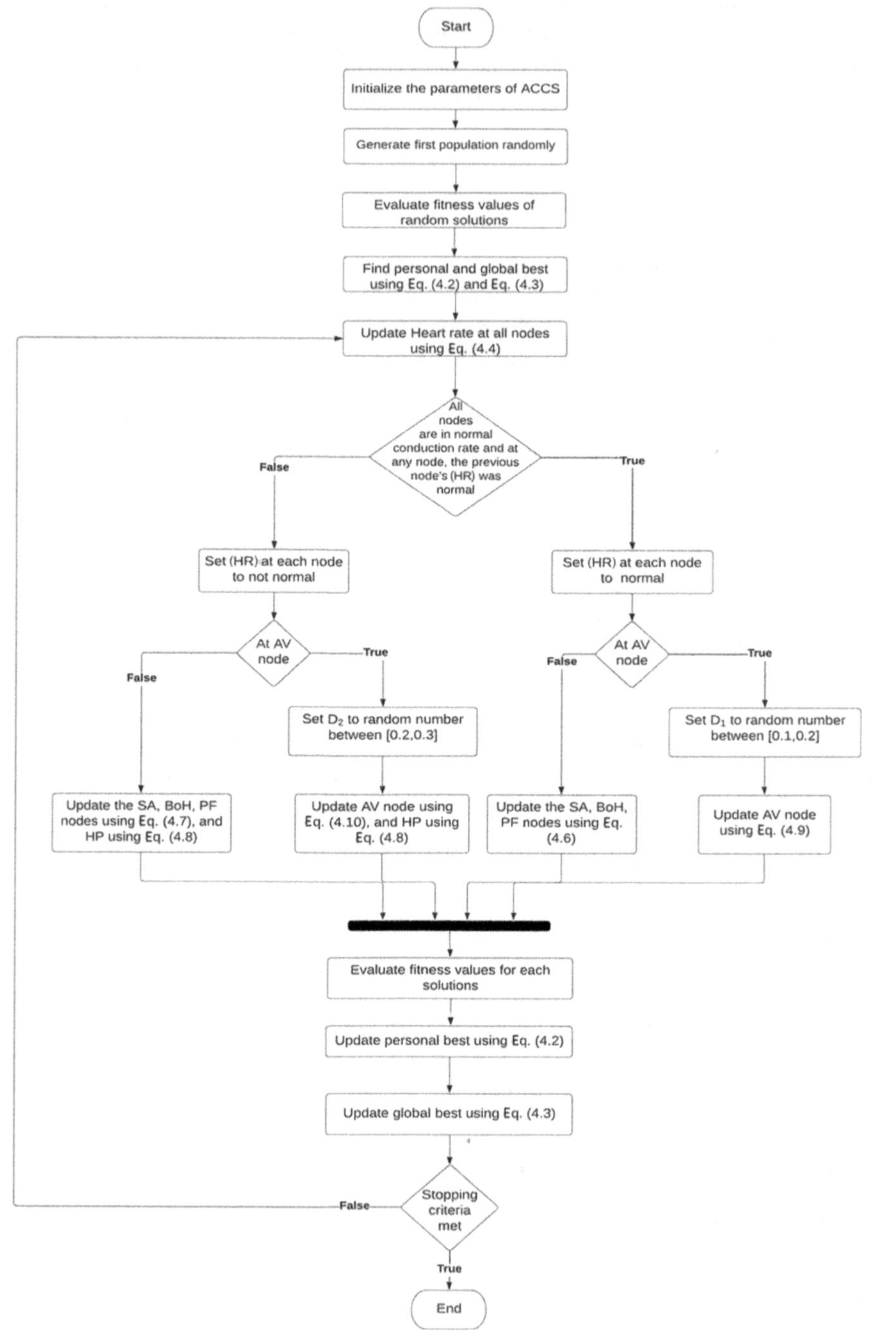

FIGURE 16.2 Flowchart for the execution of ACCS.

16.5 EXPERIMENTAL PERFORMANCE EVALUATION

To test the performance of the ACCS algorithm, several standard benchmark functions that have been conducted to evaluate the performance of existing algorithms in the literature are used. Additionally, the experimentally obtained results are demonstrated and these results are compared to five well-known optimization algorithms: WOA, PSO, GSA, DE, and Fast Evolutionary Programming (FEP). It is worth mentioning that the outcome developed in this study for ACCS was tested by solving 19 mathematical optimization problems. All 19 solved problems are classical benchmark test functions tested on all five mentioned algorithms, along with ACCS [10]. Note that *dim* indicates the dimension, *range* is the lower and upper boundary of the standard test functions' search space, and F_{min} is the minimum objective function value in Tables 16.2, 16,3, and 16.4. Many researchers have used these classical standard test functions in their studies to evaluate their algorithm performance [3–14]. The characteristics of the benchmark functions, numerical experiments, parameter settings, and performance analysis of ACCS are evaluated, and the results are demonstrated in the following sections.

16.5.1 BENCHMARK FUNCTIONS

16.5.1.1 Classical Test Functions

Generally, the ACCS algorithm is tested using two sets of classical test functions to determine the algorithm's performance. The benchmark functions can be divided into several categories based on different characteristics: unimodal test functions, multimodal test functions, and fixed-dimensions multimodal test functions [33]. Figures. 16.4, 16.5, and 16.6 represent the typical two-dimensional (2D) version plots of the benchmark mathematical test functions for some mathematical test cases conducted in this paper. However, in the benchmark test functions, whenever the number of dimensions gets high, the local minima increase exponentially, affecting the complexity of the problem to find the optimal value [34]. Therefore, each set of the considered test functions is used to compare the particular perspective of the ACCS algorithm. For instance, unimodal benchmark test functions, as indicated by their name, have a single optimum, and they are used to check performance in terms of the exploitation level and convergence of the algorithms. At the same time, multimodal benchmark test functions and fixed-dimension multimodal test functions, as their names might imply, are types of test functions that have multi-optimal solutions. They are used for testing the exploration and local optima avoidance of the algorithm. In addition, in multimodal algorithms, there are more than several optimum solutions composed of one global optimum solution and several local optimum solutions. Hence, ACCS must avoid the local optimum and converge toward the globally optimum solution.

(F1) (F2) (F3) (F4)

(F5) (F6) (F7)

(F8) (F9) (F10) (F11)

```
Initialization
        Initialize Population Members of Heartbeats Randomly
        Initialize ACC's Parameters Heart Power, delay
Repeat1 (until stopping criteria is met)
        Evaluate random solution(Finding fitness value using fitness function)
        Set the personal best, and global best
End repeat1
Repeat2 (until stopping criteria is met)
        Updating heart impulse rate at all nodes using equatio n(4.4)
        Randomly set index to update the heart rates
        Set conduction rate flag for SA _node
                //for SA, BoH, PF nodes
                If1 (Heart rate < nodes_impulse rate or conduction rate is not flagged)
                        Update Heart beat using equation (4.6)
                        Set conduction rate flag to normal using equation (4.5)
                else if1
                        Update Heart beat using equation (4.7)
                        Set conduction rate flag to normal using equation (4.5)
                End if1
                //AV node
                If2 (Heart rate < AV_node rate or conduction rate is not flagged)
                        Update Heart beat using equation (4.9)
                        Set conduction rate flag to normal using equation (4.5)
                        Set delay to D1
                else if2
                        Update Heart beat using equation (4.10)
                        Set conduction rate flag to normal using equation (4.5)
                        Set delay to D2
                end if2
        Update the local best Heart using equation (4.2)
        Update the global best (X*) if there is a better solution using equation (4.3)
end repeat 2
return X* that holds the best found solution
```

FIGURE 16.3 ACCS algorithm pseudocode illustration.

$$(F12)(F13)$$

$$(F14)\ (F15)\ (F16)\ (F17)$$

$$(F18)\ (F19)$$

16.5.1.2 CEC-C06 2019 Benchmark Test Functions

Under certain conditions, securing precision in the solution is more important than achieving rapid outcomes. Additionally, various individuals have the capability to enhance an algorithm and iterate its execution. Users arrange the algorithm that best

TABLE 16.2
Description of unimodal benchmark test functions

Function	Dim	Range	F_{min}
$f_1(x)=\sum_{i=1}^{n} x_i^2$	30	[-100,100]	0
$f_2(x)=\sum_{i=1}^{n} \lvert x_i \rvert + \prod_{i=1}^{n} \lvert x_i \rvert$	30	[-10,10]	0
$f_3(x)=\sum_{i=1}^{n}\left(\sum_{j-1}^{i} x_j\right)^2$	30	[-100,100]	0
$f_4(x)=max_i\left\{\lvert x_i\rvert,\ 1\le \lvert i\rvert \le n\right\}$	30	[-100,100]	0
$f_5(x)=\sum_{i=1}^{n-1}\left[100(x_{i+1}-x_i^2)^2+(x_i-1)^2\right]$	30	[-30,30]	0
$f_6(x)=\sum_{i=1}^{n}\left(\left[x_i+0.5\right]\right)^2$	30	[-100,100]	0
$f_7(x)=\sum_{i=1}^{n} ix_i^4+random\left[0,1\right]$	30	[-1.28,1.28]	0

aligns with their specific requirements, irrespective of the time asset involved. Within this contemporary benchmark compilation, the CEC-2019 conference introduced ten test functions [35]. The performance of these functions has been assessed using the ACSS algorithm. Referred to as "The 100-Digit Challenge", these test functions are designed for utilization in annual optimization competitions, as illustrated in Table 16.5. The functions unveiled during CEC-2019 have gained widespread popularity and are recognized as state-of-the-art benchmarks for appraising algorithmic performance in tackling real-world issues.

16.5.2 Numerical Experiment

The ACCS algorithm was coded in MATLAB R2017b and implemented on a MacBook Pro (Retina, 15-inch, Late 2013), Graphic Intel Iris Pro 1536 MB, Processor 2.6 GHz Intel Core i7, Memory 16 GB 1600 MHz DDR3. By contrast, the numerical results of WOA, PSO, GSA, DE, and FEP algorithms were taken from [1].

16.5.3 Parameter Setting

In ACCS, WOA, PSO, GSA, DE, FEP, DA, Leo, and ANA algorithms, the parameters were set as follows:

1. The number of search agents (solution candidate) was equal to 30, except in fixed-dimension multimodal functions, where it was according to the range of each function.

TABLE 16.3
Description of multimodal benchmark test functions

Function	Dim	Range	F_{min}
$f_8(x)=\sum_{i=1}^{n} -x_i \sin\left(\sqrt{\lvert x_i \rvert}\right)$	30	[-500,500]	-418.9829x5
$f_9(x)=\sum_{i=1}^{n}\left[x_i^2-10\cos(2\pi x_i)+10\right]$	30	[-5.12,5.12]	0
$f_{10}(x)=-20\exp\left(-0.2\sqrt{\frac{1}{n}\sum_{i=1}^{n}x_i^2}\right) - \exp\left(\frac{1}{n}\sum_{i=1}^{n}\cos 2\pi x_i + 20 + e\right)$	30	[-32,32]	0
$f_{11}(x)=\frac{1}{4000}\sum_{i=1}^{n}x_i^2-\prod_{i=1}^{n-1}\cos\left(\frac{x_i}{\sqrt{i}}\right)+1$	30	[-600,600]	0
$f_{12}(x)=\frac{\pi}{n}\left\{10\sin(\pi y_1)+\sum_{i=1}^{n-1}(y_i-1)^2\right\}$ $\left[1+10sin^2(\pi y_{i+1})\right]+(y_n-1)^2\}$ $+\sum_{i=1}^{n}u(x_i,10,100,4)$ $y_i=1+\frac{x_i+1}{4} u(x_i,a,k,m)=\begin{cases}k(x_i-a)^m & x_i>a\\ 0 & -a<x_i<a\\ k(-x_i-a)^m & x_i<-a\end{cases}$	30	[-50,50]	0
$f_{13}(x)=0.1\{sin^2(3\pi x_1)+\sum_{i=1}^{n}(x_i-1)^2$ $\left[1+sin^2(3\pi x_i+1)\right]$ $+x_n-1)^2\left[1+sin^2(2\pi x_n)\right]\}$ $+\sum_{i=1}^{n}u(x_i,5,100,4)$	30	[-50,50]	0

2. The population size was equal to 30.
3. The maximum number of iterations was equal to 500.
4. Heart power factor.

16.5.4 Analysis and Discussion

In this section, the ACCS algorithm's effectiveness is evaluated and performance is determined against other recently published algorithms. Usually, testing is accomplished by solving a set of standard benchmark problems [32]. Notably, 19 classical test functions were used to compare and evaluate the performance of ACCS with other recent metaheuristic algorithms. These classical test functions are categorized into three types, which are unimodal (F_1–F_7), multimodal (F_8–F_{13}),

TABLE 16.4
Description of the fixed-dimension multimodal benchmark test functions

Function	Dim	Range	F_{min}
$f_{14}(x)=\left(\frac{1}{500}+\sum_{j=1}^{25}\frac{1}{j+\sum_{i=1}^{2}(x_i-a_{ij})^6}\right)^{-1}$	2	[-65,65]	1
$f_{15}(x)=\sum_{i=1}^{11}\left[a_i-\frac{x_1(b_i^2+b_ix_2)}{b_i^2+b_ix_3+x_4}\right]^2$	4	[-5,5]	0.00030
$f_{16}(x)=4x_1^2-2.1x_1^4+\frac{1}{3}x_1^6-x_1x_2-4x_2^2+4x_2^4$	2	[-5,5]	-1.0316
$f_{17}(x)=\left(x_2-\frac{5.1}{4\pi^2}x_1^2+\frac{5}{\pi}x_1-6\right)^2+10\left(1-\frac{1}{8\pi}\right)\cos x_1+10$	2	[-5,5]	0.398
$f_{18}(x)=\left[1+(x_1+x_2+1)^2(19-14x_1+3x_1^2-14x_2+6x_1x_2+3x_2^2)\right]\times\left[30+(2x_1-3x_2)^2\times(18-32x_1+12x_1^2+48x_2-36x_1x_2+27x_2^2)\right]$	2	[-2,2]	3
$f_{19}(x)=-\sum_{i=1}^{4}c_i\exp\left(-\sum_{j=1}^{3}a_{ij}(x_j-p_{ij})^2\right)$	3	[1,3]	-3.86

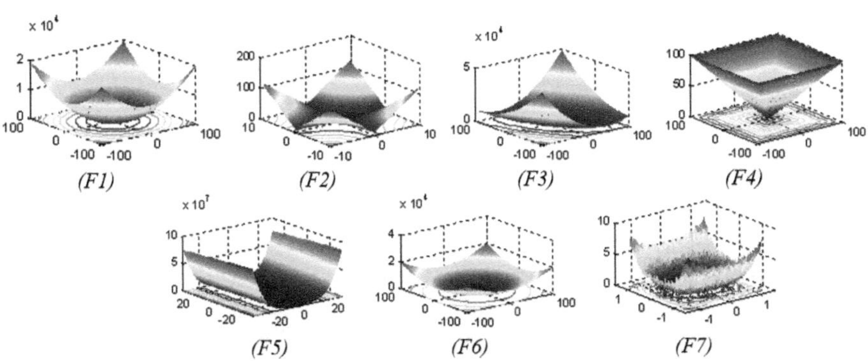

FIGURE 16.4 2D representations of unimodal benchmark functions.

and fixed-dimension multimodal (F_{14}–F_{19}), and the mathematical equations of these functions are shown in Tables 16.2–16.4, respectively. Additionally, Table 16.5 presents the mathematical cluster scripts based on the CEC-C06 2019 Test Functions.

The ACCS, WOA, PSO, GSA, DE, FEP, DA, Leo, and ANA variants were compiled 30 times on each benchmark test function, with the parameters set up as mentioned in Section 16.5.3. In each test, the algorithm was enabled to search for the best optimum solution via running the algorithm for 500 iterations. In addition, the mean and standard deviation of the optimal solutions for each function in these 30

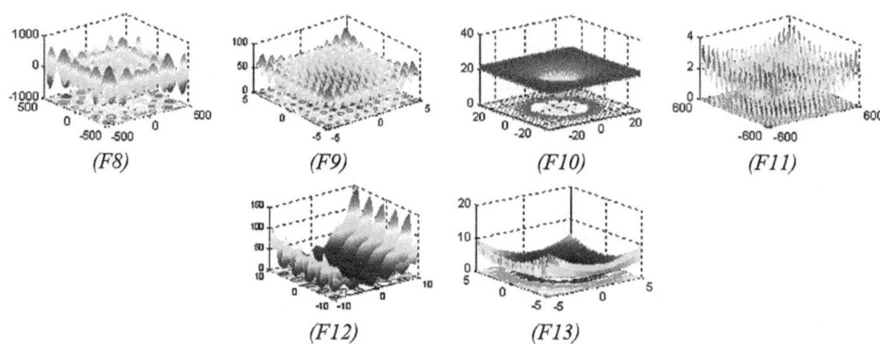

FIGURE 16.5 2D representations of multimodal benchmark functions.

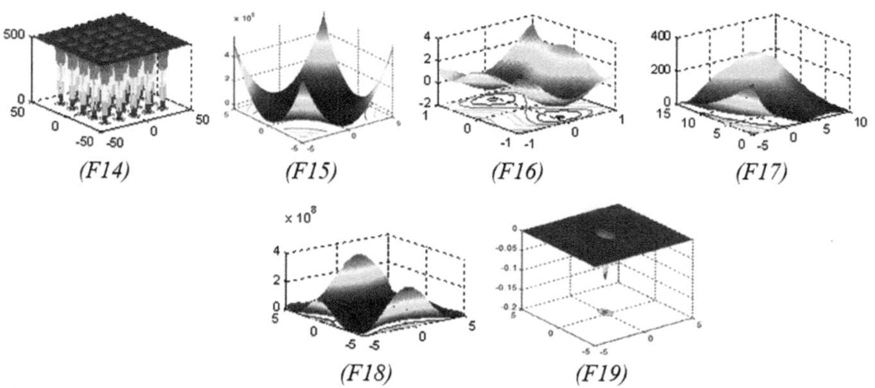

FIGURE 16.6 2D representations of fixed-dimension multimodal benchmark functions.

TABLE 16.5
CEC-2019 benchmarks "the 100-digit challenge"

f	Functions	Dim	Range	f_{min}
CEC01	Storn's Chebyshev Polynomial Fitting Problem	9	[-8192, 8192]	1
CEC02	Inverse Hilbert Matrix Problem	16	[-16384, 16384]	1
CEC03	Lennard-Jones Minimum Energy Cluster	18	[-4,4]	1
CEC04	Rastrigin's Function	10	[-100, 100]	1
CEC05	Griewangk's Function	10	[-100, 100]	1
CEC06	Weierstrass Function	10	[-100, 100]	1
CEC07	Modified Schwefel's Function	10	[-100, 100]	1
CEC08	Expanded Schaffer's F6 Function	10	[-100, 100]	1
CEC09	Happy Cat Function	10	[-100, 100]	1
CEC10	Ackley Function	10	[-100, 100]	1

Source: Brest et al., 2019.

runs were calculated, and then compared to the mean and standard deviation of the other algorithms reported in Tables 16.5–16.8. The verification of the performance of ACCS in terms of exploitation and exploration capability against the other algorithms is discussed in the following subsections.

16.5.4.1 Evaluation of the Exploitation Capability (F1–F7)

As previously discussed, functions F1–F7 represent unimodal functions since they have only one global optimal value. These functions allow testing the exploitation capability of the metaheuristic algorithms. Table 16.6 shows that ACCS obtained highly competitive solutions compared with WOA, PSO, GSA, DE, and FEP. Therefore, ACCS had a high rate of exploitation capability and was very competitive with other metaheuristic algorithms. In particular, the F1–F4 and F7 results show that ACCS was the most efficient optimizer among the other compared algorithms; however, the F5 results indicate that ACCS was better than PSO and GSA, whereas WOA, DE, and FEP were better than ACCS. ACCS in F6 showed poorer performance than the other five algorithms. Thus, the ACCS algorithm provided excellent exploitation due to the behavior of the conduction system's mathematical model, which passes through four different conditions and then updates the candidate solutions four times in each iteration.

16.5.4.2 Evaluation of the Exploration Capability (F8–F19)

Unlike unimodal standard test functions, multimodal test functions have many local optima whose number is controlled by the number of design variables (i.e., problem size). Hence, the number of local optima increases exponentially when the number of design variables increases. Therefore, this test function can be useful for evaluating the exploration capability of the ACCS algorithm. Furthermore, the statistical results of the proposed ACCS algorithms on multimodal test functions are reported in Tables 16.6 and 16.7. It can be observed that the F8–F11 results show that ACCS was the most efficient optimizer compared with the other algorithms. By contrast, in F12 and F13, ACCS provided a better experimental, numerical result than GSA, but poorer performance than the other algorithms. However, in F14–F19, the ACCS results demonstrated poorer performance than the other algorithms as they had fixed dimensions. Hence, in general, the results confirm that the ACCS algorithm had good exploration capability. In addition, the integrated mechanisms of exploration are represented by some factors inside the ACCS calculation, for instance, heart power and delay, that ensure that the search algorithm can explore the search space extensively.

Generally speaking, ACCS is the first algorithm that uses human heart behavior to model mathematical equations, and then produces a novel metaheuristic optimization algorithm. ACCS explains and shows an intense scheduling procedure that exists inside the human heart that controls the repolarization and depolarization of the heart. Unlike most algorithms, ACCS implements the cardiac conduction process in high-level procedures that match theoretical and technical details. The factors that affect ACCS are used in the algorithm's development to increase the algorithm's exploration capability. ACCS is easily tuned and implemented for various real-world

TABLE 16.6
Average of unimodal, multimodal, and fixed-dimension multimodal classical benchmark functions

	ACCS	WOA	PSO	GSA	DE	FEP
F	Avg.	Avg.	Avg.	Avg.	Avg.	Avg.
F1	0	1.41E-30	0.000136	2.53E-16	8.20E-14	5.70E-04
F2	0	1.06E-21	0.042144	0.055655	1.50E-09	0.0081
F3	0	5.39E-07	70.12562	896.5347	6.80E-11	0.016
F4	0	0.072581	1.086481	7.35487	0	0.3
F5	28.9653	27.86558	96.71832	67.54309	0	5.06
F6	7.0137405	3.116266	0.000102	2.50E-16	0	0
F7	4.82E-05	0.001425	0.122854	0.089441	0.00463	0.1415
F8	-2465.5153	-5080.76	-4841.29	-2821.07	-11080.1	-12554.5
F9	0	0	46.70423	25.96841	69.2	0.046
F10	8.88E-16	7.4043	0.276015	0.062087	9.70E-08	0.018
F11	0	0.000289	0.009215	27.70154	0	0.016
F12	1.3991	0.339676	0.006917	1.799617	7.90E-15	9.20E-06
F13	2.9919	1.889015	0.006675	8.899084	5.10E-14	0.00016
F14	10.6834	2.111973	3.627168	5.859838	0.998004	1.22
F15	0.036683	0.000572	0.000577	0.003673	4.50E-14	0.0005
F16	-0.92105	-1.03163	-1.03163	-1.03163	-1.03163	-1.03
F17	1.9425	0.397914	0.397887	0.397887	0.397887	0.398
F18	34.0941	3	3	3	3	3.02
F19	-3.1951	-3.85616	-3.86278	-3.86278	N/A	-3.86

problems due to the simple concept structure of the algorithm. ACCS can update the candidate solutions four times in a single iteration due to the behavior of the four nodes of ACCS. To sum up, all the mentioned experimental and numerical results assert that ACCS was very efficient in exploiting and exploring the test problems compared to WOA, PSO, GSA, DE, and FEP in terms of computational effort and quality of the results. However, the results of ACCS demonstrated poor performance in fixed-dimension multimodal test functions. By contrast, in unimodal and multimodal test functions, the results were highly competitive, and in most of them, ACCS was the most efficient optimizer.

16.5.4.3 Evaluation CEC-C06 2019 (CEC01–CEC10)

As previously noted, this algorithm underwent benchmarking using the CEC-C06 2019 Benchmark Test Functions. This process gauged the algorithm's capacity for exploitation and exploration, with results validated through a comparative study involving DA, WOA, PSO, Leo, and ANA.

Table 16.8 underscores algorithm performance diversity across various test functions, with the basic ACCS function demonstrating notable consistency, implying

TABLE 16.7
Standard deviation of unimodal, multimodal, and fixed-dimension multimodal classical benchmark functions

	ACCS	WOA	PSO	GSA	DE	FEP
F	Std	Std	Std	Std	Std	Std
F1	0	4.91E-30	0.000202	9.67E-17	5.90E-14	0.00013
F2	0	2.39E-21	0.045421	0.194074	9.90E-10	0.00077
F3	0	2.93E-06	22.11924	318.9559	7.40E-11	0.014
F4	0	0.39747	0.317039	1.741452	0	0.5
F5	0.037608	0.763626	60.11559	62.22534	0	5.87
F6	0.24026299	0.532429	8.28E-05	1.74E-16	0	0
F7	3.84E-05	0.001149	0.044957	0.04339	0.0012	0.3522
F8	525.5786	695.7968	1152.814	493.0375	574.7	52.6
F9	0	0	11.62938	7.470068	38.8	0.012
F10	0	9.897572	0.50901	0.23628	4.20E-08	0.0021
F11	0	0.001586	0.007724	5.040343	0	0.022
F12	0.17402	0.214864	0.026301	0.95114	8.00E-15	3.60E-06
F13	0.015473	0.266088	0.008907	7.126241	4.80E-14	0.000073
F14	3.3693	2.498594	2.560828	3.831299	3.30E-16	0.56
F15	0.029559	0.000324	0.000222	0.001647	0.00033	0.00032
F16	0.17111	4.20E-07	6.25E-16	4.88E-16	3.10E-13	4.90E-07
F17	1.5656	2.70E-05	0	0	9.90E-09	1.50E-07
F18	31.4115	4.22E-15	1.33E-15	4.17E-15	2.00E-15	0.11
F19	0.4259	0.002706	2.58E-07	2.29E-15	N/A	0.000014

TABLE 16.8
Average result of the CEC-C06 2019 benchmark

	ACCS	DA	WOA	PSO	LEO	ANA
F	Avg.	Avg.	Avg.	Avg.	Avg.	Avg.
CEC01	1.54E+05	5.43E+10	4.11E+10	1.47E+12	7294147266	8.91E+09
CEC02	19.20848	78.0368	17.3495	15183.91	17.47763	4
CEC03	12.70279	13.7026	13.7024	12.70280	12.70311	13.70240422
CEC04	25070.53	344.3561	394.6754	16.80078	69.8652733	38.50887822
CEC05	6.13997	2.5572	2.73420	1.138265	2.76024667	1.224598709
CEC06	11.32554	9.8955	10.7085	9.305312	3.01982	10.456789
CEC07	1492.699	578.9531	490.6843	160.6863	195.558303	116.5962143
CEC08	6.768143	6.8734	6.9090	5.224137	5.06228333	5.472814997
CEC09	3362.285	6.0467	5.9371	2.373279	3.26147	2.000963996
CEC10	20.28013	21.2604	21.2761	20.28063	20.0123867	2.718281828

TABLE 16.9
Standard deviation of the CEC-C06 2019 benchmark

	ACCS	DA	WOA	PSO	LEO	ANA
F	STD	STD	STD	STD	STD	STD
CEC01	84381.84	6.69E+10	5.42E+10	1.32E+12	5767198154	8.97E+09
CEC02	0.323054	87.7888	0.0045	3729.553	0.09810875	2.87E-14
CEC03	0.00155	0.0007	0	9.03E-15	0.00088954	2.01E-11
CEC04	8443.25	414.0982	248.5627	8.19908	23.7808956	10.07245727
CEC05	1.104676	0.3245	0.2917	0.08939	0.43275426	0.114632394
CEC06	0.687164	1.6404	1.0325	9.30531	0.75595651	1.5567
CEC07	284.8011	329.3983	194.8318	160.6863	236.53515	8.825046006
CEC08	0.299269	0.5015	0.4269	0.786761	0.45975194	0.429461877
CEC09	1003.938	2.871	1.6566	0.018437	0.74449295	0.00341781
CEC10	0.222432	0.1715	0.1111	0.128531	0.0285509	4.44E-16

stable performance across algorithms. This insight offers valuable perspectives on algorithms' general approaches to optimization challenges. In addition, Table 16.9 utilizes standard deviations to gauge performance variability among algorithms. Particularly for the basic ACCS function, a low standard deviation signifies consistent performance across algorithms, indicating reliability in addressing this function. This consistent trend suggests algorithms consistently exhibit either strong or weak performance, showcasing minimal variation in their approach to the basic ACCS function.

In addition, in both Tables 16.8 and 16.9, the analysis of CEC05 reveals variations in the average performance of algorithms across diverse test functions. Specifically focusing on the basic ACCS result for CEC05, a consistent trend in the average performance of algorithms is observed. On scrutiny of the standard deviation for CEC05, the variability in performance among different algorithms becomes outward. Notably, for the basic ACCS function (CEC05), standard deviations exhibit relatively low values across algorithms, indicative of consistent performance. This observation suggests that algorithms consistently exhibit either proficient or deficient performance, with diminished variability in their execution when addressing the basic ACCS function.

16.5.5 Classical Benchmark Comparative Study

As shown in Table 16.10, the global average experimental numerical result of ACCS was 3.158 on a scale of 1–6, where the consideration for the scale is to be one of the best optimizer algorithms and six of the worst optimizer algorithms. It is worth mentioning that the ranking of ACCS is ranked based on the other five compared algorithms. In this manner, the ACCS algorithm was ranked first nine times, fourth once, fifth four times, and sixth five times out of 19. However, the overall average of the ranking by the kind of benchmark test functions is as follows:

TABLE 16.10
Ranking of the optimization performance for ACCS, WOA, PSO, GSA, DE, and FEP to the classical benchmark test function

F	1st	2nd	3rd	4th	5th	6th	Rank	Subtotal
F1	ACCS	WOA	GSA	DE	FEP	PSO	1	
F2	ACCS	WOA	DE	FEP	PSO	GSA	1	
F3	ACCS	DE	WOA	FEP	PSO	GSA	1	
F4	ACCS, DE	WOA	FEP	PSO	GSA		1	
F5	DE	FEP	WOA	ACCS	GSA	PSO	4	
F6	FEP, DE	GSA	PSO	WOA		ACCS	6	
F7	ACCS	WOA	DE	GSA	PSO	FEP	1	15
F8	ACCS	GSA	PSO	WOA	DE	FEP	1	
F9	ACCS, WOA	FEP	GSA	PSO	DE		1	
F10	ACCS	DE	FEP	GSA	PSO	WOA	1	
F11	ACCS, DE	WOA	PSO	FEP	GSA		1	
F12	DE	FEP	PSO	WOA	ACCS	GSA	5	
F13	DE	FEP	PSO	WOA	ACCS	GSA	5	14
F14	DE	FEP	WOA	PSO	GSA	ACCS	6	
F15	FEP	WOA	PSO	GSA	ACCS	DE	5	
F16	WOA, PS0, GSA, DE	FEP				ACCS	6	
F17	FEP	WOA	PSO, GSA, DE			ACCS	6	
F18	WOA, PSO, GSA, DE	FEP				ACCS	6	
F19	FEP	PSO, GSA	WOA		ACCS	DE	5	34
						Total:		63
						Overall Rank:		63/19 = 3.158
						F1–F7:		15/7 = 2.143
						F8–F13:		14/6 = 2.333
						F14–F19:		34/6 = 5.667

- Unimodal test functions (F1–F7): 2.143
- Multimodal test functions (F8–F13): 2.333
- Fixed-dimensions test functions (F14–F19): 5.667

The above results show that the ACCS algorithm has a good balance between exploitation and exploration that assists this algorithm in searching and finding the global optima efficiently. Therefore, ACCS is ranked third among all six algorithms and 19 test functions.

16.6 CONCLUSION

This work proposed a novel bio-inspired metaheuristic algorithm inspired by the human heart function. The proposed method mimicked the electric generation and control behavior of the heart. Nineteen test functions were employed to benchmark the proposed algorithm's performance in terms of exploration and exploitation. The results showed that ACCS could provide highly competitive results compared to well-known metaheuristics, such as WOA, PSO, GSA, DE, and FEP. First, the results on the unimodal functions demonstrated the superior exploitation of the ACCS algorithm. Second, the results on multimodal functions confirmed the exploration ability of ACCS. Finally, the results on both types of functions confirmed that ACCS has a high convergence speed in fewer iterations and less computational time.

In future work, we will develop a multi-objective version of the ACCS algorithm, study parameters in more detail to improve the results toward better performance, adapt parameters to a much more comprehensive range of applications, test a hybrid ACCS algorithm on a vast landscape global optimization, and present a real-world application of the ACCS algorithm in various fields.

REFERENCES

[1] S. Amarel, "The history of artificial intelligence at Rutgers," *AI Magazine*, vol. 6, no. 3, p.192, 1985. https://doi.org/10.1609/aimag.v6i3.499

[2] C. Li, H. Peng, A. Xu, and S. Wang, "Immune system and artificial immune system application," in *World Congress on Medical Physics and Biomedical Engineering 2006*. Springer Berlin Heidelberg, 2007, pp. 477–480. doi: 10.1007/978-3-540-36841-0_128.

[3] M. Dorigo, M. Birattari, and T. Stutzle, "Ant colony optimisation," *Computational Intelligence Magazine*, vol. 1, pp. 28–39, 2006.

[4] J. Kennedy and R. Eberhart, "Particle swarm optimisation," in *Neural Networks, 1995. Proceedings., IEEE International Conference on*, 1995, pp. 1942–1948.

[5] S. Mirjalili, "Dragonfly algorithm: A new meta-heuristic optimization technique for solving single-objective, discrete, and multi-objective problems," *Neural Computing and Applications*, 27, 1053–1073, 2016.

[6] Mirjalili, S., Mirjalili, S. M. and Lewis, A. (2014) "Grey wolf optimizer", *Advances in Engineering Software*, 69, pp. 46–61. doi: 10.1016/j.advengsoft.2013.12.007.

[7] S. Bandyopadhyay and R. Bhattacharya, "Discrete and continuous simulation," 2014. doi: 10.1201/b17127 [Accessed 20 August 2022].

[8] D. H. Wolpert and W. G. Macready, "No free lunch theorems for optimisation," *Evolutionary Computation, IEEE Transactions on,* vol. 1, pp. 67–82, 1997.

[9] F. Glover, "Future paths for integer programming and links to artificial intelligence," *Computers & Operations Research*, vol. 13, no. 5. Elsevier BV, pp. 533–549, Jan. 1986. doi: 10.1016/0305-0548(86)90048-1.

[10] S. Mirjalili and A. Lewis, "The whale optimization algorithm," *Advances in Engineering Software*, vol. 95 pp. Elsevier BV, pp. 51–67, May 2016. doi: 10.1016/j.advengsoft.2016.01.008.

[11] J. H. Dorigo, "Genetic algorithms," *Scientific American,* vol. 267, pp. 66–72, 1992.

[12] B. Basturk and D. Karaboga, "An artificial bee colony (ABC) algorithm for numeric function optimisation," in *IEEE Swarm Intelligence Symposium*, 2006, pp. 12–14.

[13] A. M. Aladdin and T. A. Rashid, "Leo: Lagrange elementary optimization." *arXiv preprint arXiv:2304.05346,* 2023.

[14] D. N. Hama Rashid, T. A. Rashid, and S. Mirjalili, "ANA: Ant nesting algorithm for optimizing real-world problems," *Mathematics*, vol. 9, no. 23, p. 3111, 2021.

[15] X. S. Yang, *Nature-inspired metaheuristic algorithms.* Luniver Press, 2011.
[16] S. Kirkpatrick, Jr. and M. P. Vecchi, "Optimisation by simulated annealing," *Science,* vol. 220, pp. 671–680, 1983.
[17] Z. W. Geem, J. H. Kim, and G. Loganathan, "A new heuristic optimization algorithm: Harmony search," *Simulation*, vol. 76, no. 2, pp. 60–68, 2001. Available: 10.1177/003754970107600201.
[18] E. Rashedi, H. Nezamabadi-Pour, and S. Saryazdi, "GSA: A gravitational search algorithm," *Information Sciences,* vol. 179, pp. 2232–2248, 2009.
[19] O. K. Erol and I. Eksin, "A new optimisation method: Big bang–big crunch," *Advances in Engineering Software,* vol. 37, pp. 106–111, 2006.
[20] A. Hatamlou, "Black hole: A new heuristic optimisation approach for data clustering," *Information Sciences,* 2012.
[21] B. Hajek, "A tutorial survey of theory and applications of simulated annealing," *1985 24th IEEE Conference on Decision and Control.* IEEE, Dec. 1985. doi: 10.1109/cdc.1985.268599.
[22] X.-S. Yang, "Firefly algorithm, stochastic test functions and design optimisation," *International Journal of Bio-Inspired Computation,* vol. 2, pp. 78–84, 2010.
[23] "Biomimicry of bacterial foraging for distributed optimisation and control," *IEEE Control Systems*, vol. 22, no. 3. Institute of Electrical and Electronics Engineers (IEEE), pp. 52–67, Jun. 2002. doi: 10.1109/mcs.2002.1004010.
[24] R. Storn and K. Price, "Differential evolution–a simple and efficient heuristic for global optimisation over continuous spaces," *Journal of Global Optimisation,* vol. 11, pp. 341–359, 1997.
[25] S. Madappady and E. Maben, *Learn ECG in a day*. New Delhi, India: Jaypee Brothers Medical Publishers, 2013.
[26] R. Schmidt and G. Thews, *Human physiology*. Berlin, Heidelberg: Springer Berlin Heidelberg, 1989.
[27] K. Comerford, *Anatomy & physiology made incredibly easy!*. Philadelphia: Wolters Kluwer Health/Lippincott Williams & Wilkins, 2009.
[28] M. Thaler, *The only EKG book you'll ever need.* 2015.
[29] J. Postlethwait and J. Hopson, *Modern biology*. Orlando: Holt, Rinehart and Winston, 2006.
[30] B. Zaret, M. Moser, and L. Cohen, *Heart book.* New York: Hearst Book, 1992.
[31] M. McKinley and V. O'Loughlin, *Human anatomy*. New York: McGraw-Hill, 2012.
[32] N. Singh and S. Singh, "A modified mean gray wolf optimization approach for benchmark and biomedical problems," *Evolutionary Bioinformatics*, vol. 13, p. 117693431772941, 2017.doi: 10.1177/1176934317729413
[33] J. M. Abdullah and T. Ahmed, "Fitness dependent optimizer: Inspired by the bee swarming reproductive process," *IEEE Access*, vol. 7. Institute of Electrical and Electronics Engineers (IEEE), pp. 43473–43486, 2019. doi: 10.1109/access.2019.2907012.
[34] E. Cortés-Toro, B. Crawford, J. Gómez-Pulido, R. Soto, and J. Lanza-Gutiérrez, "A new metaheuristic inspired by the vapour-liquid equilibrium for continuous optimisation," *Applied Sciences*, vol. 8, no. 11. MDPI AG, p. 2080, Oct. 28, 2018. doi: 10.3390/app8112080.
[35] J. Brest, M. S. Maučec, and B. Bošković, "The 100-digit challenge: Algorithm jDE100" in *2019 IEEE Congress on Evolutionary Computation (CEC),* 2019, June, pp. 19–26. IEEE.

Index

For Product Safety Concerns and Information please contact our EU representative GPSR@taylorandfrancis.com Taylor & Francis Verlag GmbH, Kaufingerstraße 24, 80331 München, Germany

Batch number: 10397790

Printed by Printforce, the Netherlands